POLYHEDRA

POLYHEDRA

"One of the most charming chapters of geometry"

PETER R. CROMWELL

CAMBRIDGE
UNIVERSITY PRESS

CAMBRIDGE UNIVERSITY PRESS
Cambridge, New York, Melbourne, Madrid, Cape Town, Singapore, São Paulo

Cambridge University Press
The Edinburgh Building, Cambridge CB2 8RU, UK

Published in the United States of America by Cambridge University Press, New York

www.cambridge.org
Information on this title: www.cambridge.org/9780521554329

First published 1997
First paperback edition 1999
Reprinted 2001, 2004

A catalogue record for this publication is available from the British Library

Library of Congress Cataloguing in Publication data
Cromwell, P. (Peter), 1964–
Polyhedra / P. Cromwell
p.   cm.
Includes bibliographical references (p.   ) and indexes.
ISBN 0-521-55432-2
1. Polyhedra. I. Title
QA491.C76   1996
516′.15–dc20   96–9420   CIP

ISBN 978-0-521-55432-9 hardback
ISBN 978-0-521-66405-9 paperback

Transferred to digital printing (with corrections) 2008

The original colour version of the plate section (found between pages 210 and 211)
has been reproduced in black and white for this digital reprinting. These colour plates
can now be found at www.cambridge.org/9780521664059

Contents

5. Surfaces, Solids and Spheres   181

6. Equality, Rigidity and Flexibility   219

7. Stars, Stellations and Skeletons   249

Preface

Polyhedra have been a part of the fabric of mathematics for two thousand years and have been the inspiration for contributions to many branches of the subject. So it seems remarkable that information concerning their history and mathematical properties is quite difficult to find. The study of polyhedra is still an active area of research and, along with other parts of geometry, is currently enjoying something of a renaissance. However, it is still possible (even probable) that students can complete their education to graduate level and not meet such fundamental objects as the five Platonic solids. The lack of adequate sources of information may contribute to this state of affairs. This book is my response to this vacuum. It tells the story of what people have thought about polyhedra over the ages, and explains some of the mathematics that has been developed to study them.

I started the project that developed into this book by chance over seven years ago. I had just completed my Ph.D. studies at Liverpool University, where the mathematics department had recently begun to build up a collection of mathematical exhibits. These included models of all the polyhedra labelled regular: the five Platonic solids, the four star polyhedra, and five compounds. Each week, one departmental seminar was set aside for general interest talks given by members of staff and research students so that people could find out what their colleagues were thinking about. In one of these sessions I decided to explain some of the mathematical properties of the new polyhedral models. Naively expecting to find all the information I needed presented in several easily available books, I visited the library. The books I found were of three types. Some contained what is often referred to as 'recreational mathematics'. These books often had a chapter or two on the basic properties of a few kinds of polyhedra. The more advanced of such books mentioned Euler's formula. The second category of books fell at the other end of the spectrum. They concentrated on polytopes in arbitrary dimensions and occasionally dealt with three-dimensional examples. The other books I found were guides for model-making. Even large volumes on the history of mathematics contained few comments on the subject of polyhedra after the Greeks, presumably because it is not seen as part of 'mainstream mathematics'.

I gleaned all the information I could from these sources, followed up their references to journal articles and gradually built up enough material to present a survey of the area. I supplemented the display models with some from my own collection and talked on 'The history of polyhedra'. After the seminar several colleagues asked for my sources of information. On hearing of the trouble I had had they suggested that I write up the talk. Since then I have searched for information to fill in some of the details of my original talk and this book is the result.

I am not a historian and I doubt whether the book should be regarded as a history of polyhedra. I have simply collected together the things that I found interesting and arranged them by theme in approximate chronological order. My aim was to present more than a catalogue of definitions and theorems. The dry predigested style commonly used in textbooks tends to ignore the motivation that led to the results being described and leaves readers confused by details they see no need for. I have tried to place results in context, to trace the development of the underlying ideas, and find their influences on, and connections with, other subjects both within mathematics and further afield.

My selection of topics is, of course, a personal choice. Although the work is certainly not encyclopaedic, I think the coverage is fairly complete up to the turn of the century. I have chosen to concentrate on the three-dimensional, geometric aspects of the subject since a large part of the attraction of polyhedra derives from making and experimenting with models. Very few topics that could be recast as graph theory have been included. This means that many developments made this century on the combinatorial properties of polyhedra are not covered. There are three major omissions of this kind that I am aware of. The first is Steinitz' theorem, which plays a central role in connecting the geometric and combinatorial sides of the subject. A presentation of this theorem is easily available in Branko Grünbaum's book *Convex Polytopes*. A discussion of Eberhard-type theorems and Alexandrov's theorem is also missing. Perhaps more surprising to some readers will be the omission of references to duality. Although we can retrospectively find the concept in much early work, it is not clear whether the authors were aware of it. In any case, what do we mean by duality? A few passing remarks on the reciprocal properties of the Platonic solids could cause more confusion than insight: projective and combinatorial duality have been muddled for ages. Furthermore, we do not always find duality in places where it might be expected. For instance, there is no way to construct the vertex-transitive polyhedra from the face-transitive ones 'by duality'. To discuss the topic fully would take a chapter in itself.

Other topics that are not mentioned are space-filling or neighbourly polyhedra, connections with higher-dimensional polytopes or tilings, woven polyhedra of various kinds, polyhedral embeddings of mathematically interesting surfaces, or applications to linear programming and computational geometry.

Skimming through the book you will see it contains many illustrations. I think these are essential when discussing a subject so many of whose underlying intuitions derive from visual experience. The book also contains proofs. I have tried to explain these simply but fully. For the most part they are self-contained. One exception to this rule is the application of some group theory in the last two chapters. The theorem which is appealed to is discussed in Appendix 2.

Acknowledgements

I am glad to acknowledge the help and encouragement I have received whilst undertaking this project.

I am very grateful to Branko Grünbaum. He has been enthusiastic about the project and encouraged me ever since he became aware of it when I sent him drafts of the first few chapters. Since then, he has commented on the entire manuscript and supplied extra information on a variety of topics. H. S. M. Coxeter has also been very helpful in this respect. Others have also read manuscript versions and I am grateful for their suggestions: Anna Aiston, Elisabetta Beltrami, David Cromwell (all of whom studied the whole work), Richard Cromwell, Maria Dedò, Peter Giblin, Joe Malkevitch and Ian Nutt. I had helpful conversations with Robert Connelly, Peggy Kidwell and David Piggins.

I would also like to thank the Geology Department of Liverpool Museum for the loan of some pyrite crystals, Tony Hoffman of Springer Verlag for the portrait of Cauchy, Paul Bien and Donald Crowe for bringing the Tipu Sultan icosahedron and the Corfe Castle rhomb-cub-octahedron to my attention, respectively. I also thank the curator of the Rare Manuscripts Collection at University College London. Some of the plates that form the chapter frontispieces and other figures are reproduced from their copy of Wenzel Jamnitzer's *Perspectiva Corporum Regularium* (ref: Graves 148.f.13).

My parents, my brother, and friends have encouraged me throughout.

The author and publisher are grateful to the following people and institutions for giving their permission to reproduce some of the pictures: Benoit Mandelbrot, John Robinson, the American Council of Learned Societies, the Bayerische Staatsgemäldesammlungen in Munich, the British Museum, the Buckminster Fuller Institute, Cordon Art, the Deutsches Museum in Munich, the Art Museum of Princeton University, the Pushkin Museum of Fine Arts in Moscow, the Ministero per i Beni Culturali e Ambientali in both Naples and Urbino, the Collection of Rare Manuscripts at University College London, the Church of Santa Maria in Organo in Verona.

Peter Cromwell   Liverpool, 1996

Introduction

*Geometry is a skill of the eyes and
the hands as well as of the mind.*

J. Pederson

Models of polyhedra adorn the personal spaces of people with a broad range of
mathematical experience. In the university office of the professional mathematician, the teacher's classroom, and the child's bedroom, these attractive geometrical objects have a universal appeal. Their popularity has endured for centuries.
This book explores how the study of polyhedra has developed, the ways people
have used them and thought about them over the ages, and how their ideas have
evolved.

No science advances in a smooth and continuous way and the study of polyhedra is no exception. We follow some of the searches for explanations of observed
or conjectured properties of polyhedra. Sometimes this is a frustrating struggle
to understand the foundations and limitations of a new concept. What we might
retrospectively regard as a significant milestone is often the result of the accumulated effort of many people over a long period. Progress can also be made in
rapid leaps as a fresh mind brings a brilliant insight to an old problem which then
throws open a whole new region for exploration.

We shall see that the study of polyhedra has contributed to several areas of
mathematics and has connections with many others. However, polyhedra are not
confined to mathematics. Anything which is bounded by flat surfaces and which
has well-defined corners has a polyhedral form. It is easy to think of examples
from architecture. Many more can be found in the three kingdoms of nature—
mineral, vegetable and animal. Polyhedra have been used in philosophical or
scientific explanations of the world around us. They have even found their way
into art, literature and theological debate. We will begin with a brief look at
some of these.

Polyhedra in architecture

Examples of polyhedra in architecture are easy to find. The ancient pyramids
at Giza in Egypt, built over four and a half thousand years ago, are probably

the most simple in design. Modern office blocks are often prismatic structures of steel and glass. Other buildings combine both elements, placing a pyramidal roof structure on a prismatic living space. An octagonal version of this principle underlies the Baptistry at Florence in Italy.

More unusual polyhedral constructions have been used for apartments in various parts of the world. A complex of cubes with their diagonals placed vertically so that each rests on a corner has been erected in Rotterdam in Holland. In Israel, apartments have been made from a complex of dodecahedra. The geodesic domes invented by R. Buckminster Fuller in the 1940's are some of the most remarkable forms of polyhedral architecture. These first impressed the world at the *Expo 67* world fair held at Montreal in Canada. They are now used to protect astronomical telescopes and radio antennae from the elements. On a smaller scale they are used as the frameworks for hemispherical glasshouses and tents, and children's climbing frames.

Polyhedra in art

Polyhedra became popular motifs in art when linear perspective was introduced by the Italians of the fifteenth century. The flat faces and hard edges of polyhedral forms make them very good exercises for those wanting to practice perspective constructions, and many painters' manuals written in the Renaissance include instructions for foreshortening the Platonic solids. Finished paintings sometimes included polyhedra as ornaments but they were usually disguised as pavilions, architecture or headwear.

Polyhedra also appear in twentieth-century art. The *Sacrament of the Last Supper* by Salvador Dali contains a skeletal outline of part of a regular dodecahedron—the Platonic symbol of the universe. Several works by the Dutch graphic artist M. C. Escher contain polyhedra, often star polyhedra or compounds. Op-art designs include many apparently flat-faced objects which cannot be given a consistent three-dimensional interpretation. These can be regarded as 'impossible polyhedra'. A well-known object of this kind is the Penrose tribar shown in Figure I.1.

Figure I.1. The Penrose tribar—an 'impossible' polyhedron.

The same properties of polyhedra which made them attractive to the first artists who used perspective techniques (the fact the positions of a few vertices give enough information to completely describe the solid) make them appealing as computer graphics. Computers can manipulate and draw pictures of simple polyhedra very quickly. More complex polyhedral meshes are used by computer-aided design packages to assist engineers create the bodies of new cars and aircraft.

In the science-fiction film *Tron* a man finds himself transported into a computer where he meets various programs and other elements of computer architecture. In one of the more friendly encounters, he has a conversation with a floating polyhedron that continually changes shape. After having all his questions answered either 'yes' (when the polyhedron metamorphoses into a large yellow regular octahedron) or 'no' (as it becomes an orange stellation of the icosahedron) he realises that this object is the most basic element inside the machine—a bit.

Much modern abstract sculpture has a polyhedral form. Sometimes this is as simple as a cube with one corner embedded in the ground. Other sculptors use tetrahedra stuck together face to face to make vertical columns or sprawling snakelike creatures. One of the largest polyhedral sculptures is in Vegreville, Alberta, Canada. It is a huge Easter egg over seven metres high designed and built by Ronald Dale Resch. It is constructed from 2732 bronze, silver and gold tiles some eighty percent of which are equilateral triangles. The others are non-convex equilateral hexagons in the shape of three-pointed stars. The variations in the egg's curvature are achieved by changing the angles in the stars.

Polyhedra in ornament

Many ornaments have polyhedral forms. Vases and decorative containers are an obvious source. A commonly used shape is the cub-octahedron, probably because it is so easy to make. Earrings of this shape dating from 450AD have been found in Germany. Cub-octahedra are also found in the decorations on Japanese shrines. A large cub-octahedron embellished with chrysanthemums, the emblem of the emperor, sits on top of a tea house in the Shugakuin Imperial Villa in Kyoto. Sacred lanterns of this shape have been made since the thirteenth century and are still used in ceremonies to commemorate the dead. The Koreans use rhomb-cub-octahedral lanterns.

Around fifty bronze ornaments or charms of dodecahedral shape have survived from Roman times and can be seen in the museums around Europe. Most are hollow with circular holes of various sizes cut in their faces and small balls attached at their vertices. Some older examples of Etruscan origin are also known. A dodecahedron recently discovered in Switzerland has a lead core covered with silver and the names of the signs of the zodiac inscribed on its faces. The association of the twelve months with the dodecahedron still continues. One issue

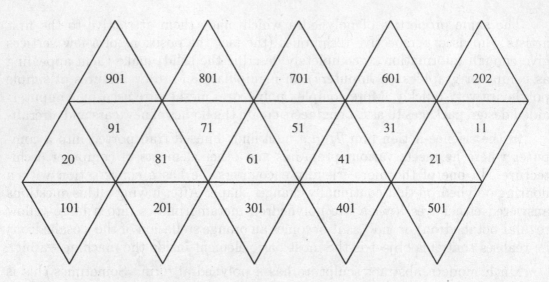

Figure I.2. Net for the Tipu Sultan icosahedron.

each year of *Mathematics in School*, a magazine for teachers, prints the net of a dodecahedron with a calendar printed on it—one month per face.

Dice are another common form of polyhedral object. All the regular solids have been used as dice. A curious icosahedral die was found in the treasure of Tipu Sultan in India when he was overthrown by the British in 1799. It is made of gold and has an unusual distribution of numbers on it. A net is shown in Figure I.2. To obtain a die with ten faces both dipyramids and prisms have been used. A very unusual die in the shape of a rhomb-cub-octahedron was unearthed at Corfe Castle, southern England, in 1973. It is made from a local black marble and is thought to be between two and three hundred years old. Only the square faces have markings on them. Pairs of letters are incised in six of them and the other squares contain patterns of circles representing the first twelve integers. Figure I.3 contains a net for this solid. Its use is unknown.

Polyhedra in nature

The precious gemstones cut and set into rings are sparkling examples of polyhedra but their facets are produced artificially. Natural processes can produce equally striking results: crystals (Plate 2). Bounded by flat reflective planes, their obvious geometric features contrast strongly with the rounded, soft, flexible or irregular qualities more frequently found in natural forms. Because of this distinctiveness, they have always attracted attention. In the nineteenth century, the study of polyhedra and crystals led to the geometric analysis of symmetry. Symmetry theory, together with the assumption that crystals are built up as repeating arrays of atoms, implies the crystallographic restriction: crystals can only have two-fold,

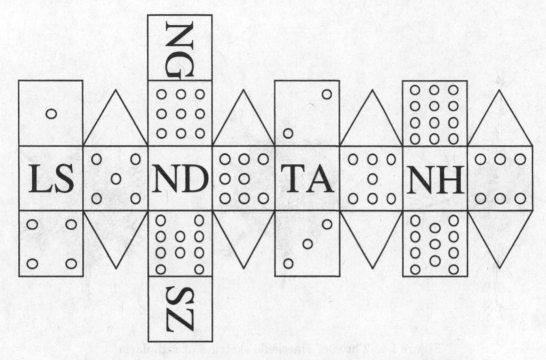

Figure I.3. Net for the Corfe Castle rhomb-cub-octahedral die.

three-fold, four-fold or six-fold rotational symmetry. For this reason, the discovery
in 1984 of a crystalline-looking substance with five-fold symmetry caused great
excitement. These objects are now called *quasi*crystals.

The kernels of some nuts and fruits contain many small seeds which grow in a
restricted space. Pomegranates are one example. As each seed grows it presses up
against its neighbours. The seeds prevent each other from expanding uniformly
and they grow to fill the available space producing flat-faced seeds with sharp
corners. If the seeds had a perfectly uniform distribution before they began to
grow and were subjected to isotropic compression forces they would end up as
rhombic dodecahedra.

The principal of economy—maximising volume from given materials—leads
to the construction of roughly spherical organisms. These sometimes have polyhedral substructures. Ernst Haeckel on his voyage on H.M.S. Challenger, in the
1880's, drew many pictures of microscopic single-celled creatures called radiolaria.
A radiolarian has a spherical skeleton that is polyhedral in character. Haeckel
named three of them *circoporus octahedrus*, *circorrhegma dodecahedra* and *circo-
gonia icosahedra* because he thought they resembled the Platonic solids. His
illustrations are shown in Figure I.4.

Spherical cages also form part of simple viruses such as that responsible for
polio. Viruses reproduce themselves by taking over the protein synthesising equip-

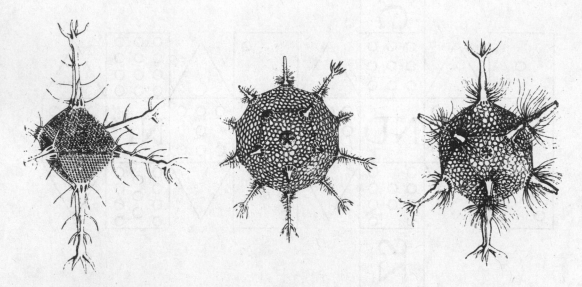

Figure I.4. Three of Haeckel's sketches of radiolaria.

ment of living cells. The viral nucleic acid introduced into the cell causes the cell's machinery to produce parts for new protein cages which protect the replicated RNA in the new viruses. The simplest virus cages are built up from repeating units that clump together in groups of five (pentamers) or six (hexamers). These pentagons and hexagons then fit together to form spherical capsules with approximate icosahedral symmetry.

The recently discovered allotrope of carbon also forms polyhedral spheres, ellipsoids and tubes. In the smallest example, C_{60}, the sixty atoms are arranged in the same pattern as the vertices of a truncated icosahedron—familiar as a soccer ball. These carbon cages have been named Fullerenes in honour of Buckminster Fuller but they are colloquially known as 'Bucky balls'.

Polyhedral molecules have been known for some time. Organic chemists have made carbon–hydrogen structures such as cubane, C_8H_8, whose carbon atoms lie at the corners of a cube. Many more examples occur in inorganic chemistry, particularly with compounds involving the transition metals. In a molybdenum chloride ion ($Mo_6Cl_8^{4+}$) the chlorine atoms form a cubic cage around an octahedron composed of metal atoms. Halides of platinum and zirconium provide further examples of molecular polyhedra. Compounds of boron and hydrogen, called *boranes*, have triangular-faced polyhedral structures which include some of the deltahedra. Molecules of the borane B_8H_8 oscillate back and forth between the forms of a Siamese dodecahedron and a square antiprism.

Polyhedra in cartography

Making maps of the world has been a problem ever since we discovered that the Earth is not flat. A globe is spherical and can be used to represent the world accurately but it provides a limited view: we cannot study the whole of the Earth's surface at the same time. Transferring data from a sphere to a flat surface presents great difficulties and always results in some distortion. In the commonly seen Mercator projection, invented in the sixteenth century, a cylinder is placed around the globe so that the two surfaces touch along the equator. The features on the surface of the sphere are then projected outwards until they meet the cylinder. The map is an exact representation along the line of contact, but away from the equator the map is less precise. The distortion is worse nearest the poles; the poles themselves cannot be represented.

Buckminster Fuller was frustrated by this inaccurate view of the world in which Greenland appears three times as large as South America when, in reality, the opposite is the case. Distortion must appear in any flat map of the Earth but it might be possible to smear it out evenly so that it is less noticeable. Fuller sought to show the shape, distribution, and relative sizes of the Earth's landmasses while containing the worst distortion to the seventy-five percent of the Earth's surface covered by water.

He took a regular solid, the icosahedron, composed of twenty equilateral triangles and subdivided each face into smaller triangles. Calculating from a similar grid superimposed on the Earth he could transfer the data from the sphere to the polyhedron. This was all done between the two world wars before computers were available to assist with the calculations. The resulting map is unique in its lack of visible distortion. Fuller also had to choose the position of the icosahedron carefully so that it could be cut along its edges and opened out flat without creating unnatural breaks in any landmass. The result, known as the *Dymaxion map*, is shown in Figure I.5.

In February 1943, *Life* magazine included a colour, cut-out version of Fuller's map. Before it was published, the editors had the map examined by a panel of experts to certify that it was an accurate representation of the Earth and that it was a new discovery. The panel, which included the Chief Geographer of the U.S. State department and two mathematicians, could not find geographical or mathematical flaws in Fuller's map but were still uncertain as to how it had been created. In terms filled with negative connotations, their report concluded that it was 'pure invention'. When Fuller applied for a patent for his map he found a ruling had been issued which decreed all possible projection methods used in cartography had been exhausted. Hence his application was rejected. He presented the Patent Office with the *Life* report. The testimony of the experts could not be argued with and he was granted the first patent this century for innovation in map-making.

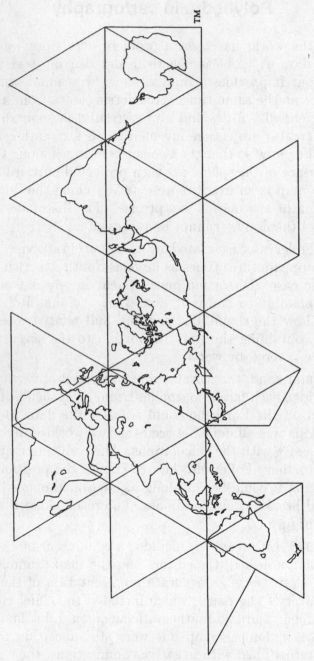

Figure I.5. Fuller's Dymaxion map

Polyhedra in philosophy and literature

Polyhedra have played roles in theories of the universe. Both Plato and Kepler made use of the five regular solids. For Kepler, the polyhedra determined the size of the known universe, the number and relative distances of the planets. Plato associated the solids with the four Empedoclean elements and the heavens, and tried to explain the properties of matter.

Following Isaac Newton's description of the 'clockwork universe' the Design Argument for the existence of God became popular: since we see evidence of design all around us, there must be a designer. The *Natural Theology* of William Paley (1743–1805) argues this case. In his argument against the existence of an inherent, universal ordering principle, Paley used the Platonic solids as the archetypal example of order.

> *Order is not universal; which it would be if it issued from a constant and necessary principle* ⋯. *Where order is wanted we find it; where order is not wanted, i.e. where, if it prevailed, it would be useless, there we do not find it.* ⋯ *No useful purpose would have arisen from moulding rocks and mountains into regular solids, or bounding the channel of the ocean by geometrical curves.* [a]

In what must be the earliest work of science-fiction, *L'Autre Monde ou Les États et Empires de la Lune et du Soleil*, Cyrano de Bergerac (1619–1655) writes about a space flight. He was imprisoned in a tower room and hatched a plan to escape. A friend brought him materials and he constructed a flying machine which he hoped would carry him to his friend's estate. The craft was light and strong, a crystal globe having many facets, a ball like a blazing mirror, in the shape of an icosahedron! It worked by catching as much light as possible to create a void which sucked in air and thereby carried the machine upwards. However, the power of the sun's rays was greater than he anticipated and instead of landing safely outside his prison, he was carried beyond the Earth's atmosphere and towards the sun. After four months he landed on a sunspot and began an adventure.

About this book

The chapters of this book are a series of related essays, each of which explores a particular theme. They are arranged in approximate chronological order but are largely independent units that can be read in any sequence. The only exception to this is Chapter 8 which develops the ideas and notation for describing symmetry. This notation is used in the following two chapters.

As in a piece of music, some themes appear several times in the book, but each entry is slightly different from the last. The subject is modified and developed or treated from a new viewpoint which brings previously hidden elements to the fore. One such topic is regularity. Regular polyhedra were introduced by the Greeks more than two thousand years ago as part of their study of solid geometry. They knew the five solids named after Plato. When Johannes Kepler experimented with new ways of constructing polyhedra, he found two more which could be labelled regular. Two centuries later, Louis Poinsot rediscovered them and found another two. The mathematical development of symmetry led to a new way of interpreting regularity which had wider applications to compound polyhedra.

Other themes which recur throughout the book are a part of every branch of mathematics—the problems of definition, classification and enumeration. What objects are we talking about? When are two objects to be regarded as the same? What kinds of object are there and how many are there? Is there any pattern or structure in such a diverse collection?

Even though the study of polyhedra is one of oldest branches of mathematics, the theory is still being advanced. However, this does not mean that it is necessary to scale high barriers of intimidating mathematics erected over the past two thousand years to attain some understanding of what has been achieved and what is being investigated today. Because of the geometrical nature of the subject, many of the results are accessible to non-specialists. Geometry has a strong visual component. It is this which gives the aesthetic appeal to the subject and which means many ideas can be communicated with the aid of pictures or models. Even though there are many illustrations in this book, sometimes they may not be sufficient. At a few places I have suggested that you make your own models to enhance your understanding of a particular point. Hands-on experience with actual three-dimensional polyhedra is the best way to find out what is going on. Once intuition is developed the pictures may suffice as reminders.

The inclusion of proofs

Although this book tells a story, it also includes proofs of the theorems. Not everyone will want to delve into the fine details of every proof, but the ideas and arguments presented in them are a part of the story. It is for the understanding provided by these explanations that people have worked so hard.

Creating a proof is like taking something apart to see what makes it work. A good proof well explained shows not only *that* something is true but also *why* it has to be true. Unfortunately, the language of rigorous argument can be intimidating. The reasons for some of the pedantry are often lost on the reader, and a mass of details sometimes obscures the flow of ideas. Remember that a proof is just an argument designed to convince its intended audience of the truth of some statement. As audiences have become more demanding over time, so the standards required have been raised. When the oversights in yesterday's proofs are taken into account, and the errors are removed, today's level of proof can be reached. Tomorrow they will be revised again when the conditions are interpreted in new and unforeseen ways and new counter-examples slip round the restrictions.

Although new ideas are often needed to create an acceptable proof, they do not grow out of logical deductions. Behind every proof there is an intuitive feeling of what should happen. Mathematicians seek a deeper understanding than a mere chain of syllogisms. This intuition is sometimes more convincing and reassuring than the proof itself. But proof is essential. The function of argument is to be critical and to focus attention on points that have been overlooked. To advance mathematics in an original direction requires imagination. Proofs point out where limitations and restrictions need to be imposed when our imaginations jump to conclusions which are too general. A proof is the check that we have not deceived ourselves.

Approaches to the book

This is both a mathematics book and a book about how mathematics is done. It can be approached on several levels. For readers wishing to trace the evolution of ideas over a long period of time, the details in the proofs may be unnecessary. Others who want to concentrate on the theory will hopefully find the information they desire. I expect most readers will vary their approach according to topic, dipping more deeply into the subjects they find attractive.

Whatever your approach, there are a few things you should note. Unfamiliar mathematical ideas cannot be fully understood by just reading about them, no matter how good the exposition. Learning mathematics is an active process not a passive one. There will be places where you cannot see everything in your mind and will need to reach for a pencil or a model to find out why something works. You learn through experience—making and handling models, solving problems, working through arguments. When you come to the statement of a theorem make sure you understand what it means. Identify the conditions and the conclusion. Check it out on a few examples to verify it.

Reading a proof is a skill which has to be learned through practice. If you get lost in a proof, try to take a step back and locate the underlying essentials and the way they fit together. If you get truly stuck then skip the proof for a while and return to it later. (The end of a proof is marked with the symbol ■.) A period of incubation allows the new ideas to settle and when you reread the arguments you may not see where you had a problem. On the other hand you may still be stuck. If this happens you can either omit the proof, work through the whole argument on a specific example and follow what happens, or try to create your own argument or counter-example. Remember that, ultimately, a proof appeals to your imagination and previous experiences. It is an argument to convince an audience. It is subjective and what matters is whether it convinces *you*. Many previously overlooked flaws are discovered because people are skeptical about some point of a demonstration.

Some final remarks on the layout of the text. A few isolated paragraphs are marked with a dagger (†) symbol. These clarify or expand some point in the preceding text but assume some prior knowledge which not all readers will possess. They provide supplementary information only, so if you do not understand them skip over them. The sources of the quotations are collected together at the back of the book. They are listed by chapter and are indexed by small letters at the end of the quotations. Figures and tables are numbered in a single sequence.

Basic concepts

There is one rather awkward problem to be overcome in writing a book about the history of polyhedra and that is to decide what is meant by the term 'polyhedron'. A glance at the figures in this book will give you some idea of the variety of objects that have been described as polyhedra. Trying to make a catch-all definition is impossible as different writers have applied the same term to several different ideas, some of which are mutually exclusive. At the most elementary level we can ask whether a polyhedron is a solid object or a hollow surface. The answers to such questions depend to a large extent on the period in which the geometers lived and the problems that they studied. To a classical Greek geometer a polyhedron was solid. Over the past 200 years, it has become more convenient to think of polyhedra as surfaces. Today, some mathematicians regard polyhedra as frameworks.

It has been said that the only thing all polyhedra have in common is the name. However, there is some common ground to be found. The polyhedra in the illustrations clearly share some characteristics. Their most obvious property is that they are made of (or bounded by) polygons. This fundamental property constituted a definition of 'polyhedron' for many centuries, even if it was not stated explicitly. As we shall see, such an open-ended definition can be interpreted in many ways. It does not supply any restriction on how the polygons are to be put together or on what kinds of polygon we can use. This ambiguity has been extremely fruitful, allowing the term to evolve in several directions and leading to the study of different kinds of polyhedral objects. Because of this, I shall leave 'polyhedron' as this vaguely defined term. Throughout the course of the book, we shall see how its meaning has been refined and altered at various times. We shall, however, need some basic terminology to refer to pieces of polyhedra— terminology which can be used however 'polyhedron' is interpreted.

I have tried to use the language appropriate to the period under discussion but occasionally it is more convenient to use modern terminology. This means that some basic terms must be introduced before their correct place in the story. These are listed below and illustrated in Figure I.6.

- Each polygon is called a *face* of the polyhedron.
- A line segment along which two faces come together is called an *edge*.
- A point where several edges and faces come together is called a *vertex*.

I shall make a distinction between the constituent parts of polygons and those of polyhedra. Thus a polygon has *sides* and *corners* whereas a polyhedron has faces, edges and vertices. Each edge of a polyhedron is formed from the sides of two faces. Faces which have two sides joined to form an edge are said to be *adjacent*.

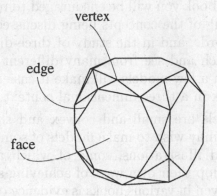

Figure I.6. Basic terminology.

There are also several kinds of angle in a polyhedron.

- The angle in the corner of a polygonal face is called a *plane angle*.
- In a solid polyhedron, the region of the polyhedron near a vertex is called a *solid angle*. It is a chunk of the corner and is bounded by three or more plane angles.
- The angle between two adjacent faces is called a *dihedral angle*. To find a dihedral angle mark a point on the shared edge and a line perpendicular to the edge in each of the two faces starting at the chosen point. The dihedral angle is the angle between the two lines (see Figure I.7).

You will probably be familiar with the names of a few basic polyhedra. A *pyramid* is formed by connecting all the corners of a planar polygon to another point not lying in the plane. The polygon is called the *base* and the extra vertex is the *apex*. All the faces surrounding the apex will be triangular. A *prism* is formed from two congruent polygons lying in parallel planes connected by a ring of rectangles. An *antiprism* is similar except that the connecting ring is composed of isosceles triangles. See Figure 2.25 for examples.

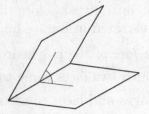

Figure I.7. A dihedral angle.

Making models

At various places in this book you will be encouraged to make a few models to enhance your understanding of the concepts being discussed. In geometry, pictures are often better than words, and in the study of three-dimensional geometry, an object that you can touch and see from many different angles is better than a picture. Experience with a few models will make it easier for you to perceive the many figures in the book in a three-dimensional context.

The suggested models are small and convex and should not present much difficulty. However, you may wish to make models of some of the other polyhedra shown in the plates and illustrations, some of which are quite intricate. All model-makers soon develop their own ways of achieving satisfactory results. The variety of instructions given in various books is evidence of this fact. For guidance purposes, I shall offer my own techniques used to construct the models shown in the colour plates.

The most important part of any model is the pattern. It is essential that the pattern pieces are constructed accurately otherwise they will not meet correctly and the model will not close up properly. The kind of pattern pieces needed will depend on how you intend to decorate the finished model. If you want to use materials of different colours then you will probably need to cut out many faces individually. If you are happy with a plain model or want to paint it when completed then you can design larger pattern pieces which have several faces in each piece. Some polyhedra can be folded up from a single piece. This reduces the number of joins that need to be made.

Having drawn the pattern pieces, the next step is to transfer the designs to the material to be used for the model. The models shown in the plates are made from thin index card. Several sheets can be marked out simultaneously by stapling them together in two opposite corners. The corners of each pattern can then be pricked through all the layers of card at once using a pin or other pointed implement. Do this on a sturdy flat surface covered with something soft to receive the pin. A carpeted floor or a folded towel on a table are good for this.

When you have pricked out the number of pieces required, separate the sheets of card and lightly sketch on the outlines of each face with a pencil. Decide where you need tabs and folds then score along these lines using a steel ruler as a guide. An empty ballpoint pen makes an excellent scoring tool. To cut out the pieces you can use a craft knife against a steel ruler for accuracy but scissors are usually adequate.

The choice of where to put tabs will depend on several factors. There are two methods to make joins: you can put a tab on both pieces to be joined and glue the tabs together, or you can put a tab on only one piece and glue the other piece onto it. The first method produces a rib inside the model along each edge which gives the model added strength. This method is particularly useful when the dihedral angle along the join is greater than 180°. The second method is the one I usually use. It is a good idea to try to alternate tabs and free sides around each face as far as possible as this helps to prevent misalignment. You may also find a pair of needle-nose pliers helpful to apply pressure along a tab while the glue dries, especially as the model progresses and the interior becomes increasingly difficult to get at.

Models which have both mountain and valley folds are often unstable. They can sometimes be deformed unless struts are added in the interior to provide extra rigidity. Some kind of internal support is also needed for models in which several faces are supposed to be coplanar (star polyhedra and compounds). Deformities in such models are usually clearly visible.

I feel I should offer a word of warning here. Model-making can become addictive. Polyhedra tend to come in natural groups and once you have made one member of a set you can feel a strong desire to make all the others. With large complex models taking many hours, this can become a very time consuming obsession. On the other hand, it is cheap, safe, and produces unusual decorative ornaments.

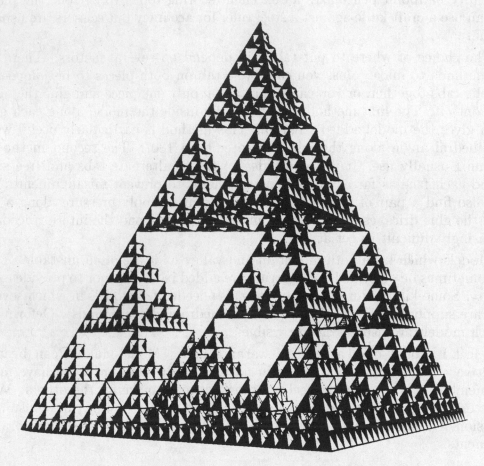

From *The Fractal Geometry of Nature* by Benoit B. Mandelbrot,
Freeman 1982, by permission.

1

Indivisible, Inexpressible and Unavoidable

> *Infinities and indivisibles transcend our finite understanding, the former on account of their magnitude, the latter because of their smallness; imagine what they are when combined.*[a]
>
> Galileo

One of the most basic applications of mathematics is the determination of area and volume. This branch of mathematics, known as *mensuration*, originated in the practical problems of everyday life. Measuring, counting, and pattern-making are probably the oldest forms of mathematical activity.

The earliest writing on geometrical problems consists of sets of rules to be followed when solving a particular type of problem: How to calculate the area of a field, or the quantity of grain that can be stored in a barn, for example. We do not know how or when these techniques were developed, just as we can say nothing of the origins of language. The early texts that have survived record no more than sets of instructions.

Later generations sought to explain the formulae they used and to find out why they worked. This led to several unexpected difficulties. Even the simplest of geometrical figures caused problems. In particular, the formula for the volume of a pyramid proved hard to establish.

Castles of eternity

Planted on the fringe of the Egyptian desert, stark and austere, the pyramids dominate the landscape. Rising from a square base, their four triangular faces slope inwards to meet at a single point. The clean lines of their geometrical form contrast with the amorphous irregularity of their surroundings; their simple outward appearance conceals an intricate design within.

Yet these huge structures are more than mere monuments: they are tombs built to protect mummified pharaohs from disturbance and destruction. The Egyptians went to such great lengths to preserve the king's body because they believed that the attainment of immortality depended on the continuing existence of his Earthly form. Thus the king's tomb had to be virtually indestructible—it was his 'castle of eternity'.

The era of pyramid building spans the Old Kingdom (2686–2181BC), the period of Egyptian history covered by the third to the sixth dynasties of pharaohs. The first pyramid was built around 2650BC. It was a step pyramid produced by successive heightening of a flat-topped tomb, a stairway to heaven. As the sun god Re became the dominant deity in Egypt some of the pharaohs proclaimed their divinity by adjoining the name of the sun god to their own: Khaef-Re and Menkau-Re for example. Also about this time the tombs evolved into the smooth-faced, geometrically true pyramid. It has been suggested that the pyramid was a material representation of the sun's rays along which the spirit of the deceased pharaoh ascended to join the gods. A more recent theory explains the locations of the pyramids along the Nile and the positions of their internal shafts by reference to the stars.

The transition from the step to the true pyramidal form occurred in the reign of Seneferu. The three pyramids that he erected show the various stages of development. The first structure started out as a step pyramid but at some stage the steps were filled in with rubble and the whole cased in limestone dressed to form a true pyramid. On another site Seneferu built a second pyramid which was intended to have the true form from the outset. However, the ground underneath the pyramid was unable to withstand the pressure of the stones piled upon it, and before the pyramid was completed it started to crack. The building work continued but, in an attempt to lessen the weight of the remainder, the angle of slope was reduced from 54° to 43° producing an abrupt change of shape midway up the pyramid. Having failed, Seneferu tried again, building a third pyramid next to the disfigured one. This time he was successful and produced the pyramid with the gentlest incline of any now known: 43° compared to the more usual 52° of later pyramids.

Seneferu's son and successor was Khufu (or Cheops in Greek) who is famous for erecting the Great Pyramid at Giza. The two other pyramids in the Giza group were built by Khaefre (Chephren) and Menkaure (Mycerinus). These edifices are colossal. While on his expedition to Egypt at the end of the eighteenth century, Napoleon Bonaparte is said to have calculated that these three mountains of stone contain enough material to build a wall along the entire French frontier ten feet high and one foot thick. The mathematician Gaspard Monge, who accompanied him, is said to have confirmed his calculation.

Work on the construction of the pyramids was probably carried out during the summer months. The cycle of life in Egypt was dictated by the river Nile. It provided an artery for commerce, and made cultivation possible along a narrow strip of land. Each year, between August and October, the river would rise and flood the surrounding area. After the water had receded, it became possible to grow crops, but by March, the ground had dried to a hard-baked mud. When farming became infeasible, the people who worked the land were available to quarry, transport and position the many tons of stone required to build each pyramid. Without the existence of such a large labour force of peasants and enslaved foreigners it seems unlikely that such huge structures could have been erected. The very existence of the pyramids is a testament to the immense power exercised by the pharaohs.

The shape of a pyramid is extremely helpful in reducing the effort required to raise stone for fewer blocks are required at each course. Over 87 percent of the volume of the pyramid is in the bottom half. Even so, the pyramids remain a remarkable achievement. They display a high standard of craftsmanship. The stones could be cut with great precision. The Egyptologist Sir Flinders Petrie found that the average gap along joints between the casing stones of the Great Pyramid is one-fiftieth of an inch. The bases of the pyramids are levelled to very high accuracy, and they are almost perfect squares.

What level of mathematics did the Egyptians possess to plan and construct such structures? What did they know of the geometry of a pyramid?

Egyptian geometry

Our knowledge of Egyptian mathematics comes largely from papyri that have been preserved by the dryness of the desert climate. They contain various problems of a practical nature: the distribution of wages, the conversion of units, the calculation of areas and volumes; together with exercises in the use of fractions. The principal difficulty in all the problems is the calculation itself.

In plane geometry the Egyptians knew formulae to find the areas of basic polygons: rectangles (length × breadth) and triangles ($\frac{1}{2}$ base × height). To calculate the area of a circle they squared $\frac{8}{9}$ of its diameter. This rule is equivalent to using a value of approximately 3.16 for π—a contrast to other early civilisations who took π to be 3. They knew how to calculate the area of a trapezium and seem to have applied a similar idea to quadrilaterals in general: they took the area to be the product of the averages of the lengths of the two pairs of opposite sides (see Figure 1.1). The result produced by this procedure is usually too large. And yet, 200 years after Euclid taught in Alexandria, this inaccurate formula was still in use. It appears among the inscriptions on the Temple of Horus at Edfu which records events in the first century BC.

In solid geometry, the volumes of cylinders, barns, and beams were evaluated

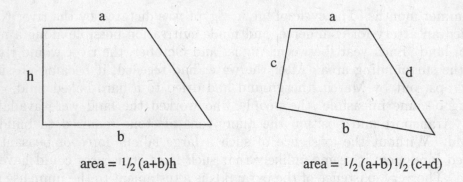

area = $\frac{1}{2}$ (a+b)h   area = $\frac{1}{2}$ (a+b)$\frac{1}{2}$ (c+d)

Figure 1.1. Egyptian formulae for finding the area of a trapezium and a general quadrilateral. The second is incorrect.

as the product of the area of a cross-section and the length (or height). The difficulty of the calculation lies in the conversion of units—from volume to grain measures, for instance. It is generally accepted (on indirect evidence) that the Egyptians could find the volume of a pyramid, and that they also had a rudimentary notion of trigonometry which allowed them to express the degree of slope, or gradient, of a pyramid.

Since the Egyptians did not possess a symbolic notation in which to express their formulae, they had to give a series of worked examples—special cases with particular numbers rather than the general case. The same example would be used in different contexts. For instance, problems 57 and 58 of the Rhind mathematical papyrus both deal with the same pyramid. In the former problem, the height and base are the data and the slope is calculated; in the latter, the height is the unknown quantity. Problem 56 also concerns the slope of a pyramid. It reads as follows:

> *Example of reckoning a pyramid.*
> *height 250, base 360 cubits.*
> *What is its slope?*
>
> *Find half of 360: 180*
> *Divide 180 by 250: $\frac{1}{2} + \frac{1}{5} + \frac{1}{50}$ cubits*
> *Now a cubit is 7 palms.*
> *Then multiply 7 by $\frac{1}{2} + \frac{1}{5} + \frac{1}{50}$*
>
> $\vdots$
>
> [calculation omitted]
>
> $\vdots$
>
> $5\frac{1}{25}$ *palms. This is its slope.*[b]

The result means that the pyramid rises 1 cubit for every $5\frac{1}{25}$ palms horizontally.

The calculation of $\frac{180}{250}$ results in the reciprocal of what today is called the *gradient*. Alternatively, it can be thought of as the cotangent of the dihedral angle between a face and the base. Notice that they only used fractions whose numerators were 1. The final part of the calculation alters the units of the measurement.

The measurements used in the problem are realistic. In the following table they are compared to those of the Giza pyramids.

	Height	Base	Slope
Khufu	147m	230m	51°52'
Khaefre	144m	216m	52°20'
Menkaure	66m	108m	50°47'
Problem 56	131m	188m	54°14'

Another papyrus also contains a problem relating to the geometry of a pyramid. The fourteenth problem of the Moscow mathematical papyrus is concerned with the volume of a truncated pyramid. This solid is sometimes called the *frustum* of a pyramid (from the Latin word for 'fragment'). The papyrus itself is about 8cm wide and over 5m long. It contains 25 problems, many of which are unclear because the papyrus is damaged. Figure 1.2 shows the portion containing problem 14. The text is written in *hieratic* script, a simplified form of hieroglyphics. In the hieroglyphic transcription shown beneath, the numbers in the problem can be picked out: the symbols ∩ and | stand for tens and units respectively. The text is read from right to left. An English translation reads as follows:

> *Method of calculating a truncated pyramid.*
> *If you are told: a truncated pyramid of 6 cubits in height,*
> *of 4 cubits on the base, by 2 on top*
> *You are to square this 4: result 16*
> *You are to double this 4: result 8*
> *You are to square 2: result 4*
> *Add together this 16, the 8 and the 4: result 28*
> *Take $\frac{1}{3}$ of 6: result 2*
> *Take 28 twice: result 56.   See, it is 56—you have found it right.*[c]

By following the instructions we obtain two intermediate results. Firstly

$$4 \times 4 + 2 \times 4 + 2 \times 2$$
$$= 16 + 8 + 4$$
$$= 28$$

and then $1/3 \times 6 = 2$. Combining these gives the volume as $2 \times 28 = 56$ cubic cubits.

Replacing the specific measurements by the symbols h, b and a for the height, base and top, respectively, we find that the algorithm evaluates the sum

$$\tfrac{1}{3}\,h\,\left(b^2 + b\,a + a^2\right)$$

which is the standard formula for the volume of a truncated pyramid.

As a corollary of this example, one is led to believe that the Egyptians must also have known the formula for the volume of a complete pyramid $\left(\tfrac{1}{3}\,hb^2\right)$ but there is no specific example where it has been used.

How did the Egyptians discover their formulae and algorithmic procedures? In general, we cannot answer this question since the papyri are not concerned with such things. They contain neither motivation, nor any justification to validate the methods they describe. Their principal function is to communicate knowledge of a technique or the rules of calculation to be followed in particular circumstances. The Egyptians did not possess a notation in which they could express general formulations of their results:

> *They were a people who had no plus, minus, multiplication, or division signs, no equals or square-root signs, no zero, no decimal point, no coinage, no indices, and no means of writing the common fraction* p/q.[d]

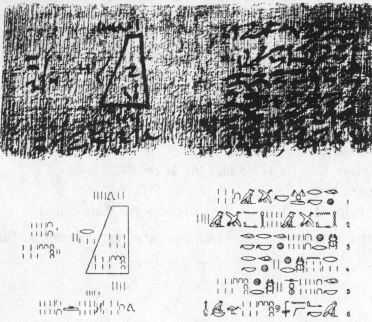

Courtesy of the State Pushkin Museum of Fine Arts, Moscow.

Figure 1.2. Problem 14 from the Moscow papyrus.

This lack of notation was overcome by giving several worked examples of a similar nature using different numbers to illustrate the method of solving a particular type of problem. They expected that the general procedure would be abstracted from sufficiently many specific examples. In these non-symbolic descriptions, the values were taken to be typical; other values can be substituted so that the general case becomes apparent. However, descriptions of procedures cannot establish the universal validity of a formula. Moreover, in some cases, a systematic derivation of a procedure is not possible since the resulting formula is either an approximation (as in the case of the area of a circle) or wrong (the area of a general quadrilateral).

Babylonian geometry

An ancient civilisation also developed in Mesopotamia, the fertile plain between the Tigris and Euphrates rivers, now part of Iraq. The mathematics of the Babylonians, along with other records, is preserved in the form of clay tablets. These tablets were inscribed with wedge-shaped signs by pressing a stylus into the surface of the soft clay. The script is now called *cuneiform* from 'cuneus', the Latin word for wedge.

The extant tablets come from two distinct periods: most date from 1800–1600BC, a few from 300BC–1BC. The mathematical tablets can be separated into problem texts and tables. The latter include multiplication tables, lists of squares, cubes, square roots, reciprocals, and so on. Some of the problem texts contain examples of elementary geometry.

In plane geometry, the Babylonians knew formulae for finding the areas of rectangles, triangles and trapeziums. In many of the problems concerning circles a bad approximation is used: for instance, the area is taken to be one-twelfth of the square of the circumference, which is equivalent to using 3 for π. It is obvious that this value for π is too small since the perimeter of a regular hexagon is three times the diameter of its circumcircle. One tablet containing a list of numbers has been interpreted as a list of approximations to the ratios of area to length-of-side for regular polygons of up to seven sides. It also contains a comparison of the perimeters of a hexagon and its circumcircle which leads to a value of $3\frac{1}{8}$ for π.

In solid geometry, the Babylonians considered similar problems to the Egyptians. Volumes of prisms and cylinders were calculated as the product of base-area and height. There are also problems dealing with cones, pyramids, and their frustums, although in these cases the wrong formulae seem to have been used. The method applied in most of these problems concerning frustums is to average the areas of the top and base, then multiply by the height. This is certainly wrong (even if we assume π to be 3) but it is analogous to the formula for the area of a trapezium. In one problem, the volume of a truncated square-based pyramid is calculated as

$$\left(\left(\frac{a+b}{2}\right)^2 + ? \right) h$$

where the second term in the calculation is unclear, the text being damaged. This term has been interpreted in various ways: possibly it reads

$$\frac{1}{3} \left(\frac{a-b}{2}\right)^2$$

in which case the formula is correct, but a more likely reading is

$$\left(\frac{a-b}{2}\right)^2$$

which is equivalent to the averaging procedure used in other problems.

Chinese geometry

The Chinese are another ancient people who developed mathematics at an early date. The oldest surviving and most influential of the ancient Chinese mathematics texts is the *Chiu-Chang Suan-Shu* (Nine Chapters on the Mathematical Art), also known as *Chiu-Chang Suan-Ching* (Mathematical Manual in Nine Chapters). It is a collection of problems that probably appeared in its current form in the first century AD but which contains older material that some believe to date from the third century BC. Even by the third century AD, the origins of the work were uncertain. In the preface to his *Commentary on the Nine Chapters* (263AD) Liu Hui says that the classics were either lost or destroyed when the emperor Ch'in Shih Huang (221–209BC) ordered the burning of all books in 213BC and the *Nine Chapters* was compiled from the remains of works that survived. The fact that some problems refer to contemporary events, ranks, or official titles which belong to particular periods helps to date the work and shows that different parts were written at different times.

The *Nine Chapters* contains a series of 246 problems together with their solutions. The problems are connected with day-to-day life and it seems that the book was intended as a handbook to be used by engineers, architects, and tradesmen. As in the Egyptian and Babylonian texts, the problems are stated with definite numbers. The solutions, however, are presented as general procedures. Two of the nine chapters (I and V) treat geometric subjects. The other chapters deal with proportions, the distribution of wealth, taxation, and the solution of simultaneous linear equations.

The first chapter, on surveying, gives rules for the calculation of areas: it deals with triangles, rectangles, trapeziums, circles and their sectors, and annuli.

The ancient Chinese evaluated the area of a circle as either $3/4$ the square of the diameter, or $1/12$ the square of the circumference. Both of these formulae correspond to using the value 3 for π. Later Chinese mathematicians improved this approximation considerably. Chang Heng (78–139AD) used $\sqrt{10} \approx 3.16$. Liu Hui described an iterative method to approximate a circle by inscribed regular polygons. Starting with a regular hexagon, and repeatedly doubling the number of sides so as to produce a regular polygon each time, he compared the perimeter of the resulting polygon with the diameter of its circumcircle (see Figure1.3). The more times the process is repéated, the more accurate the resulting value of π. After four iterations, the polygon has $6 \times 2^4 = 96$ sides. This gives the value of π as 3.14, which is good enough for practical purposes.

The fifth chapter 'Consultations on engineering works' concerns earthworks (walls, dykes, canals) and the labour required to construct them. It also contains the formulae giving the volumes of solids: prisms, cylinders, pyramids, cones, truncated pyramids and cones, and a particular case of a tetrahedron.

Although the solutions to the problems are given as general procedures, not as particular worked examples, there is no attempt to justify their correctness. In his *Commentary on the Nine Chapters*, Liu Hui supplements the brief discussions in the original work with arguments to demonstrate the validity of the algorithms. His explanations have a high standard of logical rigour but he does not state explicitly all the assumptions on which his derivations are based. Consequently, his proofs retain the intuitive element which is often excised by the pedantic axiomatic style used in many modern textbooks.

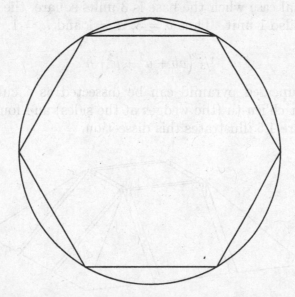

Figure 1.3. Approximations to the value of π can be calculated by comparing the perimeter of a regular polygon with the diameter of its circumcircle.

To illustrate Liu's style of proof, the frustum of a square-based pyramid will be used as an example. In his derivation of the formulae for the volumes of solids Liu uses dissection arguments. This involves splitting up the solid in question into pieces of known volume, then combining the formulae of the constituents to arrive at a formula for the whole. In order to describe the dissection of a solid he uses a set of four standard blocks, each one unit in height, width and length. These are illustrated in Figure 1.4. Liu clearly possessed a set of such blocks and expected his reader to be familiar with them. He calls them *ch'i*, a name usually used for the pieces in a board game such as chess, so it is possible that they were part of a game or puzzle with a fairly wide circulation.

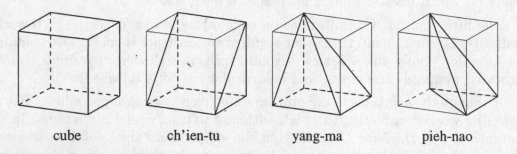

| cube | ch'ien-tu | yang-ma | pieh-nao |

Figure 1.4. The four Chinese blocks, known as ch'i, used by Liu Hui.

By restricting his dissections to be composed of these blocks, Liu can treat only particular cases of the problems. In the case of a truncated pyramid, he considers the special case when the base is 3 units square, the top 1 unit square, and the height is also 1 unit. Thus $b = 3$, $a = 1$ and $h = 1$. He needs to show that its volume is

$$\tfrac{1}{3} \left(ab + a^2 + b^2 \right) h.$$

This particular truncated pyramid can be dissected as a cube (in the centre) surrounded by four ch'ien-tu (the wedges at the sides) and four yang-ma (one at each corner). Figure 1.5 illustrates this dissection.

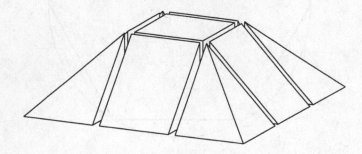

Figure 1.5. Liu's dissection of a truncated pyramid into ch'i.

Liu considers each of the terms abh, b^2h and a^2h in turn finding collections of ch'i that have these volumes. The demonstration is completed by observing that these same pieces can be arranged to form three frustums.

Firstly, to construct a figure whose volume is equal to the product of the top and base dimensions together with the height, Liu takes the central cube and the four ch'ien-tu. These blocks can be rearranged to form a rectangular prism whose dimensions are a, b and h (see Figure 1.6).

To form a figure of volume b^2h Liu builds up the truncated pyramid with extra blocks to make a cuboid. A ch'ien-tu matched with each one in the dissection fills in each side; and two yang-ma can be placed at each corner to complete the prism. (That three yang-ma can be fitted together in this situation to form a cube is fortuitous—a general cuboid cannot be dissected into three congruent pyramids.) The figure is then composed of 12 yang-ma, 8 ch'ien-tu and a cube.

The central cube in the dissection provides a figure whose volume is a^2h.

We now see that the expression $abh + b^2h + a^2h$ is the volume of

$$\text{cube} + 4\,\text{ch'ien-tu}$$
$$+ \quad \text{cube} + 8\,\text{ch'ien-tu} + 12\,\text{yang-ma}$$
$$+ \quad \text{cube}$$
$$= 3\,(\text{cube} + 4\,\text{ch'ien-tu} + 4\,\text{yang-ma})$$

or, in other words, three times the volume of the original figure—the truncated pyramid. ∎

Liu has only dealt with the particular case when $b = 3$, $a = 1$ and $h = 1$. In the general case when a, b, and h are arbitrary the dissection of the frustum can still be performed: the central block becomes a square-based prism; the other blocks are also stretched or squashed. Rearranging the pieces of the dissection is now not quite so easy. The ch'ien-tu can still be fitted together. The yang-ma, however, cannot. Liu shows elsewhere that the volume of a yang-ma is one-third

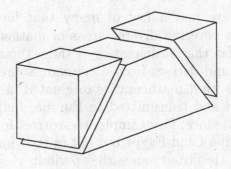

Figure 1.6.

of the product of its dimensions. In general, this is not easy to prove but, in the special case of the dissection into standard blocks, it is simple to demonstrate since three of the yang-ma blocks can be placed together to form a cube as shown in Figure 1.7.

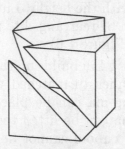

Figure 1.7. Dissection of a cube into three yang-ma pyramids

A common origin for oriental mathematics

Even though the early civilisations were separated by vast distances, their mathematics has much that is common. Examples of these similarities appeared in the geometry discussed above. The Egyptians, the Babylonians, and the Chinese all treat similar problems with a similar style of approach. Their mathematics texts take the form of problems with solutions. The volumes of elementary solids such as beams and cylinders are a common feature; perhaps more surprising is that they all treat frustums of pyramids or cones.

Some of the similarities seem to be more than just coincidences. One Babylonian tablet has the same contents as the fifth of the Chinese *Nine Chapters*: both texts start with problems on dams, walls and the number of men required for constructions, and then treat the volumes of solids. Another link between the Babylonians and the Chinese is that both use the formula '$1/12$ the square of the circumference' to find the area of a circle.

These similarities are just a few of many that have been catalogued by B. L. van der Waerden covering diverse areas of mathematics. These parallels led him to conclude that the mathematics of these three cultures, and of India as well, is interrelated and derives from a common source. Traditionally, scholars have taken the view that mathematics originated in the Near East and was subsequently developed and transmitted to Europe, India and China. Van der Waerden maintains that since, for example, the correct formula for the volume of a frustum appears in China and Egypt but not in Babylonia, the common source cannot be Babylonian. He favours an earlier period:

I have ventured a tentative reconstruction of a mathematical science

> *which must have existed in the Neolithic Age, say between 3000 and*
> *2500BC, and spread from Central Europe to Great Britain, to the*
> *Near East, to India, and to China.*[e]

He suggests that the mathematical knowledge which existed at this time was transmitted by means of problems with solutions and worked examples, that this tradition is preserved in the texts of the early cultures, and that the Chinese have kept the most faithful copy of this early mathematics for they have retained the geometrical imagery and many details whereas the others have recorded only the rules and procedures. However, these conclusions are still the subject of debate.

Greek mathematics and the discovery of incommensurability

The characteristic of Greek mathematics which distinguishes it from that of earlier cultures is the notion of proof. It is uncertain whether early civilisations could even formulate propositions in a general context, and there are no traces of deductive arguments being used to justify methods in any pre-Hellenic culture. In all ancient mathematics there is just a description of a process, often given as a sequence of worked examples. The Greeks not only stated general propositions but furnished them with rational arguments to demonstrate their validity.

Why did they find it necessary to provide statements with proofs? One possibility is that it started as a way of judging between the various results they collected abroad. The Greeks themselves wrote that the orient was a major source of material for their mathematics and other sciences. There are tales that many of the influential Greek mathematicians travelled widely: Thales, Pythagoras, Democritus, and Eudoxus are all reputed to have visited Babylon, Egypt, and possibly as far afield as India. Comparing the methods used by different cultures dealing with similar problems would have shown up discrepancies and inconsistencies. Which of the formulae for the volume of a truncated pyramid is correct: the Egyptian or the Babylonian? There was also the need to distinguish between formulae that gave exact results and those giving approximations. Given two approximations it would also be helpful to determine the relative accuracy of the different techniques.

The knowledge gained through travels in foreign lands would have consisted solely of the procedures to be followed in particular circumstances. Any arguments that had been used in the past to derive the formulae or establish their correctness would have been long forgotten.

The earliest proofs probably consisted in drawing diagrams which exhibited the desired result and reasoning from them. The Greek term for 'to prove' can be translated as 'to show', 'to point out' or 'to explain'. Thus, one pointed out reasons for believing a statement to be true. Initially, the arguments would have made a large use of visual evidence either in dissection arguments where a figure was subdivided to illustrate interrelationships, or by the use of superposition to show the equality of different parts. The word 'theorem' is derived from the Greek for 'to look at'.

Later proofs were semantic in nature relying on the meanings of words: 'odd', 'even' and 'add' for example. Although particular cases of propositions involving odd or even numbers can be illustrated by arranging pebbles in rows, the arguments used to justify the general case need to be visualised in the mind. This helped to develop the ability to reason about abstract objects.

Without abstract thought, such counter-intuitive results as the existence of *incommensurable* line segments could not have been discovered. Common sense tells us that any two lines we draw will be integral multiples of some common unit segment. But quite early on in Greek history, certainly by the end of the fifth century BC, people had found that some pairs of lines have no common measure—a discovery which pushed mathematics further away from empiricism and towards abstraction.

The Pythagoreans had a doctrine that 'number is the essence of all things', by which they meant that everything could be explained using ratios of whole numbers. This belief was strengthened by (or perhaps derived from) the discovery that consonant musical intervals fit into such a pattern. The problem of incommensurability becomes apparent when this theory is applied to geometry, for the lines in some basic figures do not conform to such a rule. The ratio of the lengths of the side and diagonal of a square (or, equivalently, the sides of a right-angled isosceles triangle) cannot be expressed using integers.

A very early proof, if not the original demonstration, of the incommensurability of the side and diagonal of a square has been preserved in an appendix to the tenth book of Euclid's *Elements*. It uses an arithmetic argument based on the properties of odd and even numbers. It shows that if the two lengths were commensurable then a number would exist which is both odd and even—an impossibility. This indirect method of proof requires strict logical reasoning and it is only when the contradiction is reached that the hypothesis is established. This has led some scholars to suggest that the initial discovery of incommensurability was made in a different context. Kurt von Fritz writes

> *It is difficult to believe that the early Greek mathematicians should have discovered the incommensurability of the diameter of a square with its side by a process of reasoning which was obviously so laborious for them if they had no previous suspicion that any such thing as incommensurability existed at all. If, on the other hand, they had already discovered the fact in a simpler way, it is perfectly in keeping with what we know of their methods to assume that they at once made every effort to find out whether there were other cases of incommensurability. The isosceles right-angled triangle in that case was the natural first object of their further investigations.*[f]

The problem is that the proof is an 'all or nothing' affair. There is no intuitive basis to guide the reasoning, no picture to suggest the result is true.

In general, the way a mathematical result is discovered and the form in which it is presented are not the same. A formal proof is constructed 'after the event'. Prior to this, the result must have been understood on an intuitive level. There is nearly always a picture of what is going on. It may have suggested the result, and then guided attempts to find a proof. Unfortunately, in the final proof, this initial insight is often lost, smothered in a mass of detail. The question is: What image aroused the suspicion that some line segments may not have a common measure?

The side and diagonal of a square are used as the archetypal example of incommensurable lines throughout Greek writings to such an extent that they were probably also the first example. However, von Fritz and others have proposed that the initial discovery could have involved the golden ratio. Line segments in this ratio can be proved incommensurable by geometric arguments. Furthermore, the geometry of the regular pentagon leads to the kind of suggestive image we seek. A pentagram is the star-shaped figure formed by the set of diagonals of a regular pentagon. In the middle of a pentagram, the lines bound another pentagon. The possibility of repeatedly inscribing pentagrams inside one another is the key element in the proof of the following theorem.

Theorem. The side and diagonal of a pentagon are incommensurable.

PROOF: Let s_1 and d_1 denote the side and diagonal respectively of the pentagon $ABCDE$ (see Figure 1.8). By using elementary geometry, in particular the properties of isosceles triangles, it is easy to show that $AB = AL$ and $LC = LN$.

Let s_2 and d_2 denote the side and diagonal of the inner pentagon $JKLMN$. Then, since

$$AC \quad = \quad AL + LC \quad = \quad AB + LN,$$

we have

$$d_1 \quad = \quad s_1 + d_2.$$

Rearranging, we get that

$$d_1 - s_1 \quad = \quad d_2.$$

Hence, if d_1 and s_1 have a common measure then this unit will also measure d_2.

We also have that $AK = LC = LN$ and $AL = AB$. So

$$AL = AK + KL$$
$$\Leftrightarrow \qquad AB = LN + KL$$
$$\Leftrightarrow \qquad s_1 = d_2 + s_2$$
$$\Leftrightarrow \quad s_1 - d_2 = s_2.$$

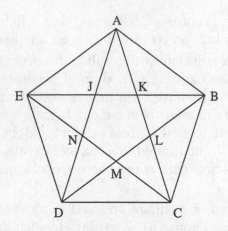

Figure 1.8.

Since we know that a common measure for s_1 and d_1 would also measure d_2, this last equality shows that it would measure s_2 as well. Therefore, s_2 and d_2 can be measured with the same unit as s_1 and d_1.

This process can be repeated to produce a ceaseless stream of smaller and smaller pentagons. A line segment which is a common measure for the side and diagonal of the initial pentagon $ABCDE$ would also measure the sides and diagonals of all the smaller pentagons. Since these can be made as small as desired, this is clearly impossible. Hence such a measuring segment cannot exist and s_1 and d_1 are incommensurable.[1] ■

Tradition assigns the discovery of incommensurability to the Pythagorean Hippasus of Metapontum (early fifth century BC). The Pythagoreans were familiar with regular polygons such as the pentagon, and also with the pentagram, or star pentagon, which they used as a badge of recognition.

An argument such as that used to prove the preceding theorem could have been understood by the Pythagoreans. It uses only the basic propositions of geometry: such fundamental results as the angle sum of a triangle and the properties of isosceles triangles. These results would have been known in Hippasus' time even if the arguments used to justify them were primitive and not up to the standards demanded by later mathematicians. All that was necessary was to present an argument which would convince its intended audience.

That Hippasus managed to demonstrate the existence of incommensurables to his contemporaries is supported by the legend that he perished at sea. Some versions say that he was punished by the gods for having done violence to the

---

[1]A similar argument for showing that the side and diagonal of a square are incommensurable is given in T. L. Heath, *The Thirteen Books of Euclid's Elements*, Cambridge Univ. Press 1908, volume 3, pp19–20.

Pythagorean dream with the introduction of 'inexpressible' ratios, others that he attempted to take glory in his discovery when traditionally all results were ascribed to Pythagoras himself, or that he betrayed the secret knowledge outside the brotherhood.

The discovery of incommensurability must have been a huge shock to the Pythagoreans for it demolished their belief that everything could be expressed in integer ratios. It is remarkable that this one result sufficed to end Pythagorean numerology, especially when it is realised that incommensurability only makes sense as a part of *pure* geometry, as an abstract notion in the realm of ideas. Incommensurability can only be justified by theoretical constructions, never by empirical methods or visual evidence. In a practical sense, the process of constructing ever smaller pentagons must terminate because, at some stage, it becomes physically impossible to construct any more. When this happens, the side and diagonal of the pentagon can be considered equal for any practical purpose. However, when considered in a mathematical context, the construction is ideal, and it is theoretically possible to continue the process indefinitely, constructing a never-ending sequence of ever-smaller pentagons.

This confrontation between practical experience and mathematical idealism pushed mathematics further towards the abstract. The visual evidence contained in diagrams could no longer be relied upon; empirical methods were rejected as people sought a more intellectual approach.

The nature of space

The discovery that some geometrical figures contain lines which are incommensurable focused attention on the properties of space. In particular, the problem of whether space can be continually subdivided or whether it ultimately becomes indivisible occupied the philosophers.

The notion that a line is composed of points underlies our understanding of the terms 'line' and 'point'. Their meanings are derived from the experience of drawing diagrams. The dot which we make to represent a point has a certain size. But is the mathematical idealisation abstracted from it extended in space? Early Greek philosophers seem to have regarded it so. This imparts a granular structure to space. If space is quantised into discrete units then a line contains only finitely many points. Assuming such points to be uniform in size, all lines would be commensurable, and the calculation of area or volume would be reduced to counting the number of atomic constituents.

The hypothesis that space is composed of extended indivisible units is called the *discrete* theory. The alternative, called the *continuous* theory, is to assume that space can be subdivided indefinitely. A consequence of this is that any line segment, no matter how small, contains infinitely many points.

Early Chinese philosophers seem to have recognised the two alternative theories. The following passage from the *Mo Ching*, a text believed to date from *c*.330BC, attempts to define a point in the discrete theory of space:

> *If you cut a length continually in half, you go on forward until you reach the position that the middle [of the fragment] is not big enough to be separated any more into halves; and then it is a point. · · · Or, if you keep on cutting into half, you will come to a stage in which there is an 'almost nothing', and since nothing cannot be halved, this can no more be cut.*[g]

Another writer describes the continuous theory:

> *A one foot-long stick, though half of it is taken away each day, cannot be exhausted in ten thousand generations.*[h]

The famous paradoxes of the Greek Zeno of Elea (*c*.490–*c*.425BC) concern the problem of the divisibility of space and time. It is difficult to identify the original thrust of the thought experiments Zeno described since they only survive in abbreviated accounts by later writers. He seems to parody both the discrete and continuous theories and deduces that the concept of motion is inconceivable.

In the 'paradox of the runner' Zeno argues against the continuous theory of space. In order to run from A to B, the runner must first reach the point halfway between. He must then reach the midpoint of the remaining distance, and so on passing a never-ending sequence of midpoints. Thus, to complete the course, the runner has to cover infinitely many distinct distances. Zeno felt that if something involved infinity then he had reached a contradiction.

In the 'paradox of the arrow' he argues against the discrete theory of space and time. Under this hypothesis, motion would consist of a succession of small jerks. At each instant of time, the arrow is in a particular place. What distinguishes a moving arrow from a stationary one? How does the arrow 'know' whether or not to jerk forward at the next instant?

When a thought experiment reaches a puzzling or nonsensical conclusion it often pays to examine the underlying assumptions on which it is based. If the reasoning is correct then what appears at first sight to be paradoxical may be resolved if the hypotheses are modified or replaced. The new hypotheses may seem strange; we may need to abandon a cherished belief; the conclusion may be surprising. Yet, if such changes make the argument consistent, it is an indication that our intuitive understanding may be at fault.

The Greeks found the concept of infinity disturbing. They tried to separate notions involving the infinite into 'potentially infinite' and 'actually infinite'. An example of the distinction is as follows: A line segment is bounded and of finite length. It can be extended by any finite amount an unlimited number of times.

The line can be made as long as required but, at every stage, the line always has a definite length. This is a potential infinity. To consider the 'whole line' would be using an actual or completed infinity.

In his runner paradox, Zeno expressed a finite line segment as a sum of infinitely many shorter segments—as a completed infinity. Although this may be disquieting, it is a consequence of the continuous theory of space. If we assume that any line segment, no matter how small, can be divided to produce two shorter segments then we are forced to accept that a line segment contains infinitely many points arranged in a continuum.

The alternative hypothesis (the discrete theory) leads to an even more puzzling situation. Zeno's arrow paradox leads to the conclusion that rest and motion are indistinguishable. We cannot tell whether a body is in motion or not until the next instant arrives and we can see whether it is occupying the same position.

In the realm of mathematics and ideal constructions the continuous theory is used. The peculiar nature of this mathematical myth is neatly summed up by Aristotle:

> The continuum is that which is divisible into indivisibles that are infinitely divisible.[i]

Democritus' dilemma

The problem of repeated subdivision arose not only in plane geometry with the construction of ever-smaller pentagons and line segments. It also appeared in solid geometry in the calculation of volumes.

Democritus of Abdera (late fifth century BC) took an interest in both science and mathematics. Although none of his works survives, he is known to have discussed the atomic theory of matter and contributed to the debate on the divisibility of space. He is said to have regarded a sphere as being really a polyhedron with imperceptibly small faces. This idea is probably connected with the atomic theory of matter rather than space. A material sphere composed of indivisible atoms would presumably have this property; a mathematical sphere is an ideal object with a perfectly smooth curved surface. The distinction between mathematical objects and their imperfect material counterparts was discussed by Plato (427–347BC) but was certainly known earlier. According to Archimedes, it was Democritus who first understood the problem of finding the volume of a pyramid though his arguments did not constitute a rigorous proof.

The Greeks had no word for volume and so they used the name of the object itself to denote its quantity. Thus in Greek mathematics, the pyramid formula is expressed as: 'A pyramid is one third of the prism having the same base-area and height'.

Suppose, for the moment, that the volume of a pyramid is proportional to its base-area and height. (In the light of other volume formulae, this is not unreasonable.) If we let V, A and h denote the volume, base-area, and height of a pyramid then we have assumed that $V = kAh$, where k is the constant of proportionality. Now, Ah is the volume of a prism whose base-area is A and whose height is h. There is a straightforward argument which shows that $k = \frac{1}{3}$.

Theorem. The volume of a pyramid is one-third the volume of a prism having the same base and equal height.

PROOF: Since any polygon can be split up into triangles, it is sufficient to prove the theorem for triangular-based pyramids as any other pyramid can be split up into these.

The triangular prism $ABCPQR$ can be dissected into three pyramidal pieces as shown in Figure 1.9. If we can show that all three pieces have the same volume then it follows that each pyramid has one-third the volume of the prism.

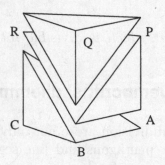

Figure 1.9. Dissection of a prism into three pyramids.

We are assuming that the volume of a pyramid is proportional to the product of its base-area and height. Thus pyramids whose bases and heights are equal will have equal volumes.

First we show that the pyramids $ABCR$ and $PQRB$ have the same volume. Regard the former as having base ABC and apex R, and the latter as having base PQR and apex B. Then, since the bases are opposite faces of the prism, they have equal area. The heights are equal since they are the height of the prism. As the bases and heights are equal, the volumes must also be equal.

To show that the third pyramid has the same volume as the others, we regard the pyramid $BPQR$ as having base BPQ and apex R, and the third pyramid as having base APB and apex R. Since the bases are each half the rectangle $ABQP$ they have equal area. And since the bases are coplanar and R is the apex of both pyramids, their heights are equal. Hence, they have equal volumes.

Thus the three pyramids have equal volumes. ■

It remains to show that the assumption on which this theorem rests is valid: that the volume of a pyramid is proportional to its base-area and height. This is not easy to justify rigorously. There is, however, an intuitive argument that makes the result seem plausible. If we imagine two pyramids having the same base and equal height, each being divided up into thin laminae, then, since the sections at equal heights will be congruent (why?), they will have the same volume (Figure 1.10). Therefore, the wholes are identical sums of equal volumes, and are therefore equal.

Figure 1.10. Layers at equal heights are congruent. Sliding the layers over one another suggests that the pyramids have equal volume.

However, in order for such a procedure to be 'accurate' the sections must be 'infinitesimally thin' and consequently the pyramid is composed of infinitely many such layers. Again we run into the problem of the continuity of space. What is the 'volume' of each elemental section which has no thickness?

Although this thought experiment does not furnish a rigorous proof, it does give an indication of what might be proved. This is important, for it is easier to supply a proof when, from previous experience or reflection, a particular result is expected to be true. Archimedes writes

> It is easier to supply the proof when we have previously acquired ···
> some knowledge of the questions than it is to find it without any previ-
> ous knowledge. That is the reason why, in the case of the theorems the
> proofs of which Eudoxus was first to discover, namely on the cone and
> the pyramid, that the cone is one-third of the cylinder and the pyra-
> mid one-third of the prism having the same base and equal height, no
> small share of the credit should be given to Democritus, who was the
> first to state the fact about the said figure[s], though without proof.[j]

And, as we shall see later, in order to apply Eudoxus' method of proof, it is necessary to know *a priori* what answer is expected.

That Democritus argued along lines similar to those outlined above is suggested in a passage in Plutarch's *De Communibus Notitiis Adversus Stoicos* (in which he argued against the common notions used by the Stoic philosophers). He records that Democritus raised the following question:

> *If a cone were cut by a plane parallel to the base, what must we think of the surfaces of the sections, that they are equal or unequal? For if they are unequal, they will make the cone irregular, as having many indentations, like steps, and unevennesses; but, if they are equal the sections will be equal, and the cone will appear to have the property of the cylinder and to be made up of equal, not unequal circles, which is very absurd.[k]*

The idea that a solid is a sum of infinitely many plane sections, all parallel and of negligible thickness, clearly underlies Democritus' argument. His dilemma again centres on the difference between atomism and continuity. Some say that in this passage Democritus is not posing a genuine dilemma but is arguing for the atomic theory of matter: a cone is obviously not a cylinder and therefore it must have 'steps'. However, his statement applies equally well to ideal mathematical cones and imperfect physical ones. This single fragment does not tell us the nature of Democritus' discussions or his conclusions, only that he was aware of the conceptual difficulties involved in one of the major problems occupying the philosophers of his time.

Liu Hui on the volume of a pyramid

Earlier we saw how Liu Hui derived the formula for the volume of a truncated pyramid by dissecting it into a central square-based prism, four ch'ien-tu and four yang-ma. The formulae for the prism and the ch'ien-tu are straightforward to demonstrate. Justifying the formula for the yang-ma proves to be more difficult. The procedure described in the *Nine Chapters* is to multiply the breadth and length together, multiply by the height, then divide by three. Liu's commentary on the problem gives an ingenious method for proving this formula is correct. It involves the use of repeated dissection which gives rise to a never-ending process. Not surprisingly, Liu has some problems trying to express what ultimately happens as the pieces become unimaginably minute. That he presents any proof at all shows him to be one of the greatest masters of empirical geometry.

The first part of the demonstration consists of fitting together a yang-ma with a pieh-nao as shown in Figure 1.11 to form a wedge (or ch'ien-tu) whose volume is known to be $1/2\,abh$. If Y denotes the volume of the yang-ma, and P denotes the volume of the pieh-nao then

$$Y + P = 1/2\,abh.$$

He wants to show that $Y = 1/3\,abh$. This would imply that $P = 1/6\,abh$ or that $Y = 2P$. It is this last equality that Liu attempts to demonstrate.

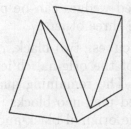

Figure 1.11. Dissection of a ch'ien-tu into a yang-ma and a pieh-nao.

He proceeds by subdividing the yang-ma and the pieh-nao into smaller components. He writes

> *To make a pieh-nao with breadth, length, and height each 2 ch'ih,[2]*
> *use two ch'ien-tu and two pieh-nao blocks, all of them red.*
> *To make a yang-ma with breadth, length, and height each 2 ch'ih, use*
> *one cubical block, two ch'ien-tu blocks, and two yang-ma blocks, all*
> *of them black.*
> *Joining together the red and black blocks to make a ch'ien-tu, the*
> *breadth, length and height are each 2 ch'ih.[1]*

These dissections are illustrated in Figure 1.12.

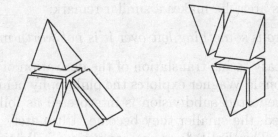

Figure 1.12. Subdivisions of the yang-ma and pieh-nao into smaller ch'ih.

To show that $Y = 2P$ we need to show that the volume of the black pieces is twice the volume of the red pieces. Liu rearranges the blocks so that the volumes of the red and black pieces can be compared.

The two red ch'ien-tu can be placed together to form a cube. Likewise, the two black ch'ien-tu form a cube. The remaining pieces comprise the black cube, two black yang-ma and two red pieh-nao. The four latter blocks can be fitted together to form a pair of ch'ien-tu coloured in the same way as the original

---

[2]One ch'ih is about 21cm.

wedge. These two black-and-red wedges can be placed together to form a cube having the same volume as the three others.

We have constructed four cubes: two black, one red, and one mixed. Thus three-quarters of the volume of the original ch'ien-tu is of known to be in the proportion black : red $= 2 : 1$. The remaining quarter of the volume comprises two black yang-ma and two red pieh-nao blocks. These fit together to produce two scaled down versions of the original black-and-red ch'ien-tu. The dissection process can be repeated on each of these wedges. This shows that three-quarters of the undetermined volume is in the ratio black : red $= 2 : 1$. Therefore, we know that $\frac{3}{4} + \frac{1}{4} \times \frac{3}{4}$ of the volume is in the desired ratio; only $\frac{1}{4} \times \frac{1}{4} = \frac{1}{16}$ is left undetermined.

By repeating this process indefinitely, continually subdividing the yang-ma and pieh-nao blocks, the unknown amount decreases, becoming arbitrarily small and tending to zero. Liu expresses this idea like this:

> To exhaust the calculation, halve the remaining breadth, length, and height; an additional three-quarters can thus be determined.
> The smaller they are halved, the finer are the remaining [parts]. The extreme of fineness is called "subtle" [or intangible]; that which is subtle is without form. When it is explained in this way, why concern oneself with the remainder?[m]

In another problem where Liu is approximating a curved figure by a polygonal one to determine its area, he makes a similar remark:

> Although there is something left over it is not worth mentioning.[n]

In the commentary to his translation of the derivation of the volume formula for the yang-ma, Donald Wagner explores the philosophy behind these ideas. The endless process of repeated subdivision is interpreted as follows. The more the pieces are subdivided, the smaller they become. Ultimately, they become intangible and formless. This limit does not comprise a collection of dimensionless, infinitely small yang-ma and pieh-nao but rather a collection of formless objects which are unimaginable and beyond description. It is meaningless to speak of their dimensions. As they cannot be examined, why bother with them?

These concepts are deeply rooted in Chinese philosophy. Wagner uses an extract from Chapter 14 of Lao Tsu's *Tao Te Ching* to illustrate the use of the ideas in another context:[3]

> Look, it cannot be seen—it is beyond form.
> Listen, it cannot be heard—it is beyond sound.

---

[3]I have chosen to use a different translation since I find this version more poetic than the one used by Wagner.

Grasp, it cannot be held—it is intangible.

These three are indefinable; therefore they are joined in one.

From above it is not bright;
From below it is not dark:
An unbroken thread beyond description.
It returns to nothingness.
The form of the formless
The image of the imageless
It is called indefinable and beyond imagination[o]

Eudoxus' method of exhaustion

The problem of how to treat the convergence of infinite processes rigorously was solved by Eudoxus of Cnidus (*c*.408–355BC). He was a contemporary of Plato and contributed to the development of many fields including geometry, medicine and astronomy. His method bears the name of 'exhaustion' but that is something of a misnomer. For although he uses the fact that, by repeated subdivision, the difference between an actual volume and a computable approximation to it can be made arbitrarily small, he does not need to pursue the calculation to its limit and 'exhaust' the volume. Rather, to prove that a solid has volume V, he shows that any value for the volume other than V can be proved incorrect in finitely many steps.

This technique was seen as an outstanding achievement. The attribution of the discovery of the method of exhaustion to Eudoxus was made by Archimedes who used it as a standard with which to compare his own work. He regarded his own work *On the Sphere and Cylinder* as his most beautiful mathematical result, so much so that he requested to have a figure illustrating the result carved on his tombstone. He wrote in the preface to this work that he had no hesitation in setting his results alongside those

> *theorems of Eudoxus on solids which are held to be most irrefragably established, namely that any pyramid is one third part of the prism which has the same base with the pyramid and equal height, and that any cone is one third part of the cylinder which has the same base with the cone and equal height.*[p]

In order to illustrate Eudoxus' method we need a dissection of a pyramid into parts of known volume and a remainder which can be further subdivided. It is sufficient to consider only pyramids on a triangular base for other pyramids can be produced as combinations of these. Figure 1.13 illustrates how a triangular-based pyramid can be dissected into two smaller pyramids, equal to each other and similar to the original one, and two prisms. The vertices of the component pieces bisect the edges of the large pyramid. The volumes of the two prisms are equal, and together they occupy more than half the volume of the pyramid. Furthermore, it is possible to show that the volumes of these prisms depend only on the base-area and the height of the pyramid: if two pyramids have bases of equal area and are of the same height, and if each of them is dissected in this manner, then the resulting prisms all have the same volume.

Figure 1.13. Dissection of a pyramid into two similar pyramids and two prisms.

As the two smaller pyramids are just scaled down versions of the original one, each of them can be subdivided to produce four more prisms and four more pyramids. This process can be continued indefinitely.

We are now set up to use the 'method of exhaustion' to show that the volumes of two pyramids, equal in height, are in the same ratio as the areas of their bases. Hence, when the base-areas are equal, the two pyramids occupy equal volumes. As in the previous theorem demonstrating the existence of incommensurable magnitudes, the indirect method of proof is used. The proof also relies on the following 'axiom of continuity':[4]

> Given two unequal quantities, U and V, where U is less than V, if we remove at least half of V, then remove at least half of the remainder, and so on repeating the process continually, eventually we will reach a quantity less than U.

In particular, suppose that a pyramid has volume V, and that W is an approximation to V which underestimates the volume. Then we can choose $U = V - W$. In the dissection shown in Figure 1.13, the two prisms occupy more than half the volume. The axiom states that by repeatedly dissecting the resulting pyramids and removing the prisms, we shall eventually arrive at a collection of pyramids whose total volume is less than U—the error in the approximation.

---

[4] Although this form of the axiom is most convenient for use here, it is worth noting that simpler but equivalent statements have been proposed as substitutes. One such is due to Archimedes:

Given two unequal quantities, the larger exceeds the smaller by an amount which if repeatedly added to itself will exceed any prescribed quantity.

Theorem. Two triangular-based pyramids of equal height have their volumes in the same ratio as the areas of their bases.

PROOF: Let P_1 and P_2 be two triangular-based pyramids of equal height whose bases have areas B_1 and B_2, respectively. Let V_1 and V_2 denote their respective volumes. Then we need to show that the ratios $B_1 : B_2$ and $V_1 : V_2$ are equal.

Suppose that this is not the case. Then there is some volume W such that

$$B_1 : B_2 \quad = \quad V_1 : W.$$

Since W is not equal to V_2, it is either less than or greater than V_2.

Suppose that W is less than V_2.

The pyramid P_2 can be dissected into two prisms and two pyramids, the latter of which can be further subdivided. The process can be repeated until the volume of the remaining pyramids is less than $V_2 - W$ (applying the axiom of continuity). Then

$$V_2 > \text{(volume of prisms in } P_2) > W.$$

Partition the pyramid P_1 in the same way repeating the subdivision process the same number of times. Now, since the volume of the prisms in such a dissection depends only on the height of the pyramid and the area of its base, and since P_1 and P_2 have equal height,

$$\text{(volume of prisms in } P_1) : \text{(volume of prisms in } P_2) \quad = \quad B_1 : B_2.$$

By hypothesis

$$B_1 : B_2 \quad = \quad V_1 : W$$

so, consequently,

$$\text{(volume of prisms in } P_1) : V_1 \quad = \quad \text{(volume of prisms in } P_2) : W.$$

But this leads to a contradiction since

$$\text{(volume of prisms in } P_1) \quad < \quad V_1$$

which implies

$$\text{(volume of prisms in } P_2) \quad < \quad W,$$

and, we know by construction that

$$\text{(volume of prisms in } P_2) \quad > \quad W.$$

Therefore, the supposition that $W < V_2$ must be false.

A similar contradiction can be derived if we assume that $W > V_2$. Since W can be neither less than nor greater than V_2, the two quantities must be equal. Hence

$$B_1 : B_2 \quad = \quad V_1 : V_2. \qquad \blacksquare$$

As a simple corollary to this theorem, we see that two triangular-based pyramids of equal height and equal base-areas have equal volumes. We saw earlier that from this key proposition, it is straightforward to show that a pyramid has one third the volume of a prism on the same base and equal height.

Eudoxus' method was an outstanding achievement. However, although the method is remarkably ingenious, and neatly avoids the problem of never-ending calculations, it suffers one major drawback: before the method can be used it is necessary to know the target formula. A proof using the method of exhaustion proceeds by eliminating all possibilities except the desired result, deriving a contradiction in a finite number of steps. Such a proof is *non-constructive*—it does not produce an answer. The answer has to be known in advance. Eudoxus' method is useless as a way of discovering new results. It can only be used to provide rigorous justification for results already suggested by other evidence. How is such prior knowledge obtained?

Thought experiments of the kind we used earlier are one source of data. Regarding a pyramid as a stack of thin laminae suggested that its volume depended only on its height and base measurements—not on its shape. We can imagine the pyramid as a stack of infinitely many planar sections so that we are no longer using an approximation but the actual volume.

Arguments such as this may not rest on such a logically firm foundation as Eudoxus' method does, but they do have the big advantage of providing an answer. Furthermore, they provide an intuitive picture of why the theorem is true, an element missing in the rigorous proof.

Archimedes used the method of exhaustion to prove many theorems concerning the areas of curved regions and the volumes of solids. And although no-one could doubt the truth of his results, it was impossible to tell how he had discovered them. He later divulged his secret source in a letter to Eratosthenes. In it he described the kind of limiting process we used on the pyramid. We see again that, in mathematics, discovery and proof are frequently two different activities: 'Democritus has found the theorem, but only Eudoxus has proved it'.

Hilbert's third problem

All the arguments seen so far which attempt to justify the validity of the formula for the volume of a pyramid have one thing in common: they all make use of infinitesimally small quantities and ability to go to a limiting case. Recall Democritus' paradox of the cone: is it a cylinder or a 'devil's staircase' whose steps are extremely thin? Liu Hui regarded the end of his dissection process as something formless, dimensionless, unimaginable and unexaminable. Consequently he asks 'Why concern oneself with it?'. Eudoxus' method of exhaustion requires the existence of a procedure by which pieces of arbitrarily small volume can be constructed. Does every proof justifying the volume formula necessarily involve such complicated concepts, or is there a subtle argument which avoids the use of limits and infinite processes?

In the geometry of plane figures such processes are not needed. If two polygons have the same area then they are *congruent by dissection* or *equidecomposable*. This means that they can be dissected into the same set of pieces; they are both solutions of a common jigsaw. Figure 1.14 shows dissections of some regular polygons which can be reassembled to form a square. In fact, to show that every pair of polygons are equidecomposable, it suffices to show that any polygon is equidecomposable with a square since polygons that are equidecomposable to the same thing are equidecomposable with each other.

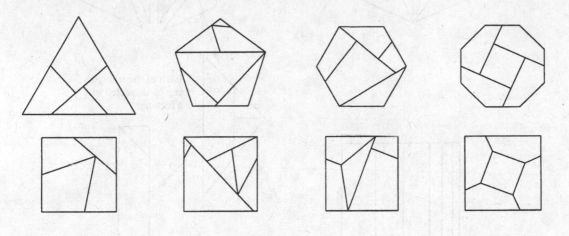

Figure 1.14.

This theorem, that any two polygons of equal area are congruent by dissection, was proved by several people working independently of one another: William Wallace discovered the essential ideas in 1807, and Farkas Bolyai (father of Janos Bolyai who did early work on non-Euclidean geometry) and P. Gerwien produced their proofs in the early 1830's. The dissections shown in Figure 1.14 are interesting in that they use small numbers of pieces. To find a jigsaw with the minimum number of pieces requires great ingenuity. However, to prove that any polygon can be dissected and the pieces rearranged to form a square requires a general strategy that can be applied in all situations. Such a process will not produce imaginative, minimal jigsaws. A description of one method for constructing a dissection is outlined in Figure 1.15.

Since any polygon can be converted into a square of equal area, and the area of a square is easily calculated, a theory of area measurement can be developed without recourse to a limiting process. Is the same true for a theory of volume?

Are any two polyhedra with the same volume equidecomposable? Can a pyramid be dissected and reassembled to form a cube? Is it just that mathematicians have not been lucky enough to find, or crafty enough to devise, a method for dissecting two given polyhedra into sets of equal pieces, or is such a task not always possible?

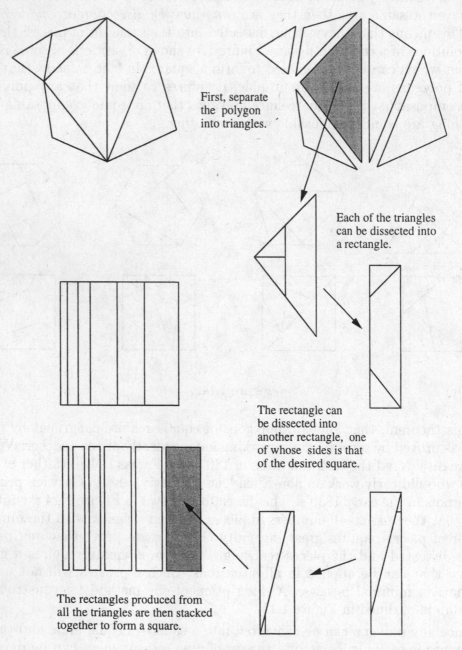

First, separate the polygon into triangles.

Each of the triangles can be dissected into a rectangle.

The rectangle can be dissected into another rectangle, one of whose sides is that of the desired square.

The rectangles produced from all the triangles are then stacked together to form a square.

Figure 1.15. How to dissect a polygon into a square of equal area.

At the beginning of the twentieth century, David Hilbert (1862–1943) compiled a list of 23 problems which he considered to be the major unsolved problems of the time and which most deserved to be worked on in the dawning century. He presented them in a famous report to the second International Congress of Mathematicians held in Paris in 1900. In his third problem he called attention to the fact that some kind of limiting process seemed to be necessary to establish a theory of volume for polyhedra. The essence of the problem was to justify the use of limits and show that without them a theory of volume is not possible. The problem reads:

> *In two letters to Gerling, Gauss expresses his regret that certain theorems of solid geometry depend upon the method of exhaustion, i.e., in modern phraseology, upon the axiom of continuity (or upon the axiom of Archimedes). Gauss mentions in particular the theorem of Euclid, that triangular pyramids of equal altitudes are to each other as their bases. Now the analogous problem in the plane has been solved. Gerling also succeeded in proving the equality of volume of symmetrical polyhedra by dividing them into congruent parts. Nevertheless, it seems to me probable that a general proof of this kind for the theorem of Euclid just mentioned is impossible, and it should be our task to give a rigorous proof of its impossibility.*[q]

Hilbert goes on to say that such a proof would be obtained as soon as a counterexample were found—when two polyhedra were discovered which could not be dissected into congruent pieces and the impossibility of such a dissection demonstrated.

Some polyhedra are equidecomposable. Any prisms of the same height whose bases have equal areas are equidecomposable (by the polygon result). In 1844, C. L. Gerling had shown that two mirror-image polyhedra are equidecomposable by dissecting them into congruent sets of acheiral pieces. (This is what Hilbert is referring to when he speaks of dividing symmetrical polyhedra into congruent parts.) Other examples of individual cases of pairs of equidecomposable polyhedra were known in 1900. In 1896 M. J. M. Hill gave examples of tetrahedra that are equidecomposable with a cube. One of these has the form of the 'pieh-nao' Chinese block. Figure 1.16 illustrates the transition from a pieh-nao to a prism with the same base and a third of the initial height. However, Hilbert felt that these were special cases, the exception rather than the rule.

This conjecture was soon confirmed. Even before Hilbert's problems had appeared in print, Max Dehn (1878–1952) announced that he had solved the problem and exhibited two polyhedra of equal volume which are not congruent by dissection. Following the discovery of an 1896 paper by the French engineer Raoul Bricard, the Russian mathematician V. F. Kagan produced a simplified and more systematic exposition of the result. Bricard's work specifies a condition

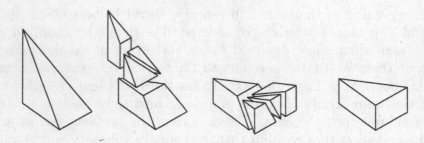

Figure 1.16. The 'pieh-nao' tetrahedron is equidecomposable with a prism
on the same base.

on the dihedral angles of two polyhedra which must be satisfied if they are equidecomposable. Unfortunately, his proof of the necessity of this condition rests on
an erroneous assumption.

The key to Dehn's proof is to associate a number (now called the Dehn invariant) to each polyhedron which is left unchanged by dissection and reassembly of
the pieces: two polyhedra which are congruent by dissection must have the same
Dehn invariant. The crux is that not all polyhedra with the same volume have
the same Dehn invariant. This happens for a regular tetrahedron and a cube, for
example, showing that these two solids are not equidecomposable. The converse,
that polyhedra with the same Dehn invariant are congruent by dissection, is also
true.

Hilbert's conjecture was correct. Unlike the case of polygons, there are polyhedra of equal volume that are not equidecomposable. Therefore, to formulate
a theory of volume we cannot use dissection arguments. To rigorously establish
the volume formulae for some polyhedral solids it is necessary to use some kind
of non-elementary method: the infinite is unavoidable.

From *Perspectiva Corporum Regularium* by Wenzel Jamnitzer, 1568.

2

Rules and Regularity

*It should be clear that the man who first intro-
duced the notion of regular solid made a significant contribution to mathematics.*[a]

W. C. Waterhouse

The oldest surviving discussion of polyhedra in a philosophical rather than a practical context occurs in one of Plato's dialogues: *Timaeus*. This is Plato's account
of the world we live in. After a brief introduction, in which the four characters set
the scene and discuss the mythical island of Atlantis, Timaeus takes the floor to
tell 'the story of the universe till the creation of man'. Through the voice of this
Pythagorean, Plato discusses the origins and workings of natural phenomena. His
enquiries range from astronomy and the motions of the heavenly bodies to the
anatomy and physiology of humans. Polyhedra appear in his detailed discussion
of the structure of matter.

The Platonic solids

The idea that matter is composed of a few elemental substances combined in
different ways had been proposed by writers of the fifth century BC. For some,
a single element sufficed: water, for example, has many forms—ice, steam, rain,
snow, dew—and is essential to life. Empedocles maintained that there were four
elements: water, earth, air and fire. He thought that the many different substances were formed by combining these in different proportions, like combining
a few letters in many ways to produce a large variety of words. (The genetic code
demonstrates the immense variety that can be expressed with a four-letter alphabet.) Leucippus and Democritus also proposed atomic theories of matter. Plato
(427–347BC) considered what makes these elements different from each other.

He suggested that they correspond to different kinds of fundamental particle. (This idea was revived only in the nineteenth century by John Dalton—one of the founders of modern chemistry.)

Plato explained that fire, earth, air, and water are bodies; bodies are solids; solids are bounded by plane surfaces, and these are composed of triangles. He chose two fundamental triangles, both right-angled: isosceles and scalene. The former is unique but of the unlimited number of scalene right triangles he selected 'the most perfect ··· that of which a pair compose an equilateral triangle.' He then proceeded to construct the four 'most perfect figures.'

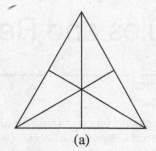
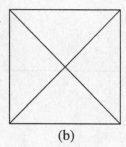

(a)   (b)

Figure 2.1.

We will begin with the construction of the simplest and smallest figure. Its basic unit is the triangle whose hypotenuse is twice the length of its shorter side. If two of these are put together with the hypotenuse as the diameter of the resulting figure, and if the process is repeated three times and the diameters and the shorter sides of the three figures are made to coincide in the same vertex, the result is a single equilateral triangle composed of six basic units [see Figure 2.1(a)]. And if four equilateral triangles are put together, three of their plane angles meet to form a single solid angle ··· and when four such angles have been formed the result is the simplest solid figure which divides the surface of the sphere circumscribing it into equal and similar parts.[b]

Prosaic accounts such as this passage by Plato are common in ancient texts. Because works did not contain illustrations, the writers' descriptions had to be sufficiently detailed to enable their audience to construct their own diagrams where necessary. The polyhedron described in this passage is shown on the left of Figure 2.2. When it rests on one of its faces it is more clearly a pyramid—it looks the same whichever face it is resting on.

If all the vertices of a polyhedron lie on a sphere then the sphere is said to *circumscribe* the polyhedron, and is called its *circumsphere*. The four vertices of Plato's pyramid lie on a sphere. Connecting these four points by arcs of great

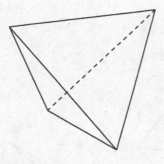
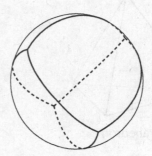

Figure 2.2.

circles divides the circumsphere into four equal parts as shown on the right of Figure 2.2.

Two of the other polyhedra that Plato describes are also constructed from equilateral triangles. One has a total of eight faces arranged so that each of the six solid angles is surrounded by four plane angles; in the other, the twenty faces form twelve solid angles, each bounded by five plane angles. See Figure 2.3. Plato continues:

> *After the production of these three figures the first of our basic units is dispensed with, and the isosceles triangle is used to produce the fourth body. Four such triangles are put together with their right angles meeting at a common vertex to form a square [see Figure 2.1(b)]. Six squares fitted together complete solid angles each composed by three plane right angles. The figure of the resulting body is the cube, having six plane square faces.*
> *There still remained a fifth construction, which the god used for embroidering the constellations on the whole heaven.*[c]

The last sentence in this passage, which seems to have been inserted as an afterthought, refers to a solid formed from twelve regular pentagons. Its inclusion reflects the fact that there are precisely five polyhedra bounded by regular polygons arranged in a regular manner. They are illustrated in Figure 2.3 and are collectively known as the *Platonic solids*, *cosmic figures*, or *regular polyhedra*.

Task. Many mathematicians have been fascinated by this group of five polyhedra and have studied their properties. They feature prominently in the following chapters, both in the theory and as a source of examples. Even if you make no other models, you should make yourself a set of Platonic solids.

Plato's two fundamental triangles cannot be put together to form a pentagon, so the last figure does not occur as one of his elementary particles. Instead

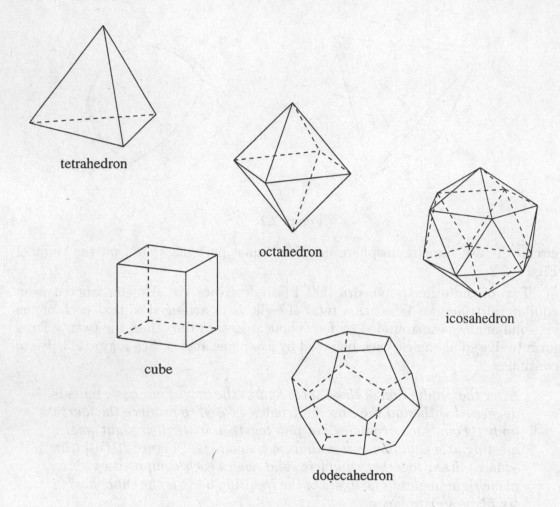

tetrahedron

octahedron

icosahedron

cube

dodecahedron

Figure 2.3. The five Platonic solids.

he employs it to hold the constellations of stars. Plutarch, in the fifth of his *Platonic Questions*, asks why Plato discarded the most perfect of figures (the sphere) and used a rectilinear figure to represent the celestial orb. He suggests that the dodecahedron can play the role of the sphere because, like the balls made from twelve pieces of leather, it is flexible and on being inflated it would become distended and spherical. He also makes the following numerological observation:

> *It has been assembled and constructed out of twelve equiangular and equilateral pentagons each of which consists of thirty of the primary triangles, and this is why it seems to represent at once the zodiac and the year in that the divisions into parts are equal in number.*[d]

Figure 2.4 shows one way to decompose a regular pentagon into 30 scalene triangles (not Plato's fundamental units). The twelve pentagons of thirty triangles can signify both the twelve houses of the zodiac each of thirty degrees, and the twelve months each of thirty days.

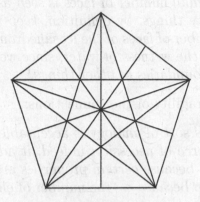

Figure 2.4.

After constructing his four fundamental particles, Plato goes on to explain how the attributes of each solid correspond to the properties of its associated element. The stability of the cube he associates with earth. The pyramid, having the fewest parts, is lightest. It also has the sharpest corners and so it is the most penetrating. These properties make the pyramid the basic unit of fire. By similar arguments, he allocates the other figures to air (octahedron) and water (icosahedron). The models shown in Figure 2.5 have been covered with tessellations of birds and fish to illustrate these relationships. A brilliant representation of the association of fire with the tetrahedron is provided by John Robinson's sculpture *Prometheus' Hearth* shown in Plate 1 but he says this is serendipitous.

Plato's associations of the elements with the regular polyhedra have inspired many illustrations. The sixteenth century plate by Wenzel Jamnitzer shown on page 50 contains a description of the octahedron in the centre panel surrounded by an assortment of things symbolising air: birds, bats, insects, windmills and wind instruments. Johannes Kepler (1571–1630) decorated his sketches (see Figure 2.6) of the five solids with symbols appropriate to each element—the cube, for example, has pictures of a tree, garden tools and a carrot. Kepler gave his own account of the connections between the solids and the elements:

> That the cube stands upright on a square base expresses stability,
> which is characteristic of terrestrial matter, whose weight tends down
> to the lowest point, while, it is commonly believed, the whole globe
> of Earth is at rest at the centre of the World. The octahedron, on
> the other hand, is viewed most appropriately suspended by opposite
> angles, as in a lathe, the square which lies exactly midway between
> these angles dividing the figure into two equal parts, just as a globe
> suspended by its poles is divided by a great circle. This is an image

of mobility, as air is the most mobile of the elements, in speed and direction.

The tetrahedron's small number of faces is seen as signifying the dryness of fire, since dry things, by definition, keep within their boundaries. The large number of faces of the icosahedron, on the other hand, is seen as signifying the wetness of water, since wetness, by definition, is held within the boundaries of other things.[e]

But he also realised the fragility of the connections:

Although, I say, this sort of analogy is acceptable, yet framed in this manner it has no force of necessity; indeed, it admits of other interpretations, not only because certain properties are at variance within the analogy, but also because · · · the number of elements and whether the Earth is at rest are matters much more open to dispute than is the number of figures.[f]

The fact that the number of elements did not match the number of solids posed a problem. Plato's evasion of the difficulty by assigning the spare solid to the heavens did not satisfy his followers and the subject was a source of much

Figure 2.5. Two models decorated to illustrate Plato's associations of elements with polyhedra.

debate at the Academy, even after Plato's death. In accounts of the theory written by later Platonists a fifth element, ether, is postulated.

Plato did not seek merely to describe nature; he attempted to explain how the four elements combine to produce the variety of matter we observe, and how the different substances change and interact. He sought a kind of 'physical chemistry'. By allowing the size of his fundamental triangles to vary he thought that the quality of the element would alter—his elements did not constitute 'being a thing' but rather 'having a quality'. Thus water was a generic form of liquid, earth and air corresponded to solid and gaseous phases of matter. He tried to explain how substances were transformed into one another. For instance, if water is heated by fire, the sharp corners of the fire particles break down the water particles into their constituent triangles and the fragments recombine to produce two air particles (steam) and a fire particle. To arrive at this 'chemical equation' Plato has balanced the number of triangles on both sides:

$$<\text{liquid}> \xrightarrow{heat} 2 <\text{gas}> + <\text{fire}>$$
$$20 = 2 \times 8 + 4$$

Even though this was extremely ingenious, it did not go uncriticised. Plato treats his elements both as solid particles and as hollow geometric shells that can be broken down. His fundamental particles disintegrate, and his elements transmute. Nevertheless, Plato's belief that mathematics can be used to understand nature has had far-reaching consequences.

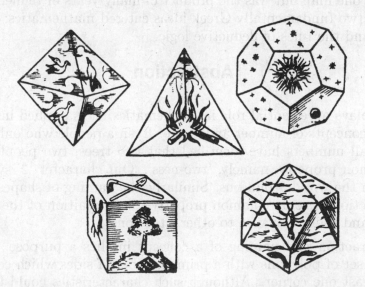

Figure 2.6. Kepler's sketches of the Platonic solids showing their associations with the elements.

The mathematical paradigm

Although Plato influenced the development of mathematics, this was not due to any significant advances of his own discovery but rather because of his enthusiasm for the subject. He encouraged his pupils to study mathematics to discipline their minds, to teach them to reason logically and provide sound arguments, as a prelude to the study of philosophy. That Plato could use mathematics in this way was part of the fall-out of the discovery of incommensurability by the Pythagoreans. This made a big impact on the development of Greek mathematics. It led to a distrust of experience and sense data, and to a greater reliance on argument to establish propositions. The degree of justification required to prove results changed with time. As the corpus of mathematical knowledge grew, mathematicians sought to verify things that had once seemed undeniable. Mathematics grew roots as well as branches. After two centuries of development, logical argument had reached the standard we find fossilised in what is probably the world's best known textbook: the *Elements* of geometry compiled by Euclid (*c*.300BC). This book was to dominate the teaching of mathematics for two millennia. Unfortunately, its very success means that we cannot trace its evolution—Euclid's *Elements* overshadowed all the previous texts to such an extent that none has survived.

Euclid's style of presentation set the paradigm for mathematical arguments that has been followed ever since: propositions are established by a series of logical deductions from explicitly stated hypotheses. This format is not the momentary inspiration of one man but was the product of many years of refinement. During its gestation, two fundamentally Greek ideas entered mathematics: the power of abstraction, and the rules of deductive logic.

Abstraction

Abstraction plays an essential role in mathematics. It is applied instinctively to generate the concepts of number and shape. Even a people who only have words for a few small numbers have identified that two trees, two people, two hands share a common property, namely, 'two-ness'. Our character '2' symbolises the many pairs in the world around us. Similarly, the naming of shapes corresponds to the abstraction of some common property—a recognition of the key qualities of an object and their relevance to other situations.

The abstraction and naming of a concept implies a purpose. There is no name for the set of polygons with a prime number of sides which contain a right angle in at least one corner. Although such characteristics could be abstracted to define a class of shapes, it has not been done because there is nothing to be gained by doing so, no role in any current theory for such polygons to occupy.

Only the need to refer to a collection gives the motivation to name it.

Abstraction can take the form of idealism. Any physical example of a cube, whether it be a natural crystal or of artificial construction, will exhibit imperfections. The mathematical notion of a cube is a purified form devoid of the inevitable defects encountered in the material world. For Plato, the idealised forms were the reality: permanent, ageless and incorruptible. The material world contained only distorted images, approximations to the genuine objects, merely representations of reality, but the mind could perceive the true unpolluted archetypal forms.

For all their preoccupation with the Platonic world, the Greeks did not dispense with sense data. They still used diagrams to assist with their arguments. Plato writes in the *Republic*:

> *Do you not know also that although they further make use of the visible forms and reason about them, they are not thinking of these but of the ideals which they resemble.* ⋯ *They are really seeking to behold the things themselves which can only be seen with the eye of the mind.*[9]

The diagrams served as reminders of the true figures. Curiously, experiments in perceptual psychology have shown that the mind 'cleans up' the data it receives. Even a roughly drawn figure can help to visualise a problem because the mind smooths out the irregularities and tries to simplify the image. 'The eye perceives but the mind will obligingly overlook.'

For a mathematician, abstraction is also an active process: it is the search for the crucial aspects of a situation. When you reach the heart of a problem, it becomes clear which data are relevant and which are redundant. The critical points can be studied in isolation without distraction. It is far easier to concentrate on particular features rather than become submerged in a mass of multifarious details. This approach has another benefit: any other object that shares the studied features will also share the properties derived from them. Discarding the inessential is the essence of mathematics.

Primitive objects and unproved theorems

The identification of properties shared by several objects leads to definitions: a name is given to the collection of all objects with particular properties in common, and to any member of it. For example, a triangle is a plane figure bounded by three lines. The definition describes how the term 'triangle' is to be interpreted, it states what the essential characteristics of a triangle are. But this is only useful if we understand the terms used in the description. We need to know something of plane figures and lines before the definition makes any sense. An attempt could be made to define lines and planes but this would only defer the problem. Eventually, if the process is not to result in an unending chain of definitions, and we are to avoid circular definitions (in which an object is defined in terms of itself), some primitive concepts must be taken as understood. Natural candidates for such primitive undefined terms are 'point', 'line' and 'space'.

In addition to accepting a few undefined terms, it is necessary to admit some statements concerning the behaviour and interrelationships of these primitive concepts. These are required to form the basis of logical arguments. Just as a definition must not be circular or infinite, so a proof must terminate. There must be some 'unproved theorems' which are admitted as hypotheses, initial statements taken on trust before a proof can get off the ground. These postulates, or *axioms* usually state things that appear self-evident and which are accepted without much trouble. For example, one such axiom states that we can draw a straight line joining any two points.

Ideally, the numbers of primitive terms and of hypotheses should be as small as possible. If one of the primitive terms can be defined in terms of the others, it is superfluous. Similarly, any axiom that can be derived from the others is redundant. The axioms must also be consistent. This means that it is impossible to prove contradictory things: every statement must be true or false—not both.

There is no objective way to choose primitive concepts and axioms. Those used by Euclid were distilled over a long period. At the beginning of book I of the *Elements* he lists his chosen axioms and purports to derive all the results of his thirteen books from just ten statements. Not surprisingly, after 2000 years, it is possible to criticise Euclid's presentation: standards of what comprises an acceptable hypothesis have changed during this time. At the end of the nineteenth century, David Hilbert revised the foundations of Euclidean geometry. He added extra axioms and supplied proofs of some of the 'self-evident' propositions which were not so easily accepted. Some of these had been so obvious to the Greeks that the possibility of doubting them, and hence needing to add them as hypotheses, would not have occurred to them. That two lines which cross have a point in common seems obvious—because we automatically attribute the quality of *continuity* to lines. Hilbert found it necessary to include axioms of continuity to ensure this.

In spite of these faults, Euclid's great achievement was to set the standard that succeeding generations of mathematicians would aspire to attain. Even though Euclid himself missed the mark on occasions, the axiomatic method developed by the Greeks has become the cornerstone of all mathematics.

The problem of existence

The purpose of a definition is to name an object, image, idea or a collection of such things. It tells us what a thing is—what properties distinguish it from other things so that we can recognise it. A definition does not, however, assert that the defined thing actually exists. It may require that the object has incompatible properties, in which case the definition is vacuous and meaningless. An example (used by Leibniz) is that of a regular polyhedron with ten faces. Any theorems about such a polyhedron are nonsense.

Apart from the primitive concepts, whose existence is assumed and whose properties are described in the axioms, the existence of a defined object must be demonstrated. This problem can be approached in two ways: in the *direct* approach the object is constructed mathematically and shown to possess the desired characteristics; the *indirect* approach seeks to show that the assumed non-existence of the object contradicts an axiom or a previously established fact. The psychological advantage of having a concrete example to experiment on, rather than just the knowledge that such an example must exist, strongly favours the first method. The Greeks certainly preferred this constructive type of existence proof.

The constructions recorded by Euclid describe how, by drawing lines and circles, the positions of the important parts of an object can be established. The steps that Euclid allowed himself are restricted by his choice of hypotheses. He provided three primitive constructions:

(1) *A* [unique] *straight line can be drawn from any point to any other point.*

(2) *A straight line segment can be continuously extended by a finite amount to produce another straight line segment.*

(3) *A* [unique] *circle may be drawn centred on any point with any radius.*

These statements encapsulate the properties of two standard drawing instruments: compasses and a straight edge. Because their properties provide the fundamental constructs in the *Elements*, these instruments are sometimes called the *Euclidean tools*. However, the properties abstracted from these implements are idealised. The straight edge would have to be unlimited in length to be able to join any two points; the compasses would have to be able to draw a circle of any radius, however large. It is also important to notice that neither of the Euclidean tools can be used to transfer distance: the straight edge is ungraduated and so is not a ruler, the compasses collapse when lifted from the page.

Even with so few operations available, a surprising amount can be achieved. Indeed, all the geometrical propositions in the thirteen books of the *Elements* are derived using only the ability to construct lines and circles. These propositions can be classified into two kinds: theorems (which demonstrate properties) and constructions (which prove existence). A construction usually has two parts: first the construction of an object is described, then the object is shown to possess the desired characteristics. The following examples, all of which are taken from the *Elements*, illustrate the process. The first example is Euclid's opening proposition in book I.

Example 1.   To construct an equilateral triangle.

Given the straight line AB, construct a circle centred on A which passes through B (postulate 3 above ensures that this can be done). Construct a second circle centred on B which passes through A, and let C be a point where the two circles meet. Draw the straight lines AC and BC (postulate 1). These steps can be followed in Figure 2.7.

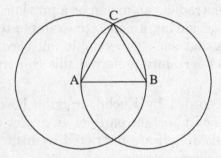

Figure 2.7. Construction of an equilateral triangle.

We now need to show that the triangle ABC is equilateral. Because AB and AC are both radii of the same circle (centred on A), their lengths are equal. Similarly the lengths of AB and BC are equal. One of Euclid's axioms is: 'things equal to the same thing are equal to each other'. Therefore $AC = AB$ and $AB = BC$ implies $AC = BC$. Hence all three sides have the same length.   ∎

Once a proposition is established it can be used in other constructions. Euclid applies his first proposition when he constructs a perpendicular to a given line.

Example 2.   To construct a perpendicular.

Suppose that AB is the given line and that the point C on AB is to be the foot of the perpendicular. The method is indicated in Figure 2.8.

Choose a point D on AC and construct a circle centred on C which passes through D. Let E be the point where the circle meets CB. Construct an equilateral triangle on DE whose third corner we shall call F. Then FC is the desired line.

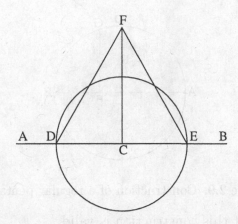

Figure 2.8. Construction of a perpendicular.

To show that FC is perpendicular to AB we must prove that angle $\angle ACF$ is a right angle. The lengths CD and CE are equal (by construction) as are the lengths FD and FE. Hence the triangles CDF and CEF have sides of the same length. In a previous proposition Euclid has shown that the angles in such triangles are also equal. In particular, angles $\angle DCF$ and $\angle ECF$ are equal. They are also adjacent to each other, and Euclid defines equal, complementary angles to be right angles. ■

This proposition forms part of a small 'tool kit' of fundamental constructions that are used many times. Other such primary constructions include finding the midpoint of a line segment and bisecting an angle.

Over the 2000 years since Euclid compiled his *Elements* people have sought to find alternatives to some of his constructions and to simplify them. The construction of a regular pentagon given below is not that given by Euclid but was found by H. W. Richmond at the end of the nineteenth century.

Example 3.   To construct a regular pentagon.

Given a line AB, let C be its midpoint. Draw a circle centred on C which passes through B. Construct the perpendicular to AB at C, and let D be the point where it meets the circle. Let E be the midpoint of CD and draw the line BE. Construct the line which bisects angle $\angle BEC$ and let F denote the point where it meets BC (see Figure 2.9).

Now construct the perpendicular to BC at F and let G be the point where it meets the circle. The points B and G are two corners of a regular pentagon inscribed in the circle. The other corners can be found by drawing a few circles. ■

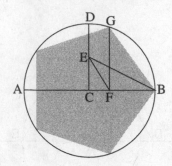

Figure 2.9. Construction of a regular pentagon.

Problem.   Verify that this construction is valid.

Although a great deal is accomplished in the *Elements*, Euclid's decision to rely on so few primitive constructions placed limitations on the kind of geometry he could perform. For example, it is impossible to construct some regular polygons using straight edge and compasses alone. A hexagon can be constructed, a heptagon cannot. An octagon can be obtained from a square by bisecting arcs on its circumcircle, but a 9-gon is impossible.

The constructibility of regular polygons with up to 25 sides is listed in Table 2.10, together with the propositions in the *Elements* for the cases covered by Euclid. Is there any pattern which separates the constructible polygons from the impossible ones? The fourth column of the table shows the decomposition of the number of sides into prime factors. The factors of 2 are isolated on the left of the column since they can be introduced or removed at will (it is easy to double the number of sides of a regular polygon, and the number of sides can be halved by joining alternate vertices). The odd primes are sorted: those in the centre of the column correspond to the constructible polygons. Are the primes correlated with constructibility? The factors 3 and 5 appear to be 'good' while 7 and 11 are not.

The pattern of which polygons are constructible, and the reasons underlying it, were unknown until 1796—two millennia after Euclid. Carl Friedrich Gauss (1777–1855) made a systematic study of the cyclotomic equation ($z^n = 1$). In geometrical terms, the solutions of this equation divide a circle into arcs of equal length, and hence locate the vertices of regular polygons. Gauss used this relationship to solve the problem of which polygons could be constructed with the Euclidean tools: it is possible to construct a regular n-sided polygon if the factorisation of n into primes has the form

$$n = 2^k \, p_1 p_2 \cdots p_r \tag{$*$}$$

where each of the odd primes p_i can be written as $2^{2^m} + 1$ for some m, and all the p_i are distinct. Prime numbers of this kind are called *Fermat primes* after the French mathematician Pierre de Fermat (1601–1665)—famous for his 'Last

Theorem'. The first few Fermat primes are

$$3 = 2^{2^0} + 1$$
$$5 = 2^{2^1} + 1$$
$$17 = 2^{2^2} + 1$$
$$257 = 2^{2^3} + 1$$
$$65537 = 2^{2^4} + 1$$

Gauss' announcement that the regular 17-gon is constructible was the first advance in this area since the *Elements* was compiled. When m is between 5 and 16, $2^{2^m} + 1$ is not a prime number so any other regular polygon which is proved constructible by this method must have an enormous number of sides ($2^{2^{16}} \approx 10^{20\,000}$). Such an object would not be constructible in any practical sense of the word.

No. sides	Constructible	*Elements*	Prime factors
3	✓	book I, 1	3
4	✓	book IV, 6	2^2
5	✓	book IV, 11	5
6	✓	book IV, 15	2 3
7			7
8	✓		2^3
9			3^2
10	✓		2 5
11			11
12	✓		2^2 3
13			13
14			2 7
15	✓	book IV, 16	3, 5
16	✓		2^4
17	✓		17
18			2 3^2
19			19
20	✓		2^2 5
21			3, 7
22			2 11
23			23
24	✓		2^3 3
25			5^2

Table 2.10. The regular polygons marked with a '✓' can be constructed with the Euclidean tools.

The converse of Gauss' result is also true: if the factorisation of n into prime factors is not of the form in equation $(*)$ then a regular n-gon cannot be constructed with straight edge and compasses. This explains why a regular heptagon cannot be constructed since seven is not a Fermat prime. The 9-gon cannot be constructed because the prime factors of nine are not distinct.

Constructing the Platonic solids

Constructions for three-dimensional objects are considerably more involved than those for simple planar figures. Indeed, Euclid does not begin his treatment of solid geometry until book XI, and the constructions of the Platonic solids are the last topic covered in the *Elements*. Few people would have progressed so far.

Euclid defines each Platonic solid by describing the number and kind of faces that contain it. He concludes the construction of each solid by showing that all its vertices lie on a sphere. This is achieved by showing that all the vertices lie on semicircles having the same diameter. It is for this reason that Euclid gives a rather unusual definition of a sphere as the surface swept out by rotating a semicircle about its diameter. From this definition it would appear that the generating axis is in some way special whereas, in fact, all the diameters of a sphere are equivalent. The definition is more a way of generating a sphere than a statement of its essential characteristic: all its points are equidistant from a given centre.

The constructions of the regular polyhedra that follow are not complete and merely sketch possible ways of approach. For most of the solids, the methods are based on those given in the *Elements*.

Construction.   To construct a tetrahedron.

First, take a semicircle of diameter NS. Divide the length NS to produce a point Q so that $NQ : QS = 2 : 1$. (The length NQ will be the height of the tetrahedron.) Construct the perpendicular to NS at Q and let P be the point where it meets the semicircle (see Figure 2.11). Draw the line NP (this will be the edge-length of the tetrahedron). This has produced what we can think of as an 'elevation' view of part of the tetrahedron. We shall now construct a 'plan' view.

Draw a circle centred on a point Q' whose radius has the same length as QP, and inscribe an equilateral triangle inside it. Denote the corners of the triangle by A, B and C. Erect a line passing through Q' perpendicular to the plane containing triangle ABC and place the points N' and S' on opposite sides of Q' so that $N'Q' = NQ$ and $S'Q' = SQ$. Now, the four points A, B, C and N' are the corners of a tetrahedron and the lines joining them determine its four triangular faces: ABC, ABN', BCN' and CAN'.

It remains to show that these triangles are all equal. The crux of the problem

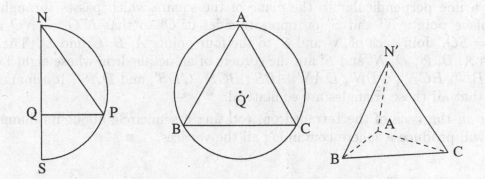

Figure 2.11. Construction of a tetrahedron.

is to show that a line such as AN' has the same length as AB—we shall not consider this further.

To construct the circumsphere, note that the triangle $N'Q'A$ equals triangle NQP so A lies on a semicircle with diameter $N'S'$. The points B and C can be similarly shown to lie on semicircles with diameter $N'S'$. Therefore, rotating such a semicircle about $N'S'$ will create a sphere that contains all the corners of the tetrahedron. ■

Construction. To construct an octahedron.

Take a semicircle with diameter NS and let Q be the point that bisects NS. Construct the perpendicular to NS at Q and let P denote the point where it meets the semicircle. Draw the line NP. This completes the 'elevation' view.

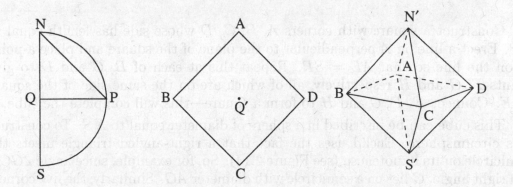

Figure 2.12. Construction of an octahedron.

Construct a square with corners A, B, C, D and with side-length equal to NP. Let Q' be the centre of the square (formed by the intersection of the diagonals).

Erect a line perpendicular to the plane of the square which passes through Q' and place points N' and S' on opposite sides of Q' so that $N'Q' = NQ$ and $S'Q' = SQ$. Join each of N' and S' to the four points A, B, C and D. The six points A, B, C, D, N' and S' are the corners of an octahedron whose eight faces are ABN', BCN', CDN', DAN', ABS', BCS', CDS', and DAS'. It remains to show that all these triangles are equilateral.

As in the case of the tetrahedron, rotating a semicircle about its diameter $N'S'$ will produce a sphere containing all the vertices.   ■

Construction.   To construct a cube.

Take a semicircle with diameter NS. Divide NS at the point Q so that $NQ:QS = 2:1$. Construct the perpendicular to NS at Q and let P denote the point where it meets the semicircle. Draw the line SP.

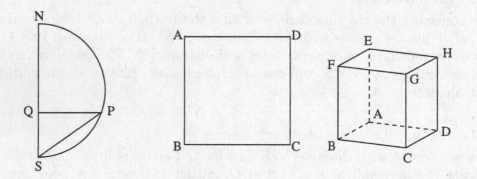

Figure 2.13. Construction of a cube.

Construct a square with corners A, B, C, D whose side has length equal to SP. Erect a line on A perpendicular to the plane of the square and place a point E on the line so that $AE = SP$. Repeat this at each of B, C and D to give points F, G and H respectively, all of which are on the same side of the square as E. Connect E, F, G and H to form a square—this will complete the cube.

This cube can be inscribed in a sphere of diameter equal to NS. To construct this circumsphere, Euclid uses the fact that a right-angled triangle meets the semicircle on its hypotenuse (see Figure 2.14). So, for example, since angle $\angle GCA$ is a right angle, C lies on a semicircle with diameter AG. Similarly, the five corners B, D, E, F and H also lie on such semicircles. The sphere swept out by rotating a semicircle about AG passes through all the vertices of the cube.   ■

A line AB can be divided by a point C into two segments AC and CB so that the ratio of the whole to the longer segment and the ratio of the longer to the shorter segment are equal: *i.e.* so that $AB:AC = AC:CB$. The Greeks called this division into extreme and mean ratio. In the Renaissance it was called the divine proportion, and it is also known as the golden section or golden ratio. This ratio is approximately $5:3$. However, the diagonal and side of a regular pentagon are in the golden ratio, and in the Chapter 1 we saw that such segments are incommensurable: they cannot be expressed exactly as the ratio of whole numbers. The precise value of the ratio is $(\sqrt{5}+1):2$.

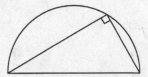

Figure 2.14. A right-angled triangle meets the semicircle on its hypotenuse.

Construction. To construct a dodecahedron.

Euclid's strategy for constructing a dodecahedron is to start with a cube and build a 'roof' shaped structure on each face (see Figure 2.15). The problem is to find the correct height and length of the ridge of such a roof.

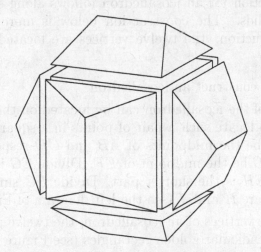

Figure 2.15. A dodecahedron can be constructed by adding a 'roof' to each face of a cube.

To construct a roof, take a square with corners A, B, C, D. Let E and F be the midpoints of AB and CD respectively and connect them by a line. Let G be the midpoint of EF. Divide EG in the golden ratio to produce H so that EH is the shorter part. Divide FG similarly to produce J. Erect lines perpendicular to the plane of the square passing through H and J and mark off points K and L so that the lengths HK and JL are equal to HG. The points K and L determine the ridge of the roof.

If this construction is applied to each square face of a cube (being careful to orient the roofs consistently) then the result is a regular dodecahedron. To prove this, it is necessary to show that the triangular end-piece of one roof and the quadrangular piece of the adjacent roof are coplanar—thus fitting together to form a pentagon. We also have to show that these pentagons are regular. ■

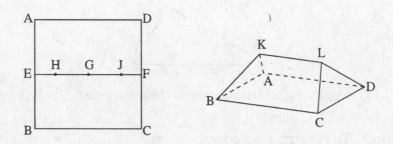

Figure 2.16. Construction of a 'roof'.

Euclid's construction for an icosahedron follows along similar lines to those of the first three solids. The construction below is more in the spirit of the dodecahedron construction: the twelve vertices are located on the surface of a cube.

Construction. To construct an icosahedron.

The twelve vertices of the icosahedron can be located on the surface of a cube—two on each face. To locate such a pair of points in a square with corners A, B, C, D, let E and F be the midpoints of AB and CD respectively and connect them by a line. Let G be the midpoint of EF. Divide EG in the golden ratio to produce H so that EH is the shorter part. Divide FG similarly to produce J. The required points are H and J (see the left diagram of Figure 2.16).

Besides being the vertices of an icosahedron, the twelve points also determine three mutually perpendicular golden rectangles (see Figure 2.17). ■

The discovery of the regular polyhedra

Almost everyone who encounters the regular solids finds something appealing about them. That the Greeks made such detailed and sustained studies of them is probably connected with their unexpected finiteness: there are only five of them, in contrast to the unlimited number of regular polygons. Fascination with

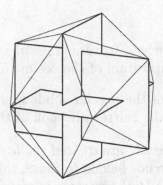

Figure 2.17.   Three mutually perpendicular golden rectangles can be inscribed in an icosahedron.

these solids led both Kepler and Plato to use them in their theories of the cosmos. Their aesthetic properties attracted Renaissance artists and craftsmen. They are also associated with ideas in many areas of modern mathematics from the algebra of group theory to the study of geometric singularities. As they have permeated so much, it seems worthwhile spending a little time trying to identify their source.

Although the oldest surviving account of the regular polyhedra is preserved in Plato's *Timaeus*, Plato should not be credited with their discovery. Solids like the cube and the pyramid are such fundamental figures that they were known very early on. Perhaps more surprisingly, the dodecahedron is also an ancient figure. Etruscan charms and ornaments of dodecahedral form dating from about 500BC have been found in Italy. Pyrite crystals possibly provided the inspiration for these ornaments. Commonly known as *fool's gold*, pyrite $(Fe\,S_2)$ is the most common sulphide and today its main use is as a source of sulphur for the production of sulphuric acid. It is often found alongside copper ores and would have been familiar to early mine workers. Its crystals are often cubic; another common shape has twelve pentagonal facets. These pentagons are not quite regular but the Platonic form is easily imagined. Plate 2 shows a group of pyrite crystals. Even though the individual crystals have grown into each other, it is still possible to identify the dodecahedral forms.

The southern part of Italy, where the Pythagorean school was situated, is particularly rich in pyrite deposits. The eye-catching crystals must have attracted the Pythagorean's attention. Indeed, the Syrian philosopher Iamblichus (*c*.250–*c*.330AD) records that Hippasus wrote about a 'sphere of twelve pentagons'.

One problem in tracing the history of Greek mathematics is that many results are attributed to Pythagoras. The Pythagorean brotherhood ascribed all their results to their founder, and later historians faced with material of unknown origin followed in this tradition. For instance, in what is known as the 'Eudemian summary'[1] we read that

> It was he [Pythagoras] who discovered the subject of incommensurable
> quantities and the composition of the cosmic figures.[h]

Fortunately, for the history of the regular solids there is an alternative source of information. A scholium to the thirteenth book of the *Elements* reads:

> In this book, the thirteenth, are treated the five so-called Platonic fig-
> ures which, however, do not belong to Plato, three of the aforesaid five
> figures being due to the Pythagoreans, namely the cube, the pyramid
> and the dodecahedron, while the octahedron and the icosahedron are
> due to Theaetetus. They are named after Plato because he mentions
> them in the Timaeus. This book also carries Euclid's name because
> he embodied it in the Elements.

These two passages are the best accounts we have. The former reports the kind of history based on legend. The latter is given more credence precisely because it goes against such traditions. Its details are not the kind of conjectured facts one would invent to fill in an unknown piece of the historical jigsaw. It records that the Pythagoreans knew the pyramid, the cube and the dodecahedron which, as we have seen, is not improbable. The discovery of the other two is credited to Theaetetus (*c*.415–369BC), a friend of Plato. The major puzzle posed by this passage is the date of the octahedron: why is it so late?

Icosahedra are not found in nature so the late discovery of this form is not unexpected. But octahedra, like cubes and almost-regular dodecahedra, occur as crystals. Surely they would have been known to the Pythagoreans.

As William Waterhouse has pointed out, this problem is solved as soon as we realise that a crucial part of the history of the regular solids has been overlooked: at some point, the notion of regularity itself had to be discovered. Until the properties that define a regular solid were isolated, the pyramid, the cube and the 'sphere of twelve pentagons' were merely useful or interesting shapes; before their common features were abstracted they were unrelated individual solids. (The reader who is unfamiliar with the definition of a regular polyhedron may like to try to find a set of properties that characterise regularity. It is not as easy as you may think.)

With the realisation that the notion of regularity had to be abstracted, the lateness of the octahedron does not seem so unnatural. Although the octahedral shape may have been familiar, it was only thought worthy of study when its

---

[1]The Eudemian summary is a passage in the prologue to Proclus' commentary on the first book of the *Elements*. It gives a brief account of the development of Greek mathematics and is thought to be based on the now lost *History of Geometry* written by Eudemus of Rhodes.

relationship to other solids was recognised. Only after the concept of regularity was invented did it become important. As Waterhouse explains:

[As T. L. Heath] *quite correctly said, the octahedron "is only a double pyramid with a square base"; and that is a very good reason why no one would have bothered with it. We can readily grant that a man who in some sense could construct a dodecahedron could in the same sense construct an octahedron—but why should he? Someone thoroughly familiar with pyramids would attach no special importance to this particular combination of them. He could assemble an octahedron, he might even admire its appearance; but mathematically he would have nothing to say about it. Only someone possessing the concept of regular solid would have reason to single it out.*

(Consider for comparison the quartz crystal. Quartz (SiO_2) is the most common mineral on earth; its crystals are large, unmistakable hexagonal pyramids and prisms; the Greek word for it has given us the very word "crystal". Yet in all of the Greek geometry there is no special study of hexagonal prisms or pyramids. The shapes were familiar enough, but there simply was nothing particular to say about them.)

The discovery of the octahedron thus was rather like the discovery of, say, the fifth perfect number: what required discovery was not so much the object itself as its significance. Some Babylonian accountant may well have written down the number 33,550,336; but he did not thereby discover the fifth perfect number because he did not observe the property which distinguishes this number from others. Similarly, the octahedron became an object of special mathematical study only when someone discovered a role for it to play.[i]

The late recognition of the octahedron as a regular solid must be closely connected with the discovery of regularity itself.

Besides the testimony of the scholium, further evidence which supports this view can be found in the etymology of the names of the five figures. The modern names for the solids derive from Greek roots: the numbers four, six, eight, twelve, and twenty, and the word 'hedr' meaning seat. Thus octahedron means 'eight faces'.

The use of this terminology is more than mere descriptive labelling. The Greeks only used these names in connection with the Platonic solids. (The hexagonal prism is never called an octahedron, for example, even though it has eight faces.) The fact that these names are sufficient to distinguish between the figures means that whoever labelled the polyhedra in this way knew several key facts: he recognised that the five figures have properties in common and, as such, form a family. Moreover, the systematic nomenclature can only have been chosen after an enumeration had been performed to find all the polyhedra sharing these properties. Only then is it apparent that no two of them have the same number of faces. This marks one of the earliest classifications in the history of mathematics.

Besides their technical names, the three solids of ancient origin also have vernacular names: pyramid, cube, 'sphere of twelve pentagons'. But it appears that the other two (the octahedron and the icosahedron) have only ever had scientific names. This suggests that the octahedron and the icosahedron are known only as part of the family, and hence their recognition or discovery occurred at about the same time as the abstraction of the notion of regular solid. On this point, then, the scholiast's account does not seem so implausible.

The scholiast's attribution of the discovery to Theaetetus is also supported by circumstantial evidence. Book XIII of the *Elements* begins with propositions on the golden ratio and inscribing regular polygons in circles—topics already covered in earlier books. If the presentation were due to Euclid, he could have shortened it considerably. This suggests that book XIII is based on an earlier treatise which was incorporated into the *Elements* in a largely unrevised and unedited state. (The scholiast also comments to this effect.)

Scholars have found strong stylistic, as well as mathematical, links between books XIII and X, connections that are sufficient to suggest that both books are based on works by the same author. And it is widely accepted that Theaetetus laid the foundations for the tenth book.

There is one last fragment of information concerning the history of the regular polyhedra. A tradition preserved up until the eleventh century AD was recorded by the Byzantine writer Suidas. He compiled an encyclopaedia, now known as the *Suda Lexicon*, which covers many topics including history, literature, philosophy, and science. It contains the following comments:

> *Theaetetus, of Athens, astronomer, philosopher, disciple of Socrates, taught at Heraclea. He first wrote on 'the five solids' as they are called.*[j]

On the whole, the traditions concerning the history of the regular solids agree. In an embryonic phase, some of the solids were studied individually for their own sake. But people were not yet aware of a family connection—they were not studying *regular* solids since the concept was unknown. The study of the regular solids begins with the abstraction of regularity. And we can be fairly confident in attributing that achievement to Theaetetus.

What is regularity?

What properties did Theaetetus abstract from the regular polyhedra when he made his definition? Unfortunately, we do not know. Book XI of the *Elements* contains definitions of the five individual solids, but Euclid did not state which of their shared properties determine their family connection. We can get some idea from the final proposition of the *Elements*, a proposition thought to have been appended by Euclid or a scholiast and not part of the original treatise that forms the remainder of the book.

Proposition. *No other figure, besides the said five figures, can be constructed which is contained by equilateral and equiangular figures equal to one another.*

PROOF: The proof proceeds by examining the different kinds of solid angle that can occur. First we note that at least three polygons must meet at any solid angle, and that the sum of all the plane angles around a solid angle must be less than four right angles.

If the polyhedron is made of equilateral triangles then a solid angle can be surrounded by three, four or five polygons. These are the solid angles of the tetrahedron, the octahedron and the icosahedron. Six equilateral triangles surround a point in the plane and so do not form a solid angle. Seven or more triangles cannot surround a point because their angle sum is larger than four right angles.

Three squares around a point form the solid angle of a cube. Four squares fit round a point in the plane, and more than four squares cannot fit around a point.

Three pentagons together form the solid angle of a dodecahedron. Four or more pentagons cannot form a solid angle since their angle sum is too large.

No other regular polygons can form a solid angle because of the angle sum restriction. ∎

From the statement of the proposition it is clear that the two following properties were associated with regularity:

(*i*) the faces must be equal, and

(*ii*) the faces must be regular polygons.

However, these conditions are insufficient and, as it stands, the statement of the proposition is false. There are other polyhedra besides the Platonic solids that are bounded by equal regular polygons. In fact, they are all bounded by triangles; together with the tetrahedron, the octahedron and the icosahedron, they form the family of eight (convex) *deltahedra* illustrated in Figure 2.18. The name 'deltahedron' was used by Martyn Cundy for any polyhedron bounded by equilateral triangles: the Greek capital letter delta ('Δ') looks like such a triangle. Two of the five non-regular deltahedra are dipyramids. Another is formed by attaching three square-based pyramids to the square faces of a triangular prism. A fourth is obtained from a square-antiprism by erecting pyramids on both of its

square faces. The names given in the figure to these last two deltahedra are due
to Norman Johnson, who developed a systematic nomenclature for polyhedra
with regular faces. We shall see more of this below. The remaining solid is
not described so easily. It has twelve faces and is sometimes called the *Siamese
dodecahedron*—a name coined by H. S. M. Coxeter. Johnson called it the *snub
disphenoid*.

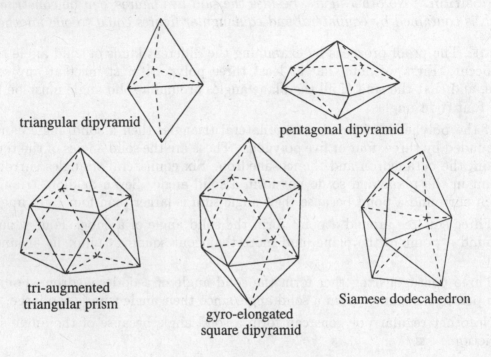

triangular dipyramid

pentagonal dipyramid

tri-augmented
triangular prism

gyro-elongated
square dipyramid

Siamese dodecahedron

Figure 2.18. Convex deltahedra.

These examples show that Euclid's characterisation of regularity is incomplete. The Platonic solids do have equal regular faces but they are not the only
polyhedra to do so. Extra conditions are required to capture precisely the aesthetic quality of regularity exhibited by the Platonic solids but which is lacking
in the deltahedra.

The lack of clarity displayed in some parts of Euclid's work has been advan-
tageous. As regularity was left incompletely defined, people felt free to propose
their own definitions. This has led to a variety of descriptions of the regular solids
which provide alternative ways to study them. The different hypotheses can be
weakened in many different ways and this gives a rich collection of ideas about
'semiregular' polyhedra. (We may note in passing that Euclid did not define
'polyhedron' either—an omission which led to a wide variety of interpretations of
that term. We shall take this up in Chapter 5.)

Let us return to our search for the original definition of regularity. Notice that

Euclid is careful to show how each solid can be constructed in a circumscribing sphere. Plato, too, notes that the tetrahedron is the simplest solid figure which divides the surface of its circumsphere into equal and similar parts. The other figures he describes are more complex solids having the same property. It seems likely that this feature was the third key property required to define regularity: the polyhedra with equal regular faces that can be inscribed in a sphere are precisely the five regular solids.

A more modern definition of a regular polyhedron is one whose faces and vertex figures are regular polygons. (In this case, the congruence of the faces does not need to be specified since it can be deduced from the other conditions.) Roughly speaking, a *vertex figure* is the polygon you see after slicing off a corner in a way that removes the same amount of each edge.[2]

Many definitions of 'regular polyhedron' require the polyhedron to have equal regular faces.[3] Some of the proposals for an extra condition to characterise regularity are collected in the following theorem. It does not matter which of the five statements is chosen to complete the definition of regularity: the theorem shows that they all lead to the same set of polyhedra.

Theorem. Let P be a convex polyhedron whose faces are congruent regular polygons. Then the following statements about P are equivalent:

(1) The vertices of P all lie on a sphere.

(2) All the dihedral angles of P are equal.

(3) All the vertex figures are regular polygons.

(4) All the solid angles are congruent.

(5) All the vertices are surrounded by the same number of faces.

PROOF: The proof consists of showing that $(1) \Rightarrow (2) \Rightarrow (3) \Rightarrow (4) \Rightarrow (5) \Rightarrow (1)$.

$(1) \Rightarrow (2)$: If two adjacent faces of a polyhedron have all their vertices on the same sphere then the dihedral angle between them depends on the radii of the circumcircles of the two faces and the length of their common edge. Since all the faces of P are congruent, their circumcircles will all be the same size. Furthermore, since all the faces are regular polygons, all the edges must have the same length. Consequently, all the dihedral angles must be equal.

---

[2]More precisely, I shall take a vertex figure to be the spherical polygon formed by the intersection of the faces surrounding a vertex with a small sphere centred on that vertex.

[3]An example of a definition which does not require this explicitly, nor even convexity, is the following proposed by H. S. M. Coxeter. A polyhedron is regular if there exist three concentric spheres one of which contains all the vertices, one contains the midpoints of all the edges, and one meets all the centres of all the faces.

$(2)\Rightarrow(3)$: The plane angles that surround a vertex determine the lengths of the sides of its vertex figure: a larger angle gives a longer side. Since all the plane angles in P are the same, all the vertex figures are equilateral. Moreover, the angles in the vertex figures are determined by the dihedral angles of P. If the dihedral angles are all equal then the vertex figures are equiangular. An equilateral, equiangular polygon is regular (by definition).

$(3)\Rightarrow(4)$: A vertex figure cuts off a solid angle from a polyhedron. If the vertex figure is a regular polygon then the solid angle has the form of a right pyramid. The angles in a vertex figure determine the dihedral angles of the polyhedron, and conversely. If all the vertex figures are regular then all the dihedral angles must be equal. This means that the vertex figures all have the same number of sides, and hence that the solid angles are all congruent.

$(4)\Rightarrow(5)$: If the solid angles are all congruent then they are surrounded by the same number of faces.

$(5)\Rightarrow(1)$: The remaining implication (as is often the case in this style of proof) is far more subtle. In fact, it seems to require the enumeration of all possible polyhedra satisfying condition (5) and then verifying that each candidate satisfies condition (1). The final proposition of the *Elements* (see page 75) shows that at most five different kinds of solid angle can be formed from regular polygons. In the preceding propositions, five polyhedra are constructed in their circumspheres, one with each type of solid angle.

This, however, is not the end of the argument. It is not immediately clear why there should be only one polyhedron for each type of solid angle. Is the shape of the complete polyhedron constrained by the local behaviour at each vertex to such an extent that a single possibility remains?

It is easy to answer this question when each vertex is surrounded by three faces. In this case the solid angles are rigid—they can take only one form. Fitting polygons together according to this rule guarantees a unique result. Therefore, the tetrahedron, the cube and the dodecahedron are the only polyhedra which can be made by fitting three congruent regular polygons around each vertex.

However, when four or more faces surround each vertex there is no such control. The contrast between the two situations is apparent in partly made models of polyhedra. Triangular vertex figures give rigidity to the structure so that a half-built dodecahedron is stiff and cannot be deformed. On the other hand, an unfinished model of an icosahedron is flexible—the local conditions on the vertices do not give it any stability until it is practically complete.

Experimenting with models will probably convince you that the icosahedron and the octahedron are unique but to prove this rigorously is not easy. To complete the proof we need to invoke the Rigidity Theorem—a topic we shall return to in Chapter 6.   ■

Bending the rules

Regular polyhedra are often defined as those having the same number of congruent regular faces meeting at each vertex. John Flinders Petrie (1907–1972) (son of the archaeologist mentioned in Chapter 1) found a new way of interpreting this condition. He and H. S. M. Coxeter had known each other since their school days and Coxeter recounts how he first heard of his friend's discovery.

> One day in 1926, J. F. Petrie told me with much excitement that he had discovered two new regular polyhedra; infinite, but without false vertices. When my incredulity had begun to subside he described them to me: one consisting of squares, six at each vertex [Figure 2.19(a)], and one consisting of hexagons, four at each vertex [Figure 2.19(b)]. It was useless to protest that there is no room for more than four squares round a vertex. The trick is to let the faces go up and down in a kind of zig-zag formation so that the faces that adjoin a given 'horizontal' face lie alternately 'above' and 'below' it. When I understood this, I pointed out a third possibility: hexagons, six at each vertex [Figure 2.19(c)].[k]

These structures are not polyhedra in the conventional sense. They do not close up, forming a structure with an natural sense of what lies inside and what lies outside. Instead, they can be extended indefinitely in any direction: the 'complete' polyhedra have infinitely many faces. Furthermore, each one separates space into two immense labyrinths, both of the same shape. These 'polyhedra' have become known as regular *honeycombs* or *sponges*.

The Archimedean solids

In the fifth book of his *Mathematical Collection*, Pappus attributes the discovery of thirteen polyhedra to Archimedes:

> Although many solid figures having all kinds of faces can be conceived, those which appear to be regularly formed are most deserving of attention. These include not only the five figures found in the godlike Plato ⋯ but also the solids, thirteen in number, which were discovered by Archimedes and are contained by equilateral and equiangular, but not similar, polygons.[l]

Pappus goes on to describe the thirteen figures. He arranges them in order according to the total number of faces, and lists the kinds of face that make up each polyhedron. These data are summarised in Table 2.20. Even though Archimedes' own account of them is now lost, the thirteen polyhedra illustrated

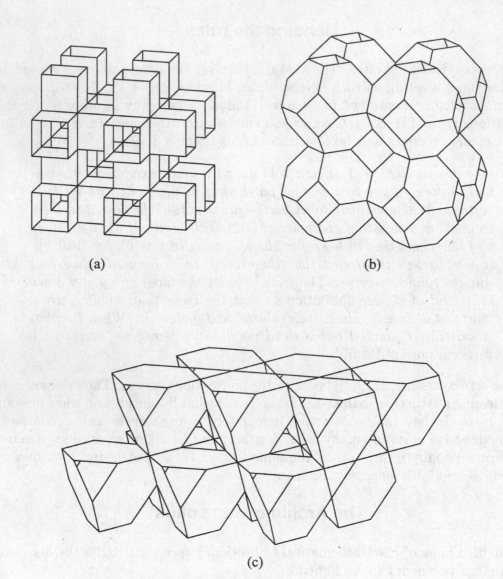

(a)

(b)

(c)

Figure 2.19. The three regular honeycombs.

in Figure 2.21 and in Plates 3 and 4 are known as the *Archimedean solids*. (They are sometimes called the *semiregular* polyhedra.)

Some of these polyhedra have been discovered many times. According to Heron, the third solid on Pappus' list, the cub-octahedron, was known to Plato. During the Renaissance, and especially after the introduction of perspective into art, painters and craftsmen made pictures of the Platonic solids. To vary their designs they sliced off the corners and edges of these solids, naturally producing some of the Archimedean solids as a result. The process of removing all the corners in a symmetrical fashion is called *truncation*.

Faces	Triangles	Squares	Pentagons	Hexagons	Octagons	Decagons
			Numbers of			
8	4			4		
14	8	6				
14		6		8		
14	8				6	
26	8	18				
26		12		8	6	
32	20		12			
32			12	20		
32	20					12
38	32	6				
62	20	30	12			
62		30		20		12
92	80		12			

Table 2.20. Composition of the Archimedean solids.

Kepler rediscovered the whole set of thirteen solids and gave them the names by which they are known today. The five Archimedean solids obtained by truncating the Platonic solids are given the obvious names: 'truncated tetrahedron' for example. For Kepler, parts of the edges are always retained in a truncated solid. If we allow deeper cuts to be made, all three of the 14-faced Archimedean solids can be obtained by slicing the corners off either a cube or an octahedron to varying degrees (see Figure 2.22). The truncated cube and the truncated octahedron are most obviously derived from the Platonic solids in their respective names. The latter polyhedron is also known as Kelvin's solid because William Thomson (Lord Kelvin) studied its space-filling properties. Kepler gave the name cub-octahedron to the solid midway between the cube and the octahedron.

An analogous situation holds for the three 32-faced solids: the truncated dodecahedron, the truncated icosahedron, and the icosi-dodecahedron. (The last two of these, especially the truncated icosahedron, may be familiar from the patterns on soccer balls.)

The solids that Kepler calls the truncated cub-octahedron and the truncated icosi-dodecahedron are not true truncations. Figure 2.23(a) shows the result of truncating a cub-octahedron: not all of its faces are regular polygons. The Archimedean figure is obtained by distorting the figure so that the rectangles become square. Kepler himself recognised this:

[· · · a polyhedron] *which I call a truncated cuboctahedron: not because it can be formed by truncation but because it is like a cuboctahedron that has been truncated.* [m]

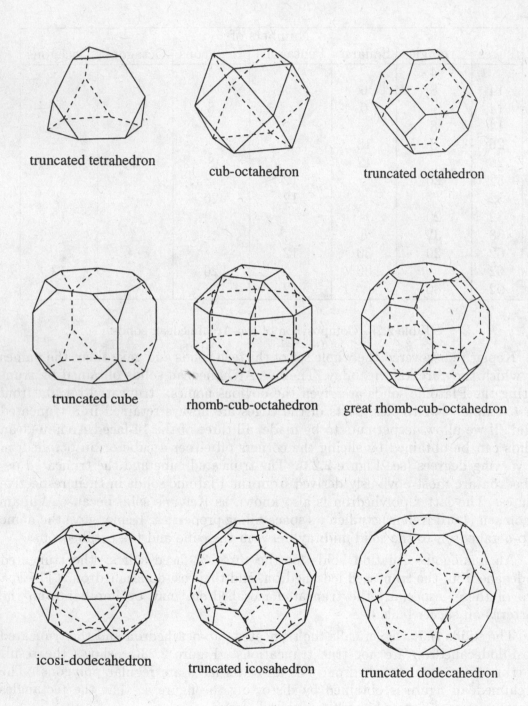

Figure 2.21. The Archimedean solids.

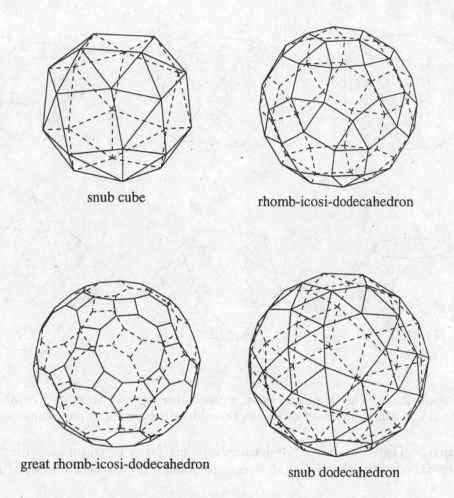

snub cube   rhomb-icosi-dodecahedron

great rhomb-icosi-dodecahedron   snub dodecahedron

Figure 2.21 (*continued*).

Because of this discrepancy, alternative names have been proposed, the most common being 'great rhomb-cub-octahedron'.

Figure 2.24 shows that the solid Kepler calls the rhomb-cub-octahedron[4] (sometimes given the prefix 'small') has faces in common with three polyhedra: a cube, an octahedron, and the rhombus-faced polyhedron shown in Figure 2.23(b). His alternative name for this Archimedean solid, *sectus rhombus cuboctaëdricus*, indicates that it is formed by slicing the corners off the rhombic polyhedron. This name is translated as 'truncated solid rhombus' but Kepler makes a distinction between this truncation and the others, for which he uses the adjective *truncus* rather than *sectus*.

---

[4]The hyphenation of these names is not standard practice but I find it easier on the eye when the names become so long.

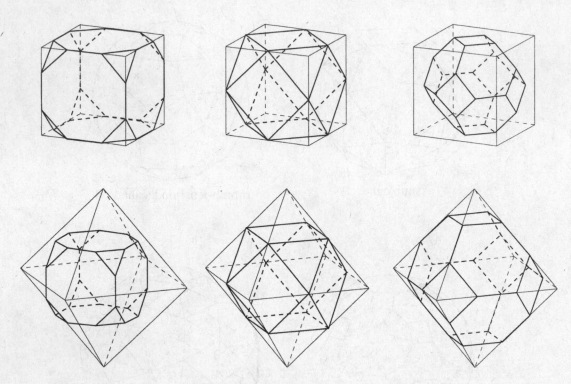

Figure 2.22. The truncated cube, cub-octahedron and truncated octahedron can all be obtained by slicing through either a cube or an octahedron.

Problem. The rhomb-icosi-dodecahedron has faces in common with another rhombus-faced solid that we shall encounter later. Try to work out what it looks like.

The two remaining Archimedean solids differ from the others in several ways. For one thing, they cannot be derived from the Platonic solids by a process of simple truncation, and consequently they have been discovered by relatively few

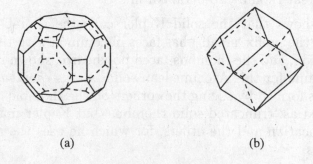

(a)   (b)

Figure 2.23.

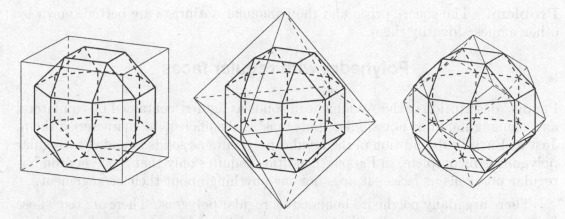

Figure 2.24. The rhomb-cub-octahedron has face-planes in common with three other polyhedra.

people. They also lack any mirror symmetry. This causes them to appear twisted, and it also means that they can exist in two forms, each a mirror image of the other—like left and right hands. Polyhedra related in this way are said to be *enantiomorphic*.

The solid Kepler called a snub cube actually has faces in common with both the cube and the octahedron. For this reason some have suggested that 'snub cub-octahedron' would be more appropriate but the name has not caught on. Similarly, the snub dodecahedron is related just as much to the icosahedron as to the dodecahedron.

Problem. Explain why there is no 'snub tetrahedron'. If you try to construct one, you should find a familiar polyhedron.

Kepler also investigated two other families of polyhedra made from regular polygons. A *prism* is formed from two n-sided polygons separated by a ring of n squares. An *antiprism* also contains two n-sided regular polygons, this time separated by a ring of $2n$ equilateral triangles. An example of each kind is shown in Figure 2.25: a pentagonal prism and a square antiprism.

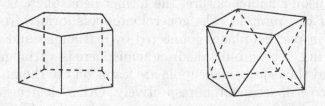

Figure 2.25. Pentagonal prism and square antiprism.

Problem. The square prism and the triangular antiprism are better known by other names. Identify them.

Polyhedra with regular faces

Pappus' description of the Archimedean solids as figures 'contained by equilateral and equiangular, but not similar, polygons' is insufficient to characterise them. Just as Euclid's description of the regular polyhedra as solids bounded by regular polygons is incomplete, so Pappus' condition requires only that a polyhedron has regular polygons as faces—it does not say anything about their arrangement.

There are many polyhedra bounded by regular polygons. There are ten whose faces are all congruent: the Platonic solids and the deltahedra. Besides the thirteen Archimedean solids, there are 87 convex polyhedra whose faces are regular polygons of more than one kind. One way to construct many of these is to dissect the Platonic and Archimedean solids into smaller pieces. The octahedron, for example, can be separated into two square-based pyramids. Pentagonal pyramids can be shaved off an icosahedron in several ways to produce five different fragments (see Figure 2.26).

The regular-faced polyhedra were enumerated empirically by Norman Johnson: some computer-assisted calculations done by Victor Abramovitch Zalgaller showed that his list was complete. Many of these polyhedra are formed by joining smaller regular-faced polyhedra together and Johnson used this property as the basis for a systematic nomenclature. The polyhedra which cannot be separated into regular-faced pieces are the building blocks and Johnson calls these *elementary* polyhedra. A regular-faced polyhedron is elementary if it cannot be separated by a plane into two smaller regular-faced polyhedra. All the other regular-faced polyhedra can be formed by sticking these basic units together in different ways.

The tetrahedron, the cube and the dodecahedron are elementary, as are nine of the Archimedean solids. The prisms, and all the antiprisms except the octahedron furnish further examples of elementary polyhedra.

The cub-octahedron and the icosi-dodecahedron can both be split into 'hemispheres'. In Johnson's nomenclature, the former hemisphere is an example of a *cupola*, the latter is a *rotunda*. The general cupola is formed from an n-gon and a $2n$-gon sitting in parallel planes connected by n triangles and n squares joined alternately in a ring. The cub-octahedron hemisphere is a triangular cupola. The square cupola and the pentagonal cupola are 'caps' off the rhomb-cub-octahedron and the rhomb-icosi-dodecahedron, respectively. Other elementary polyhedra are produced by removing such caps from these two polyhedra. The removal of opposite square cupolas from a rhomb-cub-octahedron leaves an octagonal prism. The rhomb-icosi-dodecahedron can be 'diminished' by the removal of one, two or three pentagonal cupolas. These fragments of the Archimedean solids are illustrated in Figure 2.27.

Besides the fragments of the Platonic and Archimedean solids there are a further eight elementary polyhedra. One of these is the Siamese dodecahedron— the deltahedron with twelve faces. The other seven, together with Johnson's names, are shown in Figure 2.28.

The interrelationships between the different families of regular-faced polyhedra are illustrated schematically in Figure 2.29.

The 31 elementary polyhedra fit together to produce 71 other polyhedra with regular faces. Sometimes the same set of pieces can be put together in more than one way. For example, the constituent faces of a rhomb-icosi-dodecahedron can

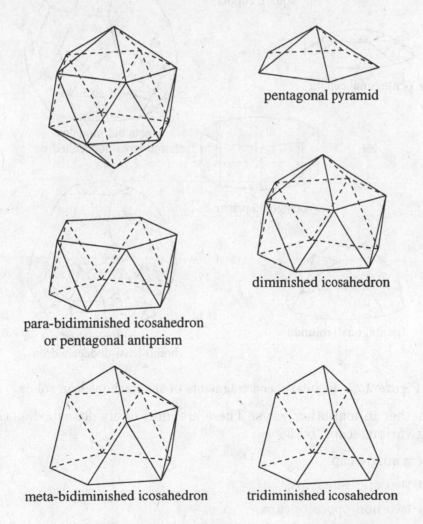

pentagonal pyramid

diminished icosahedron

para-bidiminished icosahedron
or pentagonal antiprism

meta-bidiminished icosahedron   tridiminished icosahedron

Figure 2.26. Regular-faced fragments of the icosahedron.

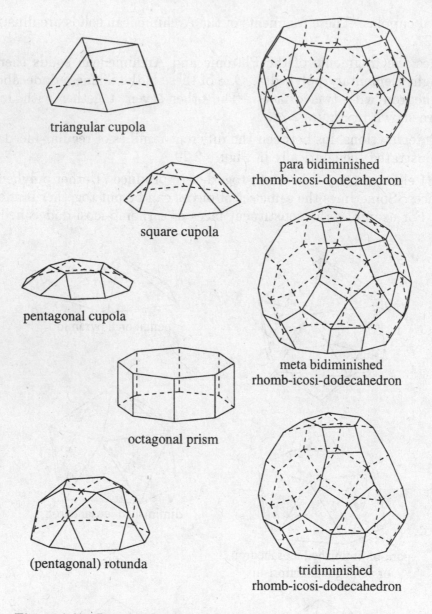

triangular cupola

para bidiminished
rhomb-icosi-dodecahedron

square cupola

pentagonal cupola

meta bidiminished
rhomb-icosi-dodecahedron

octagonal prism

(pentagonal) rotunda

tridiminished
rhomb-icosi-dodecahedron

Figure 2.27. Regular-faced fragments of the Archimedean solids.

be put together in four other ways. These are most easily described as the effect
of rotating various 'caps' by 36°:

   (i) twist a single cap,

  (ii) twist two opposite caps,

 (iii) twist two non-opposite caps,

 (iv) twist three caps.

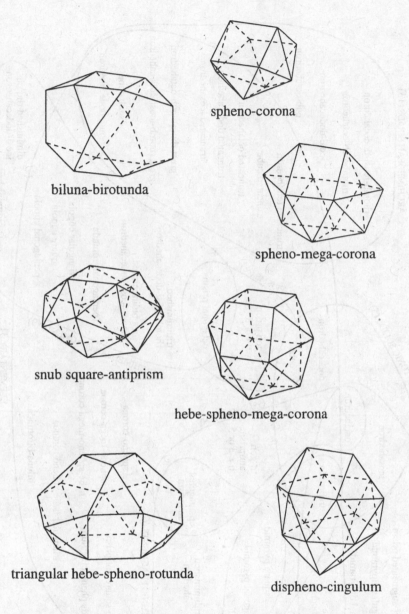

spheno-corona

biluna-birotunda

spheno-mega-corona

snub square-antiprism

hebe-spheno-mega-corona

triangular hebe-spheno-rotunda

dispheno-cingulum

Figure 2.28. More elementary regular-faced polyhedra.

Chemists use the term *isomerism* to refer to an analogous situation: molecules can differ structurally but be composed of the same set of atoms. Molecules related in this way are called *isomers* (from the Greek words meaning 'same parts'). Adopting this terminology, we can say that the rhomb-icosi-dodecahedron has five isomeric forms. Enantiomerism is a special case of isomerism.

There is one isomeric form of an Archimedean solid that is quite well-known. While trying to make a model of the (small) rhomb-cub-octahedron, J. C. P.

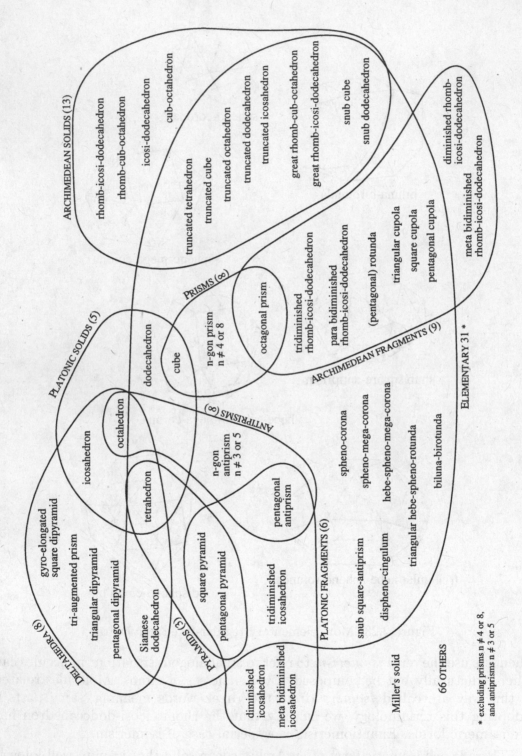

Figure 2.29. The families of convex polyhedra with regular faces.

Miller was surprised to find that he had assembled the pieces incorrectly. The polyhedron he made is shown in Figure 2.30. This polyhedron has been discovered and rediscovered many times. It is possible that Kepler was familiar with it. It has been called by a variety of names: the pseudo rhomb-cub-octahedron, Miller's solid, and elongated square gyro bicupola. The last of these names is Johnson's and it indicates how the solid is constructed from elementary polyhedra: take two square cupolas (*square bicupola*) rotated relative to one another (*gyro*) and separated by a prism (*elongated*). For brevity I shall refer to it as Miller's solid.

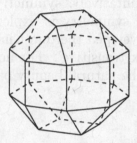

Figure 2.30. Miller's solid has congruent solid angles but they are not all equivalent.

Out of this whole menagerie of polyhedra bounded by 'equilateral and equiangular, but not similar, polygons' what is special about the Archimedean solids? What distinguishes them from the others? Pappus gives us a hint when he says that they are included among the polyhedra 'which appear to be regularly formed'. This is certainly true, but what does 'regularly formed' actually mean?

Earlier in the chapter, we saw that an extra condition was required to complete Euclid's characterisation of the regular polyhedra, and that any one of five statements ranging from the existence of a circumsphere to the congruence of the vertices proved to be adequate. How can Pappus' statement be extended so that it characterises the Archimedean solids? Simply requiring that the polyhedra have a circumsphere is not enough since the pyramids and all the isomeric forms of the Archimedean solids have this property. Surprisingly, the much stronger condition of requiring that all the solid angles are congruent is also insufficient: Miller's solid has this property. Because of this, some writers have suggested that this polyhedron should be counted as a fourteenth Archimedean solid. This, however, misses the point. The true Archimedean solids, like the Platonic solids, have an aesthetic quality which Miller's solid does not possess. This attractiveness comes from their high degree of symmetry—a property that is easily appreciated and understood on an intuitive level. It is not the congruence of the solid angles that is the important characteristic but rather the fact that the solid angles are all indistinguishable from one another. The vertices in an Archimedean solid are

surrounded by the same faces arranged in the same way, and each vertex plays
the same role in the polyhedron as a whole. For Miller's solid this is not the
case. The twist allows us to distinguish between two kinds of vertices: those
near the 'equator', and those in the 'polar regions'. This becomes clear when
the polyhedron is turned onto its side—we can detect that it has been moved.
If the true rhomb-cub-octahedron is turned on its side in the same way, it looks
untouched—if we did not observe the motion take place, we would be unable to
tell that a change had occurred.

These observations are the beginnings of a detailed investigation of symmetry.
Although easy to identify qualitatively, symmetry is quite tricky to quantify.
The mathematical analysis of symmetry is explored more fully in Chapter 8.
Applying some of the ideas developed there in Chapter 10, we shall see that the
Archimedean solids are 'vertex transitive'. For now, it suffices to recognise, as
Archimedes and Pappus did, the natural beauty of regularly formed figures.

Dodecaëdrum.

Icosaëdrum.

En ceste presente figure nous sont demonstrez les cinq Corps Reguliers de Geometrie, (lesquiels sont deduits & declarez de poinct en poinct en la su de ce present liure:) ensemble certains personnages racourciz selo ceste Art, desquels Dieu aydant, j'espere au second liure vous les deduire plus amplement.

From *Livre de Perspective* by Jean Cousin, 1568.

Decline and Rebirth of Polyhedral Geometry

③

At age twenty, Alexander succeeded his father Philip II as king of Macedonia. Two years earlier in 338BC Athens had fallen to Philip as he extended his territory southwards into Greece. Alexander continued his father's expansionist conquests. Within five years he had established an empire that extended eastwards to India and included Egypt, Syria and Persia. He chose a well-placed port on the north African coast to be the site of a new city that was to be his capital—Alexandria.

After Alexander's death in 323BC, the empire was divided and the Egyptian kingdom came under the rule of the Ptolemaic dynasty. The Ptolemies shared Alexander's vision for the new city and it became a great centre of culture and learning. To promote science and philosophy, a school and a library were built and by 300BC this 'university' had attracted a distinguished staff. It is here that the last contributions of any significance were made to the history of polyhedra for well over a thousand years. In this chapter we trace the steps of our subject as the various strands unravel, sleep-walk across vast distances in space and time, and then are reunited as the study of polyhedra emerges to flourish once more.

The Alexandrians

Alexandria was a prosperous, cosmopolitan city at a junction of several trade routes. Jews settled there as merchants and traders. Greeks arrived with Ptolemy I. Romans, Persians, Arabs, and Indians also inhabited the city. The Ptolemies recognised the importance of the schools of Pythagoras, Plato, and Aristotle and sought to provide a climate conducive to scholarship. The teachers at the university they established came from the major cultural centres and many were trained in mathematics. Besides mathematics, the students were taught literature, medicine and astronomy. The last two subjects also had a mathematical content (medicine because it made use of astrology). Although much of the syllabus was based on Greek learning, other cultures exerted their influence, most notably the Egyptian and Babylonian. Where Greek geometry had concentrated on qualitative relationships, the Alexandrians sought something more quantitative and applicable to the real world. The inductive spirit of mathematics, the 'how to' recipe approach used in the early Egyptian and Babylonian texts returned, especially in arithmetic.

Many of the great geometers taught or studied at the university. It is probably at Alexandria, with access to the great library, that Euclid compiled his *Elements* of geometry. The final book of the *Elements* is the high point of the classical study of polyhedra. It contains the fundamental properties of the regular solids and presents the mathematics of their construction and basic relationships such as comparisons of volume. Such was the prestige of the Euclidean canon that other books have been appended to it at various times. The first of these, the so-called *Fourteenth Book of the Elements*, is thought to have been written by Hypsicles in the second century BC. He was another teacher at Alexandria. The book contains further properties of the regular solids. It is based on work by Aristaeus and Apollonius. According to Pappus, these were the mathematicians who, together with Euclid, comprised the 'three geometers skilled in analysis'.

The preface to book XIV recounts a meeting between Basilides of Tyre and the author's father in Alexandria.

> *On one occasion, when looking into the tract written by Apollonius about the comparison of the dodecahedron and icosahedron inscribed in one and the same sphere, that is to say, on the question what ratio they bear to one another, they came to the conclusion that Apollonius' treatment of it in this book was not correct; accordingly, as I understood from my father, they proceeded to amend and rewrite it. But I myself afterwards came across another book published by Apollonius, containing a demonstration of the matter in question, and I was greatly attracted by his investigation of the problem. Now the book published by Apollonius is accessible to all; for it has a large circulation in a form which seems to have been the result of later careful elaboration.[a]*

It appears that Hypsicles' father saw an early version of Apollonius' work and that a complete proof was widely circulated sometime later as *Comparison of the Dodecahedron with the Icosahedron*. The theorem attributed to Apollonius states that the surface areas of the two solids are in the same ratio as their volumes.

Aristeaus' contribution to book XIV is a theorem proved in his now lost work *The Comparison of the Five Figures*. It states that if an icosahedron and a dodecahedron are inscribed in the same sphere then the same circle circumscribes a triangular face of the former solid and a pentagonal face of the latter.

Mathematics and astronomy

Alongside the development of mathematics, the Greeks laid the foundations of another science: astronomy. Many early civilisations observed the heavens in order to keep track of time. The positions of the planets and the stars established their calendar. The desire to record and map the changes in the sky naturally led to the study of the geometry of the sphere. The Pythagoreans, who studied the properties of great circles, may not have distinguished between astronomy and spherical geometry: the word 'sphaeric' applied to both. Plato's insistence that nature could be understood through mathematics, and the idea that the perfect heavenly bodies would follow paths based on the perfect figure (a circle) led to the development of the epicyclic system.

Many mathematicians spent some time studying astronomical problems. Euclid wrote about spherical geometry in his *Phaenomena*. Hypsicles and Apollonius are known as astronomers as well as mathematicians. Apollonius was very familiar with the epicyclic theory and he determined the points of orbits where planets appear stationary. In *On the Risings of the Stars* Hypsicles used the Babylonian division of the circle into 360° and divided each degree into sixty subunits, and each of these into sixty, and so on. This was the first time that degrees were mentioned in a Greek manuscript. Unlike the Greeks, who were solely interested in exact relationships, the Alexandrians also wanted to use physical measurements.

This desire to be able to calculate the positions of heavenly bodies led to the development of trigonometry—work which was done largely by the astronomer Hipparchus of Rhodes. He divided the diameter of a circle into 120 units and each of these into sixty, and so on. Using this division and the division of the circumference into degrees he made a table of chords—what would be equivalent in modern times to tabulating the sine function.

Much of the work on trigonometry and its applications to astronomy was collated by Claudius Ptolemy[1] (*c.*100–*c.*168). His *Mathematica Syntaxis* is now usually known as *Almagest* after its Arabic title. It was the dominant work in its field for many centuries and held authority similar to that of the *Elements*. The geocentric epicyclic theory that it contained was not superseded for over a thousand years.

Besides inventing trigonometry, astronomers tackled another mathematical problem: how to make a map of a round object on a flat surface. Hipparchus used orthographic (or parallel) projection to produce a map of the heavens, a task that involves mapping the hemispherical firmament onto a disc. He possibly knew the technique of stereographic projection. This was certainly known to Ptolemy, who used it in his *Planisphaerium*, where he describes the mathematics underlying the construction of an astrolabe.

---

[1]Not part of the ruling family.

Heron of Alexandria

As has been remarked above, the mathematics of the Alexandrian period had a different bias from that produced by the classical Greeks. It is concerned more with practical problems than with pure geometry. One person who exemplifies this Alexandrian outlook is Heron ($c.62$AD). Although Heron was familiar with Euclidean geometry, he was not a classical geometer. He was more practically minded, an engineer and inventor of great ingenuity. The ancient writers attribute many treatises to Heron, some of which survive intact, some only in corrupted form after revision by later editors, and others not at all. His works are classified into two categories: mathematical and mechanical. The latter class contains the *Pneumatica* which describes his well-known water and steam powered automata, *Mechanica*, and the *Dioptra* which describes a kind of theodolite used for surveying. In this last work Heron gives details of an eclipse—information that helps to determine his approximate dates.

The most important of Heron's mathematical writings is his *Metrica*. It deals with the measurement of area, progresses to volume measurement, and then considers the division of figures into parts of a given ratio. It contains the following 'Heronic formula' for the area of a triangle: if a triangle has sides of lengths a, b and c, and if $s = \frac{1}{2}(a + b + c)$ denotes half the perimeter then its area is given by

$$\text{area} = \sqrt{s(s - a)(s - b)(s - c)}.$$

It also includes a generalisation of the formula for the volume of a truncated pyramid. The solid shown in Figure 3.1 has rectangular base and top lying in parallel planes. The sides have lengths a, b, c, d as shown and the height of the solid is h. Its volume is given as

$$\text{volume} = h\left(\tfrac{1}{4}(a + c)(b + d) + \tfrac{1}{12}(a - c)(b - d)\right).$$

Problem. Verify this is correct.

In the main, the results presented in Heron's compendium are taken from works by Greek and Alexandrian mathematicians. They are far superior to the

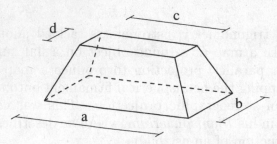

Figure 3.1.

traditions preserved by the Egyptian priests as is apparent from the inscriptions at Edfu (see Figure 1.1). Unlike the classical writers, who were satisfied to exhibit relationships between quantities, Heron was concerned with numerical answers. Intermingled with the accurate formulae, some of which are proved, are others which give only approximate solutions. The latter would have appealed to craftsmen who would have found extracting roots difficult. Like the Arabs after him, who continued the development of algebra, Heron was uninhibited by the restrictions of the Greek geometric algebra. He was happy to consider expressions equivalent to our $x^2 + y$ or $x^2 \times y^2$—things that would have appalled the Greeks, the former being an area added to a length, the other a four-dimensional quantity without a geometrical interpretation.

Pappus of Alexandria

In the 300 years from Apollonius to Claudius Ptolemy, the only mathematicians of significance to our story are Hypsicles and Heron. Other mathematicians of the period are known mainly through the writings of commentators; their original works have not survived. Pappus, a commentator of the early fourth century, has been mentioned in previous chapters because of his *Mathematical Collection*. It is only from this work, for example, that we know the discoverer of the Archimedean solids. Pappus also wrote commentaries on the *Elements*, and Ptolemy's *Almagest* and *Planisphaerium*.

The *Mathematical Collection* provides us with a lot of information. It is a handbook to the classics, a systematic account of the most important works of Greek mathematics. It includes historical comments and descriptions of the contents of many works now lost. Since he finds it necessary to include these, it is likely that many of the works were lost or inaccessible even in his own time. When an original text is easily available, Pappus gives alternative proofs or extends the results in some way. In book III, for example, he discusses the regular solids. Whereas Euclid's presentation shows how to construct each solid and then its circumsphere, Pappus shows how to inscribe each solid in a given sphere. In a later book he considers isoperimetric problems. These involve comparisons between figures with boundaries of equal size. For example, he shows that a sphere has greater volume than any regular solid with the same surface area.

Pappus precedes his proofs by an analysis which explains how his constructions arise. This is unusual for, except for the odd hint, the classical writers do not say why subjects were investigated nor how results were obtained. Their primary objective was to organise their achievements in a systematic fashion and to present the formal polished arguments needed to establish their results. Although concise, this strict deductive presentation style has some disadvantages. It can give the impression that mathematics is created in this way. We should

not forget that, before such an account can be written, many experiments have
been tried and paths explored. Conjecture comes before proof.

Another problem with the Euclidean style of presentation is that it is often
difficult to see how the proofs were discovered. Even if you can decipher the
arguments, you may not gain any insight as to how a similar problem could be
tackled. The proofs of all the propositions in the *Elements* are *synthetic*: they
work on the 'bottom-up' principle, starting from what is known and building to
the conclusion. This provides a good basis for logical arguments but rarely allows
the reader any hint of the steps that led to it. Very often the reverse process
was involved. The *analytic* approach uses the 'top-down' philosophy. In this
case we start with the conclusion and work backwards towards what we already
know. With luck, all the steps involved can then be reversed. These problems
with the classical texts may go a long way towards explaining the popularity of
commentaries.

Plato's heritage

Towards the end of the third century AD interest in Plato's philosophy began to
be revived in Alexandria. This Neoplatonism soon spread. Plotinus, its founder,
travelled to Rome and set up his own school; others took it to Greece and Constantinople. Iamblichus, who was considered a leading authority on Neoplatonism
for 200 years, introduced a mystical interpretation of some concepts and placed
an increased emphasis on religious observance.

The last great Neoplatonist was Proclus. Educated at Alexandria, and by
Plutarch at Plato's academy, he later became head of the academy whence he
received the title 'Diodochus', meaning successor. He wrote commentaries and
critiques on many aspects of Greek culture. His commentary on the first book
of Euclid's *Elements* is invaluable because of its historical passages. He had
access to many works that have not survived, including the history of geometry
written by Eudemus and the historical passages in Geminus' work. He also wrote
commentaries on many of Plato's dialogues, including the *Timaeus* and *Republic*.

The *Timaeus* became one of the most influential works written in antiquity.
The importance of the original text declined as Greek became less widely understood. The language was lost in the West after the separation of the Roman
empire; in western Asia it became replaced by Syriac. However, the first part
of Plato's dialogue passed into Europe in Chalcidius' Latin version and another
fragment preserved by Cicero. The Platonic message appealed to many thinkers
of the Dark Ages and had a strong influence on religious thought. The mixture
of rational explanation and teleology, and its creator god, fitted well with Christian notions and many Platonic ideas were introduced into Christianity. In the
West this was largely the work of Augustine but Byzantine scholars had a similar
influence in the East.

Copies of Chalcidius' work could be found in the libraries of many medieval monasteries. The monks also kept Neoplatonic works written by Boethius (c.475–524). He was a Roman, educated in both Latin and Greek, and is best known for his book *On the Consolation of Philosophy*. He also translated most of Aristotle's work on logic into Latin as well as some fragments of Euclidean geometry and Nichomachus' *Arithmetic*. The paucity of the works is illustrated by his geometry text: it contains statements of a few propositions from books I, III and IV of the *Elements* without proof. Later, these came to be considered the height of mathematical achievement.

The decline of geometry

In the sixth century, a second book became added to the Euclidean canon. Like Euclid's book XIII and Hypsicles' book XIV, this so-called book XV contains further propositions on the regular solids. It has three parts which seem to have been written at different times. The first part, which may have been around in Pappus' time, describes how some regular solids can be inscribed in others. In the second part the numbers of edges and solid angles of the five solids are given. The third part shows how to calculate the dihedral angle between the faces of any Platonic solid. This is done by constructing an isosceles triangle whose apex equals the dihedral angle. The rules for drawing these triangles are attributed to 'Isidorus our great teacher'. This refers to Isidorus of Miletus, one of the architects of the great church of Saint Sophia in Constantinople. He also had a school in the city and wrote commentaries on the *Elements*.

The fifteenth book does not reach the high standard of the work to which it is appended. Mathematics in general, and geometry in particular, had begun to decline about the beginning of the first century AD. Few new results were obtained and mathematicians were reduced to studying the works of their predecessors, finding alternative proofs, or filling in gaps where manuscripts had been lost. The Greeks had concentrated on geometry and had restricted this to deductions from the properties of lines and circles. The main reason for doing this was to solve the problem of existence: objects could be constructed using simple operations to ensure that definitions were consistent. This narrowness of vision and the insistence on completeness allowed a thorough examination of a small field to be conducted. But the Greeks' conception of mathematics had severe limitations. They valued the aesthetic beauty of abstract mathematics and wanted their knowledge to be secured on the solid foundations of obviously true axioms. This led them to reject any notion that involved the infinite whether it be infinitely large or infinitesimally small objects, or endless processes. Even the length of the diagonal of a unit square was never fully accepted as a number. Irrational quantities were limited to geometric interpretations. Thus Greek contributions were mainly in the areas of number theory and geometry. They mined the vein of line and circle geometry so thoroughly that it was hard to see how to continue. Further progress in the same direction was very difficult.

The Alexandrians, and those who followed them, had to broaden their outlook in order to proceed. Geometry continued to be studied but it did not make any major advances. Instead, people concentrated on arithmetic and algebra (subjects where the Greeks felt inhibited) and trigonometry, the mathematics motivated by astronomy. These were young subjects for investigation and could be advanced relatively easily.

With the rise of Alexandria, the Greek schools had gone into decline. The Romans, who controlled central and northern Italy and the Greek colonies in southern Italy and Sicily, conquered Greece in 146BC, and Mesopotamia in 64BC. Unable to ignore a major power on the north African coast, they attacked Alexandria. Part of the library was destroyed when they set fire to the Egyptian fleet at harbour in the city. Some of the manuscripts escaped this tragedy only to be destroyed later by Christian and then Moslem raids on the city. The first of these was a consequence of Theodocius' order in 392 to destroy Greek temples throughout the empire. The spread of Christianity had led Constantine to adopt it as the official religion, giving the Christians greatly increased influence. Opposed to any pagan learning (which included much Greek thought) they sought to eradicate it. Plato's academy lasted until 529 but this, too, came to an end when the Byzantine emperor Justinian closed all the Greek schools.

The rise of Islam

The fall of the Roman empire in the mid-fifth century signalled Europe's descent into the Dark Ages. East–West trade ceased, engineering projects lay abandoned, many skills were forgotten. The Church became a powerful body and exerted its influence on many aspects of life. Latin, the language of the Church, became the language of scholarship. A tiny remnant of the former learning was preserved in the monasteries but there was a general disinterest in the physical world.

Spiritual concerns were not confined to Europe. The seventh century witnessed intense religious fervour in Africa and western Asia. As Europe was being overrun by Huns, Goths and Vandals, the Arabs united under Mohammed and began to build the Islamic empire. The capture of Egypt and Alexandria in 640 meant many of the remaining manuscripts were burned. Most of the scholars fled the city. Many, seeking relative safety, travelled to Constantinople taking their precious manuscripts with them.

By 642 the Moslem conquest of Persia was complete. During the next hundred years Islam spread east across India to the Chinese border and west across north Africa and southern Italy to Spain and southwest France. The early political centre of this empire was at Damascus. In 755 the empire split into two kingdoms:

the Umayyad caliphate reestablished itself in Cordoba in southern Spain and the Abbasids remained in the East. The second Abbasid caliph, al-Mansur, moved his capital to Baghdad. The city was to become a flourishing cultural and intellectual centre—a new Alexandria. The ambitious programme was carried out by caliphs Harun al-Rashid and al-Mamun. They erected a library, an observatory, and an institute for translation and research known as the 'House of Wisdom'. Together these housed some of the greatest scholars of the period. Here, they had access to all the major Greek works, either in Greek or as translations into Syriac or Hebrew. The empire subsumed any Alexandrian learning that had survived. People were sent to seek out and buy manuscripts in foreign lands. In fact, manuscripts were so highly prized that al-Mamun obtained them as part of peace treaties. The classical works of science and mathematics were translated and became part of the Arabic heritage.

Moslem geometry owes most to writers like Euclid, Archimedes and Heron but other civilisations also had things to contribute. For example, an approximate construction for a regular heptagon in its circumcircle is known as the *Indian rule* which may indicate its origin. The side of the heptagon is taken to be the altitude of one of the six equilateral triangles forming a hexagon inscribed in the same circle. (Recall that an exact construction for this polygon using the Euclidean tools is impossible.)

Thabit ibn Qurra

One outstanding translator, who also made original contributions to algebra and geometry, was Thabit ibn Qurra (836–901). He belonged to the Mandaean sect whose astrology had much in common with that of the Babylonians. The Mandaeans produced many astronomers and mathematicians. Thabit's gift for languages and his mathematical ability gave Arabic some of its best translations. They also include fine illustrations. Where men less skilled in mathematics copied manuscripts they carefully left spaces for diagrams but, unfortunately, many have not been drawn in.

Thabit's *Kitab al-Mafrudat* (Book of Data) was very popular during the Middle Ages and was included in a compilation volume by Nasir al-Din al-Tusi, together with the *Elements* and the *Almagest*. It contains problems in elementary geometry, geometric algebra and some constructions. Another work, *Kitab fi Misahat al-Ashkal al-Musattaha wa'l-Mujassama* (Book on the Measurement of Plane and Solid Figures) contains rules for computing the areas of plane figures and the surface areas and volumes of solids. It contains a rule by Thabit (though his proof has not survived) for calculating the volumes of truncated pyramids and cones. If A and B are the areas of the base and the top, and h is the height of such a solid then its volume is given by

$$\text{volume} = \frac{1}{3} h \left(A + \sqrt{AB} + B \right).$$

He also wrote on the 'construction of a circumscribed solid with fourteen faces'—
the cub-octahedron.

Abu'l-Wafa

The last great representative of the Baghdad school of mathematicians was Abu'l-
Wafa (940–998). Of Persian descent, he moved to Baghdad in 959. His work
Kitab Fi Ma Yahtaj Ilayh al-Kuttab Wa'l-ummal Min Ilm al-Hisib (Book on
What Scribes and Businessmen Need from Arithmetic) was very popular. In it
he sets out the methods of calculation used by merchants and finance clerks in a
systematic manner. He also gives some methods used by surveyors and ridicules
the $\frac{1}{2}(a + b)\frac{1}{2}(c + d)$ rule for the area of quadrilaterals. He remarks that it is
obviously incorrect and rarely corresponds to the truth.

Another of his works, the *Kitab Fi Ma Yahtaj Ilayh al-Sani Min al-Amal
al-Handasiyya* (Book on What the Artisan Needs from Geometry) describes two-
and three-dimensional constructions. Some of them are original but many are
taken from Euclid, Archimedes, Heron and Pappus. He includes constructions
of Platonic and some Archimedean polyhedra in their circumspheres. Most constructions are exact and use the Euclidean tools but a few are good approximations such as the construction of a heptagon. However, Abu'l-Wafa is best
known for his constructions using straight edge and compasses of fixed opening.
These so-called *rusty compass* constructions had been studied earlier but he conducted an extensive exploration of the field solving a large number of problems.
The inclusion of these constructions in a book for craftsmen is easily explained—
in practical applications they are more accurate than the standard Euclidean
methods.

Europe rediscovers the classics

At the turn of the eleventh century, Christianity began to surge back in northern
Spain, and the East was overrun by the Seljuk Turks. The following two centuries
also saw the crusades. Through these, and the revived Mediterranean trade
routes, northern peoples came into contact with many new ideas. They also
became aware of the ancient Greek learning. As knowledge of the Greek classics
began to filter into western Europe, scholars travelled to Africa, Spain, and the
Near East to buy manuscripts. They sought both Arabic translations of Greek
texts and Moslem learning. Much of it became easily accessible when Toledo was
recaptured by the Christians in 1085. Like the ninth century, the twelfth century
became a century of translation. Latin versions of works by Euclid, Ptolemy,

Aristotle, and Archimedes soon appeared. Just as the Moslems had sought out scientific works, the Europeans also concentrated on science paying little attention to literature.

This influx of Greek learning led scholars to examine the physical world around them. Attempts at rational explanations of phenomena became increasingly more common. Although the first half of the *Timaeus* had been available in Latin since the fourth century, it had not attracted much attention. Now it was studied in the cathedral schools and the Platonic vision of understanding nature through mathematics reinforced their new way of looking at the world. Chalcidius' translation did not include Plato's own geometric theory of the universe—his atomic theory based on the regular solids. The principal source of knowledge of this theory came from Aristotle's criticism in *De Caelo et Mundo*. This was translated into Latin in the early thirteenth century and Aristotle's vigorous attack was almost universally accepted.

A part of the *Timaeus* that medieval scholars found more relevant was Plato's discussion of light and vision. The science of optics was virtually unknown in medieval Europe but by the Renaissance it had become a university subject.

Optics

Speculation on the nature of light and vision goes back to antiquity. Greek thought on the subject was collected by Euclid in the third century BC and recorded in his *Optica*. It postulates that light travels in straight lines, and that light rays connect the eye and the object being observed. The phenomenon of sight was explained in various ways by different schools of philosophy. Some proposed that the eye was a source of light rays which then interacted with the object; others that the object itself emitted an image-like essence which radiated in all directions carrying the form of the object to the eye. Euclid, however, does not concern himself with the physiology of sight, but concentrates on the geometry. Irrespective of the mechanism involved, if light travels in straight lines then the rays which convey the image of an object to the eye can be thought of as a pyramid or cone with the eye at the apex and the object as the base. The principles of geometry can be applied to this pyramid, and Euclid showed that the apparent size of an object depends on the angle at the apex. The larger the angle, the larger the object appears to be.

Claudius Ptolemy also wrote on optics. He discussed reflection and mirrors, and refraction. He was also interested in the perception of colour. He noted that light rays reflected from a surface take on its colour, and that the more distant an object is, the duller it appears. Artists of his day made use of the latter phenomenon applying bright colours for near objects and darker shades for those farther away. This technique is now called atmospheric perspective.

The Moslems investigated mirrors and lenses, spherical aberration, and the formation of rainbows. The most influential writer was Abu ali al-Hasan ibn al-Hasan ibn al-Haytham (965–1040), known in the West as Alhazen. He maintained that light was produced by the sun and candles, and that when an object is illuminated by such a source some of the light is reflected towards the eye. Despite much experimentation, he was unable to fully explain the properties of refraction. However, he used the phenomenon to give an explanation of the visual process. Drawing on the writings of Aristotle, Euclid, Ptolemy and Galen, he used the visual pyramid and the laws of optical geometry to show how an image could be conveyed into the eye from where the form of the object can be recognised by the brain. Alhazen's book *Kitab al-Manazir* was introduced into Europe in a Latin translation entitled *Perspectiva*. Around 1265 John Peckham wrote *Perspectiva Communis*, which was essentially a summary of Alhazen's work. It was very popular and achieved a wide circulation.

At this time the Latin word 'perspectiva' referred to the study of optics and visual perception. It was not concerned with what we think of as perspective: an artistic technique for representing space on a flat surface. Later, the notion that viewing a painting should be like looking through an open window was connected with the visual pyramid of light rays converging on the eye. Mathematics was then applied to justify the artistic process: if a painting is to be a cross-section of the visual pyramid then *natural* perspective (optics) explained the geometric constructions used in *artificial* perspective (art). As we shall see below, art was one of the important factors in the rediscovery of solid geometry.

Campanus' sphere

The twelfth-century translations of scientific and mathematical works were mainly from Arabic sources. Translations of the *Elements* were made by people such as Adelard of Bath and Gerard of Cremona. Gerard's translation was one of the better versions since he used one of the best Arabic sources—a translation made by Thabit. However, the most widely used edition of the *Elements* until the Renaissance was not a direct translation. It was written in the late 1250's by Campanus of Novara and is a reworking of several earlier editions. At least one of his sources seems to be based on Adelard's translation. Campanus was a competent mathematician who understood and communicated Euclid's geometry well. His clear exposition of all fifteen books became *the* version of Euclid read in the Middle Ages and, subsequently, the first to appear in print. It retained its position as the definitive version of the *Elements* until translations from Greek sources appeared in the sixteenth century.

In proposition 17 of book XII Euclid shows how to construct a polyhedron sandwiched between two concentric spheres. (He goes on to apply Eudoxus' method of exhaustion to this setup to show that the volume of a sphere varies as the cube of its diameter.) The construction produces a polyhedron with n rings, each containing $2n$ faces. So the polyhedral sphere has $2n^2$ faces in total. The 72-faced example (shown in Figure 3.2) obtained by taking n to be six was described by Campanus. During the Renaissance this polyhedron attained a status and popularity equal to that of the Platonic solids, as we shall see later. The construction of this polyhedral sphere is outlined below.

Proposition. How to construct a polyhedron which approximates a sphere.

Let N and S be the North and South poles of a sphere and let C be its equatorial circle. Inscribe in C a regular polygon with an even number of sides, say $2n$. We can now construct n circles, each of which passes through the two poles and two opposite corners of the equatorial polygon. Into each of these circles we can inscribe a regular $2n$-gon so that the points N and S appear as corners. These correspond to lines of longitude on the sphere. The polyhedron is completed by connecting up the vertices of these n polygons with lines of constant latitude in planes parallel to the equator. The result when $n = 6$ is shown in Figure 3.2. ∎

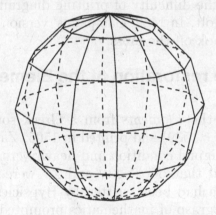

Figure 3.2. A polyhedron which approximates a sphere.

Collecting and spreading the classics

The thirteenth century saw the new universities replace monasteries as the centres of learning. The writings of Aristotle became very popular and the revived interest in mathematics was transferred to the study of scholastic physics. At this time, trade across the mediterranean increased and Italy became the gateway into Europe. Goods from Asia and north Africa entered its seaports, and the banking houses made it a great financial centre. The great wealth this produced supported the advance of learning and the arts.

The passion to rediscover the classics of antiquity led to the rapid growth of libraries in the fifteenth century. These collections, much larger than the libraries of the Middle Ages, had a strong bias towards classical works and included many Greek manuscripts. This was particularly true in the case of mathematics texts. Whereas medieval collections contained only Euclid and Archimedes in Latin translation, the humanists assembled an almost complete corpus of mathematical writings including works by Apollonius, Heron, Proclus and Pappus in the Greek language. This endeavour was greatly assisted by the collapse of the Byzantine empire. The fall of Constantinople to the Turks in 1453 led to a flood of scholars fleeing to Italy bringing their treasured manuscripts with them.

The fifteenth century also witnessed many other changes in Europe. The secrets of making paper and gunpowder arrived from China. Gunpowder radically altered the nature of warfare and fortifications had to be redesigned. The study of projectiles also became important. Paper replaced parchment and combined with the invention of printing with movable type to produce a revolution in the spread of information. Although books by both ancient and contemporary writers were printed daily in Venice, it was thirty years before a mathematics text appeared. This was partly due to the difficulty of printing diagrams but this problem was overcome by Erhard Radolt. In 1482, Campanus' version of the *Elements* became the first mathematics book off the press.

The restoration of the Elements

The first translation of the *Elements* from a Greek source was made by Bartolomeo Zamberti (1473–*c*.1539) and printed in 1505. Zamberti sought to restore the *Elements* to their original condition and heavily criticised Campanus for his inaccuracies. He realised that books XIV and XV were not part of the original corpus and thought both had been written by Hypsicles. Zamberti's attack on Campanus and his poor grasp of mathematics prompted retaliation. Campanus' edition was defended by Luca Pacioli, who revised it and republished it in 1509. Pacioli thought the source of the errors in Campanus' text did not rest with the author but was to be found in the manuscripts he had available: careless copyists introduced many mistakes, especially into diagrams.

Even though Zamberti worked from Greek sources and tried to produce a faithful translation, he failed to spot errors in the mathematics. The inadequacy of Zamberti and other translators as mathematicians led to the realisation amongst many scholars that good translations of mathematics texts were still desperately needed. Ideally, these new translations should be made by mathematicians who knew Greek. It is part of the beauty of mathematics that the restoration of mathematical texts is possible at all. Translators in other fields perpetuated the errors in corrupted texts, often without recognising the faults. This was also the case in mathematics, where few translators were expert in the field. But the logical development and internal consistency of mathematics makes it clear when something is amiss. These same features also allow the original meaning of a text to be recovered even if the original words are lost.

The desire to revive the study of mathematics and repair the ancient texts became an obsession for many Italian mathematicians. Francesco Maurolico (1494–1575) was one such man. His father had left Constantinople and settled in Sicily where he taught his son mathematics and astronomy. Maurolico's knowledge of Greek and mathematics meant he had both the linguistic and technical skills to reconstruct the classical texts. His translation of the *Elements* covered all fifteen books but, sadly, his labour contributed little to the revival of mathematics. Nearly all his work exists in manuscript form but only books XII to XV on solid geometry were published after his death. The translator whose edition of the *Elements* made the biggest impact was Federigo Commandino (1509–1575). He was one of the leading translators and editors of Greek classics in the sixteenth century. His version of the *Elements* was printed in 1572 and it dominated geometry until the nineteenth century.

A new way of seeing

The new way of looking at the world which arrived with the rediscovery of ancient knowledge influenced art. As graven images are condemned in the Bible, the decoration allowed in the early churches was strictly limited. Its purpose was to glorify God, and Biblical themes were very popular. Paintings were used to educate as well as decorate. Many of the Church's members were illiterate and the pictures served as reminders of its teachings. They had to narrate a story simply and clearly, concentrating on the essential elements. Generally, the paintings have a gold background indicating that something special is happening. The scene is not natural, rather something supernatural or miraculous. The characters which inhabit these paintings are stiff. They show no emotions and seem lifeless, even weightless—a quality which adds to their ethereal appearance. The Greek mastery of motion and expression is completely absent. Where there is variation in height it conveys the relative importance of objects or some social hierarchy: the saints are depicted larger than the people.

The infusion of classical learning into Europe during the twelth and thirteenth centuries, with its emphasis on man and the universe, stimulated artists and scholars to study their surroundings. Their new interest in nature and their attempts to record it revealed the lack of realism in their art. They became aware of the lifelessness of their pictures. Gradually the solid gold backgrounds were replaced by blue, and other features were introduced to suggest a location, a room interior for example. Shading was used to give the impression of volume.

In their early attempts to convey solidity, painters depicted objects as they were experienced rather than as they actually looked. A rectangular table would be shown as in Figure 3.3(a). The object is drawn as though seen from many sides simultaneously instead of from a single fixed viewpoint. This produces a result in which there is a diminution in size towards the observer. In contrast to the optical perspective which was to come, this tendency is sometimes called inverted, or reversed, perspective, although there is no systematic reduction in scale as there is in the later mathematically-based system.

Diagrams of this form continued to be used in mathematical and scientific treatises long after optical perspective was introduced into art. A cylinder, for example, would be illustrated as shown in Figure 3.3(b) where its most notable characteristics, circular ends and parallel sides, are combined.

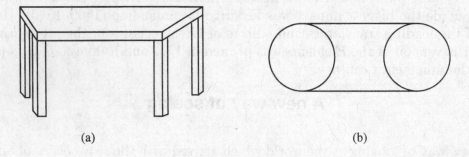

(a)　　　　　　　　　　　　　　　　　　(b)

Figure 3.3.

This portrayal of subjective experience gave way to a more objective approach. The artist became detached from the scene and observed from a single viewpoint rather than imagining himself involved in it. The compositions showed more purposeful arrangements of objects and consistency in their spatial relationships; they were no longer viewed as collections of separate objects, each dealt with independently.

By the fourteenth century, artists had achieved realistic representations of space. Objects had mass and volume, and depth was convincingly portrayed. There was an intuitive feeling of the geometrically precise perspective to come. Lines that were above eye-level sloped downwards as they receded; those below eye-level sloped upwards. Lines to the left of centre moved to the right as they went into the distance; those on the right inclined to the left. These general principles meant that lines which, in reality, were parallel to each other and skew to the picture plane, showed a tendency to converge, although not always to a single point. Despite what we would regard as a high degree of inaccuracy in these constructions, a satisfactory illusion of space can be obtained by following these simple rules of convergence. Oblique views of buildings became popular settings because the converging lines they contain produce a strong three-dimensional effect. Giotto and Duccio often achieved a good impression of space in this way. They were among the first artists to display substantial-looking buildings with convincing structures.

Over this period the idea that a painting should be like a window evolved. The observer looking at a canvas should not see a flat image representing the world, but rather see through the canvas, as through an open window, to a scene beyond. Painters aspired to produce a truly three-dimensional illusion, to portray space and not merely suggest it. The ultimate illusion was performed in public on the steps of Florence cathedral some time in the first quarter of the fifteenth century. In a now famous demonstration, Filippo Brunelleschi (1377–1446) exhibited a picture which not only looked realistic but which actually produced the same image in the eye as did the real scene. He had painted a picture of the baptistry reflected in a mirror so that his painting was a mirror image of the actual building. He made a small eye-hole in the board and invited people to look through the eye-hole from the back of the board at the real baptistry. He then held a mirror in front of the painting so that the observer saw a reflection of the painting. By removing and replacing the mirror the observer could compare the painting with the real scene. We do not know what motivated Brunelleschi nor how he created his picture. However, the results were so convincing that this event is now taken to mark the beginning of perspective theory in art.

Perspective

Up until the middle of the fifteenth century, the Latin word 'perspectiva' was not used in connection with artistic attempts to represent space. Earlier treatises on the subject dealt with *natural* perspective, that is, optics and visual perception. None were concerned with representing space on a flat surface. At some time, the notion that viewing a painting should be like looking through an open window must have been connected with the idea of a visual pyramid of light rays converging on the eye: for the eye to receive the same impression from a painting as from the original scene, the picture must be a cross-section of the visual pyramid.

As an architect, Brunelleschi would have been familiar with making ground plans and elevations of buildings. Given the desire to construct a 'realistic' picture, the information in such plans is sufficient to construct geometrically the intersection of a picture plane and visual pyramid. One merely needs to add the position of the observer and the picture plane to the plans (see Figure 3.4). Then the ground plan (the scene viewed from above) is a horizontal section through the visual pyramid, and the elevation (the side view) is a vertical section through the pyramid. The picture plane is a transverse section through the pyramid and appears as a line on both plans. The lateral position of a point in the picture can be read off from the ground plan, and its height from the elevation. Combining these two pieces of information allows the image of the point to be located precisely on the picture plane.

The completed image is dependent on the position of the observer. As the distance from the observer to the building changes, the view of the building alters in response. The distance of the picture plane from the observer does not affect the perspective construction—only the size of the image changes, not its proportions. It is this that is the essence of linear perspective. However, the distance of the picture from the observer is vital when it comes to viewing the resulting painting. If the picture is to produce the same image in the eye as the real scene then the eye has to be placed at the apex of the visual pyramid. Thus, if a realistic effect is to be achieved, the ratio of the width of the picture to its distance from the observer must be the same for both the construction and the final viewing.

After constructing several of these images some shortcuts become apparent. Any lines parallel to the picture plane are undistorted; parallel lines which are skew to the picture plane converge and meet in a single point. Furthermore, all such vanishing points lie on a single line—the horizon. Whether or not Brunelleschi went through a process such as this when he discovered artificial perspective, this kind of technique was described in later manuals for artists.

Early perspective artists

For sometime after Brunelleschi's demonstration, the techniques of perspective painting were passed from one artist to another. The systematic rules are easily learned and applied to produce realistic looking structures. However, their use does not guarantee aesthetically pleasing or artistic results. Perspective accuracy is only one factor influencing the overall impression given by a painting. Achieving a unified and balanced composition is far more important than constructing a strictly mathematical diagram. Among the first to use perspective constructions were Brunelleschi's friends Masaccio, Masolino and Donatello. Masaccio quickly found that strict perspective can give harsh results and he made instinctive adjustments to the rigid formulae to produce more comfortable compositions. Other fifteenth-century artists also felt free to accept or ignore perspective as their compositions required. It is the talent of great artists to know when and how to break the rules.

The overpowering impression of depth which can result from strict application of perspective can be opposed in a variety of ways. The sense of depth can be lessened by reversing the atmospheric perspective—applying bright colours for distant objects, dull colours for near ones. Another technique is to choose the

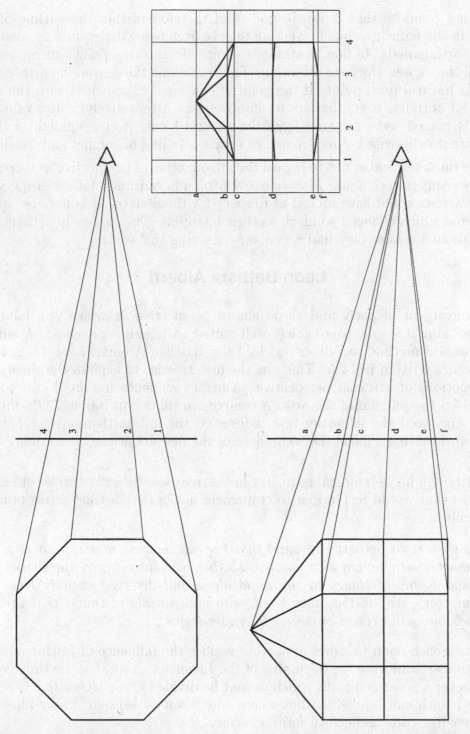

Figure 3.4. Using a plan-and-elevation diagram to construct a perspective view.

vanishing point so that it lies behind (and therefore within the outline of) an object in the painting, thus preventing the eye from being drawn into the distance by the orthogonals. In fact, a strategic choice of vanishing point can be used to control the viewer's focus of attention. The eye follows the converging orthogonals to their natural limit point. If the point of convergence coincides with the main centre of activity, it emphasises its importance. Alternatively, if the vanishing point is placed some distance from the natural focus, it can highlight a detail that might otherwise be overlooked, or create a feeling of conflict and tension.

We must remember not to regard the introduction of perspective as merely an improvement in technique. Masaccio loved to paint naturally falling drapery but earlier artists could have looked at drapery for themselves. It is not the artists' technique which changed so much as their intention. The change in artistic style came about because they had a new way of seeing the world.

Leon Battista Alberti

The conversion of plans and elevations to perspective drawings is a laborious process, and it is not particularly well suited to imaginary scenes. A simpler construction method was described by Leon Battista Alberti (1404–1472) in his *De Pictura* written in 1435. This was the first treatise to explain the theory and constructions of artificial perspective. Alberti's audience for this Latin volume comprised the patrons of the arts. A companion volume in Italian *Della Pittura*, which appeared the following year, addressed the artists themselves. Through these works Alberti made the techniques of the new art accessible to many other artists.

Although he uses optical geometry in his treatises, Alberti expresses the ideas in terms that would be familiar to craftsmen, not in the abstract terms of mathematicians.

> I wish it to be borne in mind that I speak in these matters not as a mathematician but as a painter. Mathematicians measure the shapes and forms of things in the mind alone and divorced entirely from matter. We, on the other hand, who wish to talk of things that are visible, will express ourselves in cruder terms.[b]

This approach soon becomes apparent, as does the influence of Euclid. His first definitions paraphrase the beginning of the *Elements*. A *mark* is anything visible to the eye; a *point* is a mark which cannot be divided into parts; a *line* is a mark whose length can be divided but whose width cannot be split; many lines close together like threads in cloth form a *surface*.

In order to paint as though the picture were an open window, it is essential to be able to judge the position and scale of all the objects in the painting.

Alberti established the size and location of objects on a canvas by marking out a tessellated pavement to act as a reference grid. The pavement, which is sometimes called the ground plane, is regularly subdivided and diminishes correctly as it recedes into the distance. It determines the position and proportion of the other objects in the scene. The fundamental axiom of perspective is that the image of an object is reduced linearly—its proportions remain the same, only the scale changes.

The pavement forms the heart of Alberti's method so it is fundamental to be able to construct it correctly. This is precisely the problem which earlier artists had failed to solve adequately. One method in common usage was to make constant ratios between adjacent rows of tiles—take a third off the depth each time. Alberti's geometrically-based method which replaced this became known as *costruzione legittima*—the legitimate construction.

Paolo Uccello

The new perspective geometry applied most naturally to objects composed of straight lines and plane surfaces. Although the principles of perspective are simple, applying them can have problems. As a result the kinds of physical spaces and objects which artists dared to tackle were greatly simplified. Fifteenth-century pictures contain many more right angles and straight lines than appear in nature or earlier paintings. Buildings still provided popular backdrops: external architecture and room interiors could be constructed easily and accurately. In some cases the grid described by Alberti as part of the construction process appears as a patterned floor or pavement in the finished painting.

Besides buildings, another source of rectilinear objects was provided by solid geometry. Polyhedral objects and frameworks were frequently used as exercises in perspective construction. Paolo Uccello (1397–1475) took great interest in geometry and perspective. He devoted much of his time to the study of perspective problems, constructing complex objects in great detail. In his *Lives of the Painters, Sculptors and Architects* the biographer Giorgio Vasari (1511–1574) writes of Uccello

> When Paolo showed his intimate friend, Donatello the sculptor, maz-zocchi [polyhedral tori] with projecting points and bosses, represented in perspective from different points of view, spheres with seventy-two facets like diamonds, and on each facet shavings twisted round sticks, with other oddities on which he wasted his time, the sculptor would say, "Ah, Paolo, this perspective of yours leads you to abandon the certain for the uncertain; such things are only useful for marquetry, in which chips and oddments, both round and square, and other like things are necessary."[c]

Vasari was of the opinion that Uccello spent too much time on his perspective constructions and that his paintings would be improved if he devoted himself to animals and figures in the same way. Notice also the reference to Campanus' sphere.

Between 1426 and 1431 Uccello was in Venice. It is thought that he was involved in designing the patterns in the floor of the Basilica of San Marco. Some of the designs have a striking three-dimensional quality due to the careful juxtaposition of contrasting shades of marble. In one of the doorways at the west end is a panel showing a polyhedron surrounded by a necklace of hexagonal prisms (see Plate 7). This design is often attributed to Uccello.

Polyhedra in woodcrafts

In the above quotation from Vasari, Donatello mentions perspective in association with marquetry. Wood was a popular medium in which the new art was expressed. The intarsia craftsmen excelled in the study of perspective, so much so that they were often called 'maestri di prospettiva'. Many inlaid pictures survive to show their virtuosic skill. The early panels depicted simple geometrical objects or views of buildings in perspective. Some motifs became so popular they were a kind of trademark. The relatively simple forms of polyhedra were a favourite theme. The Platonic solids and some Archimedean solids were used, as was the 72-faced sphere described by Campanus. Another common object was the *mazzocchio*—a kind of polyhedral torus. An example is included in Figure 3.5. The name is derived from a form of headwear. Other popular motifs were books, scientific instruments such as armillary spheres, and musical instruments like the lute and the organ.

Figure 3.6 shows a still-life design for a marquetry panel by a Florentine craftsman, dating from about 1470. To make the construction more difficult, thus showing greater skill and technical proficiency with the new techniques, hollow forms of the polyhedra were used so that the faces at the rear were made visible. These frameworks of edges are called *skeletal* polyhedra. Polyhedra inscribed in each other were also used. These combinations of polyhedra were taken up by carvers of wood and ivory. They placed polyhedra inside each other so that the smaller ones were imprisoned by the larger ones but rotated freely.

The best surviving examples of intarsia are in Verona and Urbino. Fra Giovanni da Verona created some magnificent panels in the choir-stalls and the sacristy in the church of Santa Maria in Organo, in Verona. Two of the scenes in the sacristy contain perspective pictures of polyhedra.[2] One, containing Campanus' sphere, an icosahedron and its truncated form, is shown in Plate 6. The

---

[2]If you want to see them, you will need to ask to be shown the sacristy.

Courtesy of the Ministero per i Beni Culturali e Ambientali, Urbino.

Figure 3.5. A panel from Duke Federigo's study in his palace at Urbino.

other panel also contains skeletal forms: an augmented cube, an augmented icosi-dodecahedron, and a cub-octahedron.

Duke Federigo's study in the palace at Urbino is lined with marquetry panels, attributed to Baccio Pontelli. Some depict open cupboards containing things such as musical and scientific instruments, and books. The one shown in Figure 3.5 contains a polyhedral torus. Other examples of these mazzocchi can be found in the choir-stalls of Modena cathedral.

There are also examples of polyhedra in the royal palace at El-Escoriel just outside Madrid. The palace was erected by Philip II (1527–1598), who is said to have excelled in mathematical studies as a young prince. The doors to the throne-room at the palace were a gift from his father-in-law, Maximilian II. They were ornately carved and inlaid by German craftsmen. The intarsia panels contain some of the typical elements (lutes, books) and some polyhedra.

Courtesy of The Art Museum, Princeton University. Gift of Frank Jewett Mather, Jr.

Figure 3.6. A fifteenth-century still-life design for a marquetry panel.

Piero della Francesca

The principles set down by Alberti which allowed artists to produce realistic representations of space are not sufficient to enable complex designs such as polyhedra to be constructed. Although the methods for constructing polyhedra accurately were not included in *Della Pittura*, it is clear that techniques for doing so were known, certainly in the second half of the century if not before. They were first written down by Piero della Francesca (*c.*1410–1492), and after this, perspective constructions of polyhedra became a standard feature of painters' manuals. Vasari wrote of him:

Piero was, as I have said, a diligent student of his art who assiduously practised perspective, and had a thorough acquaintance with Euclid, so that he understood better than anyone else all the curves in the regular bodies and we owe to him the fullest light that has been thrown on the subject.[d]

In the early 1470's Piero gave up his career as an artist and concentrated on his studies of the mathematical theory of perspective. His work *De Prospectiva Pingendi* was written some time between 1482 and 1487 but the ideas it contains were formed much earlier. He describes two methods of converting plan views of objects to perspective views in a picture plane. One of these is similar to the method described above by which Brunelleschi might have constructed perspective views of buildings. Piero's other method is an extension of Alberti's idea using a tessellated pavement of square tiles. Alberti and Piero probably met quite often at Urbino.

Figure 3.7 is Piero's diagram showing how to construct a regular pentagon lying in the ground plane. The point labelled A is the central vanishing point, and $BCED$ is a square projected into the picture plane. Imagine that this square is not an image composed of lines in the picture plane but an actual tile existing behind the picture plane. If the edge BC is regarded as a hinge then this tile can be rotated so that it hangs down under the hinge. To foreshorten a figure, in this case a regular pentagon, first draw it in this genuine square on the paper. The problem now is to transfer this figure to the image of the square. To facilitate this, the diagonal BE is added to both the square $BCED$ and its image.

To show how the image of a point in the figure may be located, the corner labelled H will be used as an example. Transferring the lateral coordinate is straightforward: construct a vertical line through H until it meets BC, then join its end point on BC to the vanishing point A. The image of H must lie somewhere on this line. The diagonal BE is used to determine the precise location. First construct a horizontal line through H and let N be the point where it crosses BE. The image of N can be found easily since its lateral position can be determined in the same way as that for H, and its image must lie on the image of the line BE. The image of the horizontal line NH is also horizontal. Thus the image of H lies at the intersection of the two lines which are known to pass through it. Proceeding in this way, all the corner points of the pentagon can be transferred to the image plane. Piero goes on to describe how to locate images of points which do not lie in the ground plane. This method of constructing an object point by point may be tedious and time consuming but it is not difficult. Given sufficient patience any point can be constructed and complicated patterns can be properly foreshortened.

Besides being an artist, Piero was also a competent mathematician of his time. His two mathematical treatises are *Trattato d'Abaco* written about 1450,

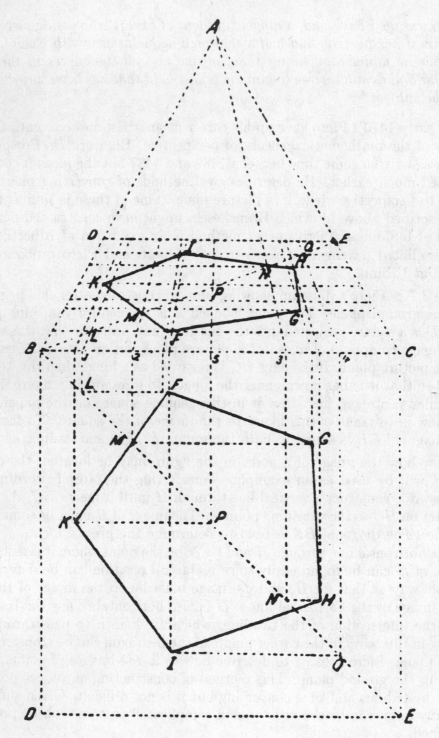

Figure 3.7. Piero's perspective construction of a regular pentagon.

and *Libellus de Quinque Corporibus Regularibus* (Book on the Five Regular
Bodies) written about thirty years later. The latter work is dedicated to Duke
Guidobaldo and is concerned with the mensuration of regular polygons and polyhedra. Piero takes the fixed proportions between the geometric solids and uses
them as arithmetic problems. For example, given a cube inscribed in a sphere of
diameter seven units, find its surface area. Again, if an octahedron is inscribed
in a cube composed of squares of side 12 units how long is the side of the octahedron? Archimedean solids also feature in the exercises: given a polyhedron
composed of four triangles and four hexagons (a truncated tetrahedron) inscribed
in a sphere of diameter 12 units, find its sides and surface area.

Piero's *Trattato d'Abaco* is one of a long line of *abaco* texts, all ultimately derived from two works by Leonardo of Pisa (*c.*1170–1250), also known as Fibonacci.
Fibonacci was educated in north Africa and travelled widely. He introduced Arabic numerals and methods of calculation into Europe in his *Liber Abbaci* (1202)
and translated some Greek and Arabic geometry and trigonometry into Latin
in his *Practica Geometriae* (1220). He lists three of the five Platonic solids and
refers interested readers to Euclid for further information. Piero must have followed this direction and gone to Campanus' version for he discussed all five solids,
and some parts of his *Quinque Corporibus Regularibus* come from book xv.

Abaco texts were primarily concerned with teaching the mathematics of commerce to future merchants. Some were of the recipe type giving rules and worked
examples; others included the theoretical basis for the rules. At this time there
were no standardised weights and measures, nor even containers of uniform size.
Each city also had its own currency. Thus merchants needed to work with proportions (for rates of exchange) and had to be able to find the volume of barrels and
other containers quickly and accurately. Graduated rules seem to have been used
to gauge sizes in northern Europe but in Italy they used geometry. Solids were
divided up into two categories: pyramidal and columnar. The volume of barrels
could be found by treating them either as short fat cylinders or as two truncated
cones stuck end to end. These skills in decomposing complex forms into simpler
ones were also used by artists. The objects whose volumes are easily calculated
are also simple to draw. This produced an affinity between the artist and his
audience for the viewers would be able to analyse a painting into its underlying
forms. They could see a tent as a cone on a cylinder, or a hat as a prism.

Like Piero's book on painting, his abaco is more advanced than similar works
in its field. He does not restrict his applications of geometry to everyday objects
but studies abstract polygons and polyhedra. He even considers the regular solids
and other polyhedra that can be inscribed in a sphere which is certainly far more
than any merchant ever needed.

Luca Pacioli

Luca Pacioli (1445–1517) grew up in Borgo San Sepolcro in Tuscany, the town where Piero had his workshop. It is uncertain whether Piero provided Pacioli's early education but the two were certainly friends in later life. They often met at the Urbino court of Duke Federigo da Montefeltro. It is through Pacioli that Piero's work achieved wide circulation. His manuscripts were kept in the Urbino library and were only printed because Pacioli incorporated parts of them into his own works.

The well-known portrait of Pacioli shown in Plate 5, originally displayed at Urbino, now hangs in the Museo di Capodimonte in Naples. Painted in 1495, it shows the master giving a geometry lesson. The organisation of the painting resembles that commonly used at the time for the title pages of mathematics books. The glass polyhedron floating in the upper left of the picture symbolises the pure and eternal truths of mathematics in the Platonic realm of ideas. The teacher sees this perfect archetype and tries to explain it to his pupil with the aid of books, diagrams and models. The identity of the second figure, who appears disinterested in the lesson, is unknown. One proposal is that it is a portrait of Duke Guidobaldo; another that the artist included himself. (Who painted the picture is another point of debate. It is usually attributed to Jacopo de' Barbari.) More recently, Nick MacKinnon has proposed that the figure is Albrecht Dürer.

Pacioli was ordained as a Franciscan in the 1470's and is shown in his friar's robes. Other objects in the picture include geometrical instruments such as compasses, a model of a dodecahedron, and a slate with 'Euclides' inscribed on the frame. The pages of the book from which he has copied a diagram onto the slate contain the words LIBER XIII, which must refer to Euclid's account of the regular polyhedra in the last book of the *Elements*. Applying perspective geometry to the circle on the slate should have produced an ellipse but the oval in the picture appears not to be a conic section. The other book in the portrait has the letters LI. R. LUC. BUR. inscribed on its spine. This can be decoded as 'Liber Reverendi Lucae Burgensis' (Book by the Reverend Luca of Burgo) identifying the work as Pacioli's encyclopaedia of mathematics *Summa de Arithmetica, Geometria, Proportioni et Proportionalità*, which was printed in 1494 and dedicated to Guidobaldo. It is a compilation which draws on many sources including Piero's *Trattato d'Abaco*.

This pattern was adopted for several other portraits of Renaissance mathematicians. The best-known work of Nicolas Neufchâtel is a portrait of the Nuremberg mathematician and calligrapher Johann Neudörfer with his son, painted in 1561 (Figure 3.8). Neudörfer is shown explaining the geometry of the dodecahedron, a model of which he is holding, while his son attentively takes notes. The Platonic realm is represented by a skeletal cube floating above them.

Courtesy of the Bayerische Staatsgemäldesammlungen, Munich.

Figure 3.8.   Johann Neudörfer and his son (1561) painted by Nicolas Neufchâtel.

In 1496 Pacioli was invited to the court of Ludovico Sforza in Milan to teach mathematics. He stayed until 1499, when the Duke lost the city to Louis XII of France. Here he met Leonardo da Vinci (1452–1519). Amongst many other things, Leonardo was interested in geometry, and in particular, the construction of regular polygons. He knew approximate constructions used by engineers and craftsmen and tried devising constructions of his own, both exact and approximate. Pacioli may have given Leonardo lessons in Euclidean geometry. Leonardo certainly studied Euclid and would have found the Latin text hard to follow. His notebooks show that he studied books I and II on regular polygons, books V and VI on the theory of proportions, and book X also on polygons. His notes are supplemented by summarised extracts from *De Proportionibus et Proportionalitatibus*—the second of the two books of Pacioli's *Summa*.

It was at Milan that Pacioli completed work on his *Divina Proportione*, a work probably in progress at the time of the portrait but which was not published until 1509. Written in Italian, the work comprises three books. The first, *Compendio de Divina Proportione*, is dedicated to Ludovico Sforza. It contains a summary of the properties of the divine proportion, or golden ratio, and a study of polyhedra. The 'best of all' properties was that the diagonals of a regular pentagon divide each other in the golden ratio. An 'almost incomprehensible effect' concerned the icosahedron. The five triangles forming a solid angle of the icosahedron can be thought of as a pentagonal-based pyramid. A rectangle can be inscribed in an icosahedron whose sides are alternately sides and diagonals of such pentagons. Thus the sides of the rectangle are in the golden ratio. Three such golden rectangles can be simultaneously inscribed in an icosahedron so that they are mutually perpendicular (see Figure 2.17).

The first book also contains studies of the regular solids and other polyhedra which can be derived from them. These include the glass polyhedron shown in the portrait, the rhomb-cub-octahedron. Pacioli mentions that this solid and Campanus' 72-faced sphere are useful for architectural purposes as they can be used to construct hemispherical structures such as the dome of the pantheon.

Pacioli uses two procedures to generate new polyhedra—truncation and augmentation. The truncation process is not applied consistently. When pyramids are sliced from the corners of the triangular-faced solids then the usual truncated tetrahedron, truncated octahedron and truncated icosahedron result. The truncated forms he gives of the other two regular solids are what we now call the cub-octahedron and the icosi-dodecahedron. Augmentation is a kind of dual process in which pyramidal pieces are stuck on to the faces of a polyhedron. Pacioli refers to augmented polyhedra as 'elevated'. Augmentation is applied to the five regular solids, the cub-octahedron, the icosi-dodecahedron and the rhomb-cub-octahedron. Illustrations of all these polyhedra were drawn by Leonardo. They are depicted in both skeletal (or vacuous as Pacioli puts it) and solid forms. Some examples are shown in Figure 3.9. The augmented octahedron can be interpreted in several ways and will reappear many times in later chapters. Apart from being an octahedron with a pyramid stuck on each face, it can also be thought of as two interpenetrating tetrahedra.

The second book of *Divina Proportione* is concerned with Vitruvian architecture. The third book *Libellus in Tres Partiales Tractatus Divisus Quinque Corporum Regularium* is an Italian translation of Piero's *Quinque Corporibus Regularibus*. The Montefeltro dukes were patrons of both Piero and Pacioli, and Pacioli would have had access to the original manuscript in the Urbino library.

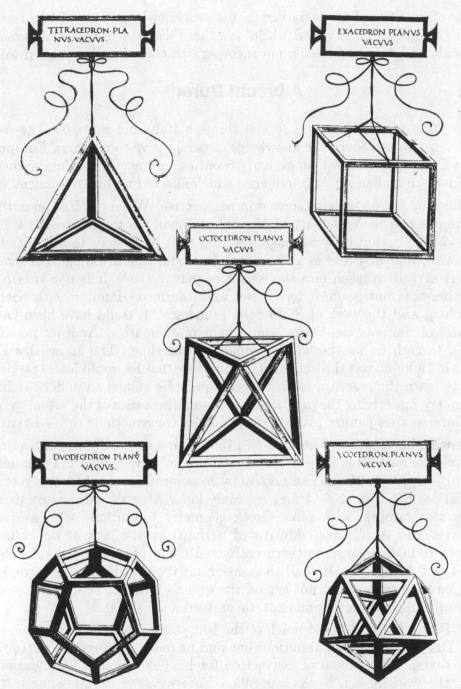

Reproduced from the *Companion Encyclopaedia of the History and Philosophy of the Mathematical Sciences*.

Figure 3.9. Some of Leonardo's drawings for Pacioli's *Divina Proportione*.

Pacioli is often accused of plagiarism for his unacknowledged use of Piero's work. However this may be, it is very doubtful whether Piero's ideas would have reached such a wide audience had Pacioli not incorporated them into his printed works.

Albrecht Durer

The new style of painting soon spread through Italy and beyond. The person who did most to introduce the knowledge of perspective to northern Europe was Albrecht Dürer (1471–1528). In his early twenties, he travelled to Venice where he was exposed to Italian art—an influence which affected all his subsequent work.

He became increasingly interested in perspective. When in 1505, an outbreak of plague reached his home city of Nuremberg, many of its citizens took flight. Dürer took the opportunity to make a second visit to Venice. In the autumn of the following year he wrote in a letter that he would ride to Bologna as someone there had agreed to teach him the 'secret of perspective'. It is not known who this teacher was but he must have been well informed—familiar with both the Milan school and the work of Piero della Francesca. It could have been Donato Bramante, or Scipione del Ferro, who taught mathematics. Another possibility is Pacioli though he was based in Florence at the time. The knowledge Dürer acquired in Bologna was theoretical. (It is unlikely that he would have travelled so far just to learn the practical tricks of technique.) He gained an understanding of the geometry underlying the practice, how the mathematics of the visual pyramid and an intersecting picture plane allowed precise constructions to be derived.

Following his return to Nuremberg in 1507, Dürer devoted much of his time to the study of perspective. Help in this direction may have come from a handbook on the subject, *De Artificiali Perspectiva* by Johannes Viator, which was reprinted by a local publisher in 1509. Dürer certainly knew Alberti's treatise on painting and was also familiar with some Greek geometry by authors such as Euclid, Archimedes and Apollonius. Motivated perhaps by the lack of books on the theory of art in languages northern craftsmen could understand, he decided to write his own handbook. He had thought about the idea before his second visit to Italy but the project did not get off the ground until his return. He planned to cover all aspects of art—practical, theoretical and ethical.

Like Piero, Dürer devoted much of the last years of his life to his theoretical studies. His project was an ambitious one and he made comprehensive studies of classical texts. The manual of instruction for his fellow artists finally appeared in 1525: the work in four books was called *Unterweysung der Messung mit dem Zirkel un Richtscheyt in Linien Ebnen unnd Gantzen Corporen* (Instruction in the Art of Measurement with Compasses and Ruler of Lines, Planes and Solid Bodies).

The first book begins by discussing the basics of geometry (point, straight line) and progresses to more complicated curves such as conchoids, epicycloids and helices. When dealing with the conic sections Dürer fails to overcome the intuitive expectation that an oblique section of a cone must be egg-shaped whereas his method of constructing an ellipse should have given the correct symmetrical form.

Book II is concerned with regular polygons for which Dürer gives both theoretically exact and also approximate constructions. This is the first time that approximate constructions had appeared in print though many of them had been known for a long time. He includes a construction used by the craftsmen of his day that produces an approximate 9-gon, and he gives the Indian rule method of inscribing an approximate heptagon in a circle. Besides a theoretically correct construction of a pentagon taken from Ptolemy, he also includes two approximations, one of which is still one of the quickest methods known.[3] Dürer goes on to show how regular polygons can be incorporated into ornaments, parquet floors and tessellated pavements. Practical problems are also considered in book III, which deals with problems in architecture and engineering.

The fourth book returns to the geometric theme and starts with solid geometry. The Platonic solids and several of the Archimedean solids are discussed together with some polyhedra of Dürer's own invention. He introduces a technique of conveying information about three-dimensional objects on a flat surface via paper-folding which in modern times is called a *net*. The method involves developing the surface of a polyhedron onto a plane sheet of paper so that the resulting figure can be cut out as a single connected piece then folded up to form a three-dimensional model of the original polyhedron. His illustration for the dodecahedron is shown in Figure 3.10.

The techniques for drawing in correct perspective are also included in this book. The notion that a painting can be thought of as the intersection of the picture plane with a visual pyramid with its apex fixed at the viewpoint is vividly illustrated by a series of woodcuts. In one of these, a man maintains a single viewpoint by placing his eye at the top of a small obelisk placed on the table and paints on a glass screen the features of a man sitting on the other side. Although artists' manuals showed how to apply perspective to simple shapes, drawing instruments of some kind were often used to render more complex designs.

The painting-on-glass technique has some limitations. The distance between the eye and the picture plane can be at most an arm's length. This causes apparent distortion when an object has a large depth of field (such as a lute with a long fingerboard, for example). The parts near to the screen appear too large. This can be cured by moving the object further away from the screen, in which case the image is reduced in size, or by moving the viewpoint further back from the screen. The second approach can be achieved by using a piece of rope to follow the path of the light rays from the object to an artificial eye. Jacob

---

[3]See Dan Pedoe, *Geometry and the Liberal Arts*, Penguin 1976, pp66–67 for details.

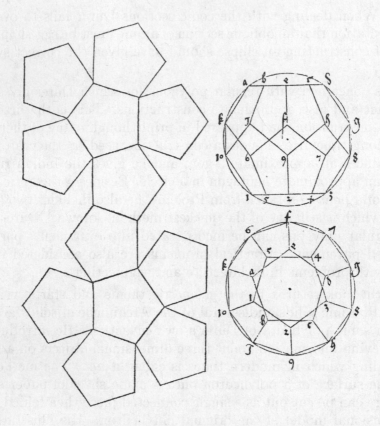

Figure 3.10. Dürer's net for a dodecahedron.

de Keyser is credited with inventing an instrument of this type. The second of Dürer's woodcuts (reproduced in Figure 3.11) shows two men using such a device to produce a perspective image of a lute—one of the set-piece motifs. The complexity of their task is clearly illustrated when we recall that the most complex of the regular solids, the dodecahedron, can be completely determined by constructing just twenty points. In contrast, Dürer's lute has over 150 points.

Wenzel Jamnitzer

Dürer's influence on the theory and practice of art was enormous. This is particularly noticeable in German graphic art. Taking their lead from Pacioli and Dürer, Nuremberg artists adopted polyhedra for perspective studies. A series of books were printed which contained perspective drawings of the Platonic and Archimedean solids and many elaborate variations derived from them. Augustin Hirschvogel's *Geometria* (1543), Lorenz Stoer's *Perspectiva* (1567), and Hans Lencker's *Perspectiva* (1571) are a few examples.

Figure 3.11. One of Dürer's woodcuts illustrating the principles of perspective.

The outstanding work of this genre is Wenzel Jamnitzer's (1508–1585) *Perspectiva Corporum Regularium* (Perspective of regular bodies) published in 1568. I have used his plates for some of the frontispieces of the chapters in this book.

Jamnitzer and his brother were accomplished goldsmiths and jewellers. Sixteenth-century Nuremberg was renowned as a centre for gold work and its goldsmiths worked closely with other local arists such as sculptors and printmakers who supplied them with patterns and models. In 1561 Jost Amman (1539–1591), one of the best graphic artists, arrived in the city. He was a printmaker and designer of jewelry and goldwork, and collaborated with Jamnitzer on several occasions. He made an engraved portrait of his friend at work in his studio involved in a geometric construction (see Figure 3.12).

It was Amman who turned Jamnitzer's perspective drawings of polyhedra into engravings suitable for publication. The high quality of Amman's work and the sheer variety of Jamnitzer's designs combined to make the *Perspectiva Corporum Regularium* a masterpiece. The title page of the work is shown in Figure 3.13. The frame includes the figures of Arithmetic, Geometry, Architecture and Perspective, all depicted with their associated symbols. Notice the visual pyramid which is Perspective's main attribute. These fields echo Dürer's belief

Courtesy of the British Museum, London.

Figure 3.12. Engraving of Wenzel Jamnitzer in his study.

that an artisan should understand geometry and proportion (arithmetic) so that his work is natural and balanced. Jamnitzer was a great admirer of Dürer and owned a collection of all his prints in their first editions.

The main part of the book is arranged in six chapters, each with its own title page. The first five show the influence of Plato's *Timaeus* and the correspondence of the regular solids with the four elements (fire, air, earth, water) and the heavens. The title page of the chapter on octahedra is shown on page 50. The format of a central description of the solid surrounded by objects indicating its associated element is repeated for the other chapters. The symbols of fire on the title page of the chapter on tetrahedral forms include candles, lanterns, dragons and canons. Earth, for chapter three, is symbolised by fruit and vegetables, rabbits, rams and farm tools. Crabs, shells, fish and sea serpents symbolise water; and stars, clouds, astronomical instruments and zodiacal figures cover the page for the chapter on dodecahedral forms.

The drawings comprising each chapter are of the Platonic solids, both solid and skeletal, and of other polyhedra derived from these fundamental forms by augmentation, truncation, cutting notches into the sides or faces, or some combination of these, all done in a regular manner. These themes with variations are arranged like ornaments on page after page. Figures 3.14–3.16 reproduce typical

Figure 3.13. Title page of Jamnitzer's *Perspectiva Corporum Regularium*.

plates showing variations based on the cube, the dodecahedron and the icosahedron. Notice the true truncation of the cub-octahedron (Figure 2.23(a)) in the
lower right of Figure 3.14.

The final chapter is concerned with mazzocchi, cones, spheres and what are
best described as polyhedral monuments. (It is this chapter that has provided
the frontispieces.) Many of the monuments could never be erected as structures.
Some balance precariously on a single vertex, others are connected only through
vertices. Some have parts which are not supported at all but which float in space.
A typical example is shown on page 288.

Although in one sense Jamnitzer's book represents the culmination of research
begun by Piero and Dürer, its purpose is different from theirs. His geometrical
fantasies are presented simply for what they are—completed designs. He does not
explain how they were constructed. We can, however, guess at his methods, for
the pictures themselves tell us something about his technique. The perspective is
very shallow as though the objects are seen from a relatively distant viewpoint. It
is probable that he used an artificial eye of the kind described by Dürer. Indeed,
Amman's portrait of Jamnitzer shows him using something similar to the Dürer–
Keyser device to sketch geometric bodies. As a manufacturer of all kinds of
scientific instruments he could easily have made such an instrument specially for
the purpose. Even so, it is unlikely that he made models of all his varied designs.
His perspective machine, along with a few models, would provide the frameworks
for his drawings. The rest is pure imagination.

Perspective and astronomy

The Platonic themes which so appealed to Pacioli and Jamnitzer were also taken
up by Daniele Barbaro (1514–1570). He wrote two treatises on perspective with
almost identical titles. One work *La Practica della Prospettiva*, which deals with
the geometry of regular and semiregular polyhedra, exists only as a manuscript.
The other, *La Practica della Perspettiva* printed in 1559, covers various aspects
of perspective. Barbaro was taught perspective geometry by Giovanni Zamberti,
brother of Bartolomeo, but he also draws on the work of Piero and Dürer. Following their lead, he gives the two standard methods of construction (plan/elevation
and Alberti's pavement) and then progresses to the study of polyhedra, a topic
he covers in more detail than earlier treatises. He deals with the construction of
regular and semiregular solids, mazzocchio, and some variations. He also includes
nets and explains the rules for shading and the construction of shadows.

Besides the projection of shadows, Barbaro also explains the optical geometry
of sundials and techniques for mapping the celestial sphere. By this time it was
apparent that perspective geometry was closely connected with the stereographic
projection used in astronomy. In fact, the two ideas are equivalent. This was made

Figure 3.14. Page from Jamnitzer's *Perspectiva Corporum Regularium* showing polyhedra derived from a cube.

Figure 3.15. Page from Jamnitzer's *Perspectiva Corporum Regularium* showing polyhedra derived from a dodecahedron.

Figure 3.16. Page from Jamnitzer's *Perspectiva Corporum Regularium* showing polyhedra derived from an icosahedron.

explicit by Commandino in the preface to his edition of Ptolemy's *Planisphaerium* published in 1558. This text attracted attention not only as a treatise by the author of the *Almagest* but also as a source of the mathematics underlying the construction of astrolabes. This involves projecting the tropics of Cancer and Capricorn and the circle of the ecliptic onto the equatorial plane using the celestial South pole as the centre of projection. These three things, the fixed eye (South pole), the picture plane, and the projected object, are the three basic elements of perspective. The technique of orthographic (or parallel) projection provides the basis for the geometry of sundials and shadows. Commandino treated this aspect in a companion volume—a commentary on Ptolemy's *Analemma*.

It is uncertain how soon the connection between perspective and astronomy was made. It has been suggested that it may even have played a part in the initial process of discovery. Brunelleschi's training as a goldsmith would have included the practical mathematics used by craftsmen and found in the abaco texts. As a maker of scientific instruments, he would also need to know the projective geometry needed to construct an astrolabe. Unfortunately, there are many threads linking perspective to other areas of science and we shall probably never know which ones Brunelleschi drew together to make his revolutionary discovery.

Polyhedra revived

The fifteenth and sixteenth centuries were a time of renewal in Europe. The humanists collected, translated and commented on the principal texts of antiquity and scholarship began to be revived. Mastery of the new ideas was essential before new steps could be taken. But this was not a simple process of assimilation or replacing obviously false notions with correct ones. The new philosophies and ideas had to be grappled with; inconsistencies in the many different sources had to be removed. Amid this uncertainty there was one body of knowledge which seemed to offer a secure foothold—the axiomatic truths of mathematics. The publishing of mathematics texts played an important role in spreading new science. This is particularly true in solid geometry. Increased trade had led to great interest in stereometry—the determination of volumes of containers.

The rediscovery of Plato in the fifteenth century introduced the Pythagorean creed 'Number is the basis of all things' and the idea that nature could be understood through mathematics.

The Neoplatonist writings of Plotinus were translated into Latin in 1492. Platonism came into vogue as the Renaissance thinkers sought to throw off medieval scholasticism: it became a major force in the fight against Aristotle. The Platonic tradition, though never entirely lost in the West, now acquired many new adherents with its attractive fusion of rational explanation with theology through the mathematical design of the Creator. Contemplating the universe and uncovering the divine plan held great appeal for the Renaissance philosophers. One of the most prolific thinkers to explore this idea is the subject of the next chapter.

Johannes Kepler.
Courtesy of the Deutsches Museum, Munich.

Fantasy, Harmony and Uniformity

> *Kepler's most notable contributions to pure mathematics were in the geometry of regular polygons and polyhedra.*[a]
>
> H. S. M. Coxeter

Johannes Kepler (1571–1630) lived in an age of transition, in a turbulent Europe with its upheavals of religious and political structures, at the frontier of the medieval and modern eras. His theology, belief in God the Creator, coupled with his Pythagorean-like mysticism led him to search for the mathematical order underlying the phenomena of nature, the Eternal Geometry behind its design. Unlike earlier philosophers, Kepler also demanded that his theories matched the facts.

Although he is remembered chiefly for his astronomical works, Kepler was a prolific author and wrote on many diverse subjects from crystals to optics. His interest in polyhedra spans his whole career. They occur in his first published treatise, *Mysterium Cosmographicum*, and also in one of his last major works, *Harmonices Mundi Libri V*. Both works exhibit his desire to expose the mathematical design of the universe, to find the harmonious and unifying scheme which the great Architect used in its creation.

A biographical sketch

Kepler was born in Weil, a small town near Stuttgart. After completing his schooling he won a scholarship to study at the university in Tübingen. The Dukes of Württemberg, having been converted to the Protestant religion, founded the universities at Tübingen and Wittenberg to train leaders for the Reformed Church, and Kepler's education was to prepare him for the Lutheran ministry.

Kepler was taught geometry and astronomy by Michael Mästlin, an astronomer with whom he maintained contact after leaving Tübingen. Although Mästlin and Kepler discussed the new Copernican philosophy in private, in his lessons Mästlin taught only the traditional astronomy. Kepler graduated from the faculty of arts to the faculty of theology, but never completed his training. A Protestant school in Graz had asked the university to recommend a candidate to fill a vacant teaching position. The university suggested Kepler and, in 1594, he withdrew from his studies and moved to the Austrian province of Styria to teach astronomy and mathematics.

While at Graz, Kepler read the Neoplatonic writing of Proclus, whose commentary on the first book of Euclid's *Elements* had been published in a Greek edition some sixty years earlier. He also studied Commandino's Latin translations of works by Archimedes, Apollonius and Pappus. It appears that he also had access to Franciscus Flussatus Candalla's Latin version of the *Elements*. This edition purported to be a translation from Greek sources but it seems to be based as much on Campanus' and Zamberti's editions as on any original manuscript. Candalla appended his own work to the canon: propositions showing how the regular polyhedra could be inscribed in one another. This may have stimulated Kepler's interest in polyhedra—an interest he was to retain for the rest of his life.

It was also at Graz that Kepler produced his first treatise. It contained his answer to the mystery of the universe. His model of the planetary orbits was an attempt to explain the proportions of the known universe and he believed that the regular polyhedra played a key role in shaping its design. Now that his future career was not to be in the church, he felt his calling was to reveal God's perfect design in nature, to fathom the mind of the Creator himself.

He sent copies of his *Mysterium Cosmographicum* to many of the leading scholars in Europe including Galileo and the Danish astronomer Tycho Brahe. Tycho thought the work original and imaginative and recognised Kepler's mathematical ability. When the Protestants were expelled from Graz, Kepler travelled to Prague and became his assistant leaving Graz on New Year's day, 1600. Their collaboration lasted only 18 months and ended with Tycho's death. Kepler succeeded him as Imperial Mathematician to Rudolph II. This gave him access to the large collection of astronomical observations Tycho had made during his life. Kepler continued to work on the problem he had been set by his late master—an analysis of the motion of Mars. Mars has the most eccentric of the planetary orbits and thus caused the most problems to the circular motion theories. Eventually these studies led him to the discovery of his first two laws—that planets follow elliptical orbits, and that a line from the sun to a planet sweeps out equal areas in equal time—thus demolishing the paradigm of circular motion at constant speed. He presented these two laws in his *Astronomia Nova* published in 1609.

The political situation deteriorated a few years later and, after the emperor's death, Kepler moved to Austria taking a position as Mathematician at Linz. Here he wrote two major treatises: *Epitome Astronomiae Copernicanae* (1618–1622) and *Harmonices Mundi* (1619). The first of these, despite its title, is not on Copernican astronomy but is a textbook presenting Kepler's own astronomical laws. The second work is almost a sequel to the *Mysterium Cosmographicum* and, indeed, its conception can be traced back to 1599, just three years later. In it Kepler attempts a grand unification of geometry, music, astrology and astronomy; he wants to expose the hidden harmonies of the universe. He aims to derive the phenomena included in his survey from the 'one and eternal Geometry'. The first two of its five books are concerned with the mathematics of regular polygons and their combinations. The second book contains nearly all Kepler's writing on polyhedra, including his explanation of why each of the Platonic solids is associated with a particular element (see Chapter 2). However, *Harmonices Mundi* is not a treatise on pure mathematics and Kepler is not investigating the properties of geometrical bodies for their own sake. He is concerned with establishing a hierarchy of polygons according to their 'sociability'—a measure of their capacity to combine with other polygons to form tessellations and polyhedra. In a later book of *Harmonices Mundi* this ranking of polygons is used to explain astrological phenomena.

When discussing music he uses the geometry of polygons to explain why certain ratios of frequencies are discordant. The Pythagoreans had discovered that strings whose lengths are in simple numerical ratios form pleasant chords. Thus the octave is 1:2, and the fifth is 2:3. But why are ratios such as 1:7 or 3:7 discordant? Kepler argues that the answer lies in geometry, not arithmetic, and is connected with the constructibility of regular polygons. A heptagon is not constructible with ruler and compasses, so it is not truly known and, therefore, cannot have been used in a harmonious design of the universe. Kepler also found harmonic ratios between the angular velocities of the planets. In amongst all this, the work contains Kepler's third law of planetary motion relating the distance of a planet from the sun to its period, the time taken for a complete orbit.

His efforts were then directed towards completing the task with which he had been charged many years earlier by Rudolph II—the compilation of new astronomical tables to replace the outdated Alphonsine tables. The great work was concluded in 1626. Kepler travelled to Germany to supervise their printing and at the beginning of 1628 he handed a copy of the *Tabulae Rudolphinae* to the emperor Ferdinand II.

Kepler lived for another three years. It is fortuitous that we still have his original papers for, after his death, his manuscripts suffered several misadventures before finding a safe resting place. They were rescued from a fire, only to be pawned and then lost. After their surprise rediscovery nearly 150 years later their new owner offered them for sale. Several universities and scientific societies were approached but none showed any interest. Eventually scholars at the Russian Academy of Science in Saint Petersburg (one of whom was Leonhard Euler— a mathematician who also figures prominently in the history of polyhedra, as we will see in Chapter 5) convinced the empress Catherine II of the value of the manuscripts and she agreed to buy them. They are now preserved in the Academy's library.

A mystery unravelled

> *Seldom in history has so wrong a book been so seminal in directing the future course of science.*[b]
>
> O. Gingerich

A precursor to Cosmographical Treatises, containing the Cosmic Mystery of the admirable proportions between the heavenly Orbits and the true and proper reasons for their Numbers, Magnitudes and Periodic Motions demonstrated by the five regular geometric solids.

So reads the full title of Kepler's treatise *Mysterium Cosmographicum*—a work in which he attempts to recreate the Architect's design for the universe. Kepler records that he began to wonder why the cosmos was as it was. Why are there six planets? Why are their distances and speeds what they are? The originality of such questions is easily overlooked. Prior to this, astronomers had merely observed the sky and recorded what they saw. Kepler sought not merely descriptions but causal explanations.

At first he tried to find numerical relationships between the sizes of the orbits but without success. He devised various schemes hoping to force sensible ratios to appear, even resorting to the addition of several planets, but these also failed. Inspiration came while he was giving a lesson. He was explaining how the conjunctions of Saturn and Jupiter progress through the houses of the zodiac. He marked on a circle the points where such conjunctions occur and connected them in sequence. This resulted in a diagram like the one in Figure 4.1. Each line spans a gap of eight signs so three lines almost form a triangle. The conjunctions progress round the zodiac precisely because the 'triangle' does not close up. When many of these lines have been drawn in they outline a smaller circle. Studying this figure, it suddenly struck him that he could use geometrical relationships rather than numerical ones, for the circumscribed circle and the inscribed circle of an equilateral triangle were in the same ratio to each other as the orbits of

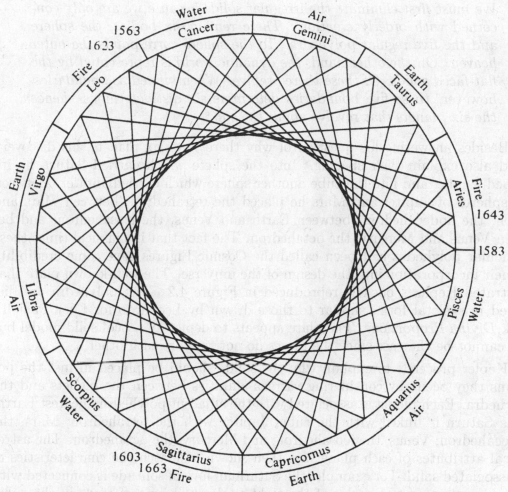

Reproduced from the *Dictionary of Scientific Biography* with permission of the
American Council of Learned Societies.

Figure 4.1. The progression of the conjunctions of Saturn and Jupiter
through the zodiac.

Saturn and Jupiter. These were the outermost planets, and the triangle is the
first figure. Almost instantaneously he reasoned that the next interval, that between Jupiter and Mars, must correspond to the next figure, the square—and so
on. But again he met with failure. He was now thinking in geometrical terms.
It dawned on him that instead of fitting two-dimensional figures to the orbits in
space, it was more natural to use three-dimensional forms. And there was the
answer. For although there are an unlimited number of regular polygons, there
are only five regular polyhedra. Thus there can be only five intervals between the
planets and hence only six planets. He writes:

*We must first eliminate the irregular solids because we are only con-
cerned with orderly creation. There remain six bodies, the sphere
and the five regular polyhedra. To the sphere corresponds the outer
heaven. On the other hand, the dynamic world is represented by the
flat-faced solids. Of these there are five. When viewed as boundaries,
however, these five boundaries determine six distinct things—hence
the six planets that revolve about the sun.*[c]

Besides answering his question of why there were six planets, his discovery
could also explain their spacing. Into the sphere which carried Saturn he inscribed a cube, and into the cube another sphere which carried Jupiter. Between
the spheres of Jupiter and Mars he placed the tetrahedron; between Mars and
Earth, the dodecahedron; between Earth and Venus, the icosahedron; and between Venus and Mercury, the octahedron. The fact that in classical times these
particular polyhedra had been called the Cosmic Figures gave some credibility
to their incorporation into the design of the universe. The well-known plate that
illustrated Kepler's book is reproduced in Figure 4.2. The polyhedra are displayed in skeletal form, similar to those drawn by Leonardo for Luca Pacioli's
book *Divina Proportione*. The plate appears to depict an actual solid model but
this cannot be the case since the pieces do not support each other.

Kepler proceeds to explain why the solids should be placed in just the positions they occur by considering various affinities between the planets and the
polyhedra. Each planet is associated with the adjacent polyhedron nearest Earth.
Thus Saturn is linked with the cube; Jupiter with the tetrahedron; Mars, the
dodecahedron; Venus, the icosahedron; and Mercury, the octahedron. The astrological attributes of each planet are then shown to match the characteristics of
its associated solid. For example, the Saturnian love of solitude is connected with
the uniqueness among angles of the right angle—the type of angle in the cube.
By contrast, Jupiter has chosen from the many acute angles.

As astrology is geocentric, the Earth has no astrological attributes. Its position among the polyhedra is explained as follows. The polyhedra fall naturally
into two classes: those whose nature is to float (the octahedron and the icosahedron) and those whose nature is to stand upright. For if the former are made
to rest on a face, or the latter on a corner then 'the eye shies from the ugliness
of such a sight'. Falling back on the medieval homocentric view of the universe,
Kepler argues that the most appropriate place for the Earth, home of Man created in God's image, is so as to separate one sort from the other. By arguments
of this kind Kepler justifies everything he believes about his model.

After all this fantasising and speculation there is an abrupt change in his presentation. He announces that the model must be checked against empirical data.
He knew that the planetary orbits are not circular but oval shaped. The spheres
which carried the planets had to be thought of as shells of sufficient thickness

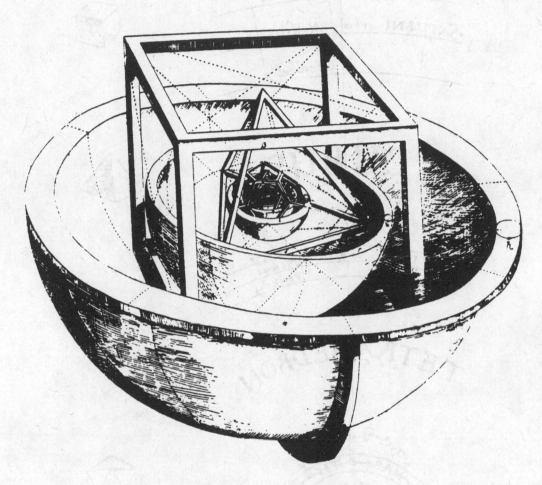

Figure 4.2. A plate from Kepler's *Mysterium Cosmographicum* illustrating his model of the universe.

to contain the oval orbits, with the innermost wall marking the planet's minimum distance from the sun (perihelion) and the outer shell its greatest distance (aphelion). These shells are marked in Figure 4.3, which illustrated *Harmonices Mundi*. In some places the model agreed quite well, in others there were significant discrepancies. The problem with Jupiter was attributed to its great distance. The shell for Mercury was too small so instead of the inscribed sphere, he used the *midsphere* of the octahedron, that is, the sphere that touches the midpoints of all the edges. This is shown in the illustration of the four innermost planets (Figure 4.4).

The Copernican data he was using gave the distances of the planets as mea-

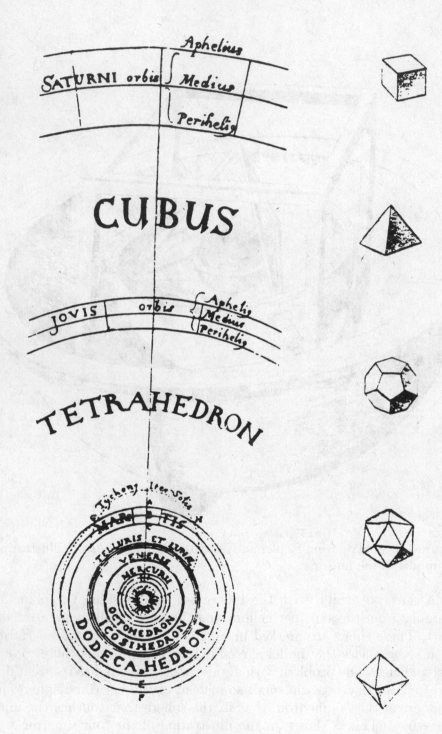

Figure 4.3.

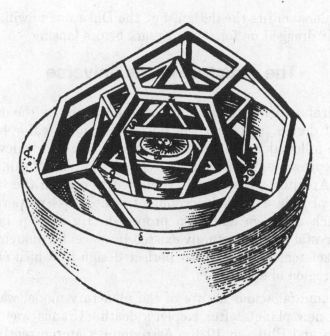

Figure 4.4.

sured from the mean sun—the centre of the Earth's orbit. This was a mathematically convenient origin for Copernicus to choose as it simplified his calculations. Kepler, hoping to improve the fit of his model, engaged Mästlin's help to recalculate the distances as measured from the actual sun. Even though this change in the data did not result in any significant improvement, Kepler continued to use his version of the data. His interest was in the actual universe and its properties, and he wanted physically meaningful data. When trying to decide whether or not the Earth's shell should include the moon's orbit, he admits that he will choose whichever solution gives the best fit. Convinced that his model has to be right, he explains away as many discrepancies as he can, and blames those that remain on Copernicus' faulty data. In his book *The Sleepwalkers*, Arthur Koestler likens it all to a game of 'Wonderland croquet through mobile celestial hoops'.

Kepler took the draft of his book back to Württemberg for publication. In his enthusiasm he persuaded Friedrich, Duke of Württemberg, to have a model of the universe made in silver showing the positions of each of the five regular polyhedra and with the planetary symbols inset in precious stones. It was to be made in parts by different silversmiths so that the Duke would be the first to see the assembled model. In order to satisfy himself that the project was worthwhile, the Duke asked to see a copper version. Kepler, not having sufficient money, made a model of coloured paper instead. The Duke asked Mästlin's opinion of the completed model; he commended it, whereupon the Duke consented to the work being done. The project was never completed. It appears that the silversmiths had difficulty understanding the design, but the Duke was unwilling to abandon the project and it dragged on for several years before lapsing.

The structure of the universe

Kepler's polyhedral model of the universe was motivated by the desire to expose its mathematical design, to reveal the plan which its Creator had used in its construction. He followed in the Pythagorean tradition and believed that such a plan would be expressible in harmonious geometrical relationships reflecting the decisions of the Architect. His model attempted to explain the structure of the universe and to provide a unified account of some of its properties. This was the first time such a system had been proposed. He did not believe that the polyhedra and crystal spheres actually existed in space; he thought of them more as an invisible skeleton, as part of the perfect design by which each planet was allotted its own region of space.

The illusory and fallacious nature of the planetary model was shown up by the discovery of new planets after Kepler's death. Uranus was found in 1781, Neptune in 1846 and Pluto in 1930. Astronomers are currently searching for a tenth planet, the so-called planet X, whose existence has been postulated to account for certain discrepancies between the predicted and observed paths of the known planets. Similar gravitational anomalies led to the discovery of Neptune.

Had the planets been discovered in Kepler's lifetime, he could possibly have accommodated them in a revised version of his model. For, although there are only five Platonic solids, under a wider definition of regularity there are four other regular polyhedra. Kepler discovered two of these polyhedra and tinkered with his model to see whether it could incorporate one of them.

Kepler's planetary model has been compared to the modern theory of particle physics which seeks to explain the number and properties of elementary particles. Just as Kepler admired the regularity of the Platonic solids and was attracted by the idea that nature must be constructed around such elegant forms, so the modern physicist idolises symmetry. Kepler intuitively recognised the high degree of geometric symmetry exhibited by the regular polyhedra; the physicist searches for more abstract symmetries in nature. Current models of particle physics are based on what is known as unitary symmetry, a notion closely related to rotational symmetry.

Unitary symmetry theory suggests that the elementary particles can be arranged in families according to their properties and that they display definite patterns. The first classification scheme proposed in the 1960's was based on representations of the special unitary group SU(3). It incorporated all the then known particles and successfully predicted the properties of new particles which were later discovered. Rather surprisingly, Kepler also managed to make a successful prediction based on his model of the universe. When he heard that Galileo had discovered four new planets with the newly invented telescope, he refused to accept that they could be planets orbiting the sun as he had proved that there could be only six such planets. Thus he deduced that these four astronomical bodies must be satellites of the known planets, like the Earth's moon. Galileo's four planets are, in fact, all moons of Jupiter.

The patterns shown by the SU(3) model led to the proposal that protons and neutrons are not fundamental particles but are composed of smaller entities. These constituents have come to be called quarks. The original quark model associated with SU(3) required three quarks which were labelled 'up', 'down' and 'strange'. These quarks combine in triplets producing baryons such as the proton and neutron, and as quark–antiquark pairs producing mesons such as the pion and the kaon.

In the mid-1970's a new kind of particle was discovered which did not fit the current model. A fourth quark labelled 'charm' was postulated to solve the problem, and the unitary symmetry theory now considered representations of the group SU(4). Unlike Kepler's polyhedral hypothesis, which was demolished with the discovery of the outer planets, the discovery of new particles does not cause such a devastating problem to particle physicists. Although the number of regular polyhedra is limited to five, there are many groups of symmetries to choose from.

Fitting things together

Kepler's writing on the mathematics of polyhedra is mostly contained in the second book of his *Harmonices Mundi*. The first two of the five books are concerned with polygons and the different ways in which they form 'congruences'. In book I Kepler defines a polygon to be *regular* if it is equilateral and has equal angles. He then defines a *half-regular*[1] polygon to be one which is equilateral, and restricts attention to those having four sides. Thus Kepler's half-regular polygons are, in fact, rhombi.

In the second book he investigates the ways in which regular and half-regular polygons can be fitted together around a point. This leads to the construction of tessellations of the plane, and of polyhedra. Kepler uses the word *congruence* meaning 'fitting together' to describe both situations—a tessellation being a congruence in the plane, and a polyhedron being a congruence in space. Here Kepler is concerned with harmony in its broad sense for the word 'harmony' is derived from the Greek for 'fitting together'. He remarks that stacking poly-

---

[1] This would normally be translated 'semiregular'. I have used 'half-regular' instead because, further on, Kepler defines half-regular polyhedra as those composed of half-regular figures. Retaining the conventional translation could be a source of confusion since, in current terminology, 'semiregular polyhedra' refers to the Archimedean and Catalan solids.

hedra together to fill space is another form of congruence. He considered such space-filling polyhedra in *De Nive Sexangula*.

He defines a polyhedron as follows:

> There is a congruence in space, and a solid figure, when the individual angles of several plane figures make up a solid angle, and regular or half-regular figures are fitted together so as to leave no gap between the sides of the figures, which join up on the opposite side of the solid figure, or, if a gap is left, it is such that it can be filled by a figure of one of the kinds already employed, or, at least, by a regular figure.[d]

To Kepler, then, a polyhedron was a three-dimensional figure composed of regular polygons or rhombi fitted together edge-to-edge. He describes polyhedra that do not close up completely as *semisolid*.

He continues his list of definitions and describes, in turn, the kinds of polyhedra that he is interested in. His classification is summarised by the diagram in Figure 4.5.

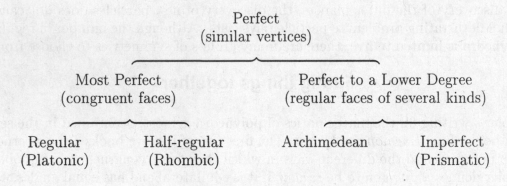

Figure 4.5. Kepler's classification of polyhedra.

Having defined a *perfect* congruence as one in which all the vertices are similarly surrounded, he subdivides this class into those that are *most perfect* and those that are *perfect to a lower degree*. The former category comprises the polyhedra whose faces are all the same shape. These are further subdivided into *regular* and *half-regular* according to whether their faces are regular or half-regular polygons. In fact, the rhombic polyhedra are not a subclass of the perfect polyhedra at all because not all the vertices are surrounded by the same number of polygons. Kepler knows this but remarks:

> There is no reason why we should not call this congruence most perfect, for its imperfection is in the faces and is not a consequence of its being solid but, rather, an accidental feature.[e]

The polyhedra that Kepler calls *perfect to a lower degree* have regular faces of several kinds. We know them as the Archimedean polyhedra and the families of prisms and antiprisms. Kepler notes that among the prismatic figures, the triangular antiprism and the square-based prism belong to the most perfect polyhedra (the former being the regular octahedron, the latter the cube). All the other prismatic solids are classified as *imperfect* being either more like plane figures, in the case of prisms, or more like parts of figures than whole ones. Kepler's diagram showing an exploded view of an icosahedron shows clearly what he had in mind (Figure 4.6). The central section is a pentagonal antiprism.

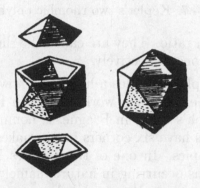

Figure 4.6. Kepler's sketch of an exploded icosahedron shows it decomposed into an antiprism and two pyramids.

Following his definitions, Kepler describes examples of the various kinds of polyhedra. He tries to find all the possibilities and then often continues to prove that there are no more. The first family that he enumerates are the regular polyhedra. He notes that this classification forms the last proposition in Euclid's *Elements* and his proof follows the Euclidean one: he tries fitting combinations of polygons around a point and eliminates all the arrangements whose angle sum is greater than 360°. The five remaining possibilities can be realised by the Platonic solids. Kepler uses this method of proof consistently in his study of polyhedra. He considers all possible ways that a vertex could be surrounded and excludes those that are impossible, either because the angle sum is too large or because the required pattern cannot be continued. This proof by exhaustion of possibilities is long-winded but effective. It has elegance in its uniformity of style rather than in any concise ingenious arguments.

Rhombic polyhedra

Kepler knew two examples of his half-regular polyhedra. They are shown in Figure 4.7. The first is bounded by twelve rhombi whose diagonals are in the ratio of $1:\sqrt{2}$. The second rhombic polyhedron is bounded by thirty rhombi whose

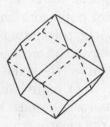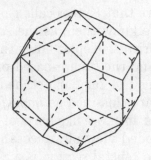

Figure 4.7. Kepler's two rhombic polyhedra.

diagonals are in the golden ratio. They are called the rhombic dodecahedron and the rhombic triacontahedron, respectively.

In an earlier work, Kepler provides a hint as to how he discovered these two polyhedra. The *De Nive Sexangula*, a work written in 1611 as a New Year's gift for a counsellor at Rudolph's court in Prague, is nominally concerned with the question of why snowflakes have six corners but it makes frequent diversions and touches on many other topics. In one of these asides Kepler considers another example of hexagonal forms occurring in nature, namely a honeycomb. He notes that a bee's cell is terminated at the base by three equal rhombi. He continues:

> *These rhombi put it into my head to embark on a problem of geom-*
> *etry: whether any body, similar to the five regular solids and to the*
> *fourteen Archimedean solids could be constructed with nothing but*
> *rhombi. I found two, one with affinities to the cube and octahedron,*
> *the other to the dodecahedron and icosahedron—the cube can itself*
> *serve to present a third, owing to its affinity with two tetrahedra cou-*
> *pled together. The first is bounded by twelve rhombi, the second by*
> *thirty.[f]*

The clue is in the reasons given for including the cube as a third possibility. Two equal tetrahedra can be placed so that the edges of one meet the edges of the other at right angles, and so that the points of intersection bisect the edges. Such a pair of coupled tetrahedra is shown in Figure 4.8(a). Kepler named this body *stella octangula* but it was known to many others including Pacioli and Jamnitzer. Pacioli called it 'octaedron elevatum'. The eight corners of this compound polyhedron coincide with the corners of a cube, the edges of the tetrahedra forming diagonals of the square faces.

An octahedron and a cube of appropriate sizes can also be coupled together so that their edges bisect each other at right angles (see Figure 4.8(b)). The relationship between the rhombic dodecahedron and this compound is analogous to that between the cube and the compound of two tetrahedra: the corners of

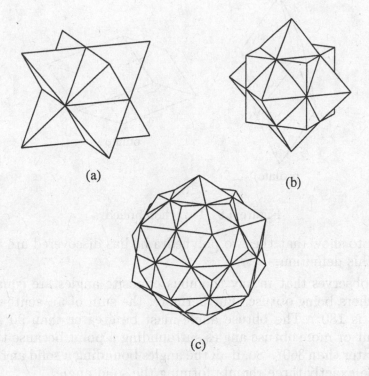

(a)   (b)

(c)

Figure 4.8. Compounds of the Platonic solids.

the rhombic polyhedron coincide with those of the cube and the octahedron, and two intersecting edges of the compound are diagonals of a rhombic face. The regular dodecahedron and the icosahedron can also be coupled together to form a compound polyhedron (Figure 4.8(c)). This compound shows a similar affinity with the rhombic triacontahedron.

These rhombic polyhedra have some resemblance to the Platonic solids. Like them, they are spherical in shape, have congruent faces, and a high degree of symmetry. Moreover, the geometry of the Platonic solids limits the number of possibilities to only five. Kepler wanted to show that the number of rhombic polyhedra is limited and that he had found all the possibilities. To enumerate something you need a definition. Kepler's was as follows:

> *The solid formed is half-regular when the plane figures are half-regular. Its solid angles are then not all the same, but differ in the number of lines they contain, though the angles are not of more than two kinds and neither are they distributed on more than two spherical surfaces, which are concentric. The number of angles of each kind must be the same as the number of angles of one of the regular solid figures.*[9]

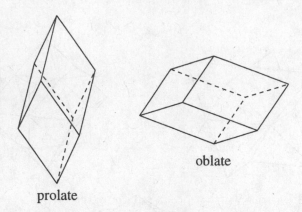

prolate

oblate

Figure 4.9. Two rhombohedra.

He then tries to show that the two polyhedra he has discovered are the only ones that satisfy this definition.

First he observes that in any rhombus opposite angles are equal, two being acute, the others being obtuse. Furthermore, the sum of an acute angle and an obtuse angle is 180°. The obtuse angle must be greater than 90°, hence there cannot be four or more obtuse angles surrounding a point because the angle sum would be greater than 360°. So, if all the angles bounding a solid angle are obtuse, there must be exactly three rhombi forming the solid angle.

The acute angles of three rhombi can be fitted together to form a solid angle, and two such sets can be joined to form a (prolate) rhombohedron (see Figure 4.9). Its eight vertices lie on two concentric spheres, six on the inner sphere, and two on the outer one. The final clause of Kepler's definition of a half-regular polyhedron excludes this case since no regular polyhedron has two vertices. They are also ruled out because the same number of polygons surround every vertex. But, Kepler argues, this polyhedron should also be excluded because the six equatorial solid angles are 'mixed' being composed of both acute and obtuse angles:

> *Each of the six obtuse solid angles is formed by two obtuse plane angles and one acute one, an irregularity which is once more contrary to the definitions.*[h]

Such 'irregularities' are not in fact excluded. Perhaps they are not in the spirit of perfection implicit in the definitions. In the remainder of the proof, he relies on this extra 'implicit' condition: that each solid angle is bounded by a single type of plane angle so that acute angles meet only acute angles, and obtuse angles meet only obtuse angles.

If the obtuse angles in the rhombus are greater than 120° then three obtuse angles cannot be brought together to form a solid angle. If the obtuse angles are exactly 120° then the rhombi can be fitted together to form a tessellation of the plane (Figure 4.10). So, to form a polyhedron the obtuse angles in a rhombus must be less than 120° and, consequently, the acute angles must be greater than 60°. If all the plane angles surrounding a solid angle are acute then there can be at most five of them, since if there were six or more, the angle sum would be greater than 360°. The case with three acute angles meeting has already been excluded, so the only possibilities remaining are for a solid angle to be bounded by four or five acute angles. These two possibilities are realised in the two rhombic polyhedra. ∎

There are two points to note arising from this discussion. Firstly, the proof is not constructive—it does not tell us what shape of rhombus we need to be able to make the polyhedra. Nor does it guarantee their existence. However, Kepler already knows that two such polyhedra do exist and only wants to show that there are no others. Even so, the question of uniqueness still needs to be considered. Is there only one of each kind?

The second point concerns the definition of half-regular polyhedra. Kepler assembled his definition by listing several properties shared by both polyhedra and hoped that these would be sufficient to characterise them. In order to exclude the rhombohedron, he restricted the number of solid angles of each kind to be the same as occurs in one of the Platonic solids. (He applied a similar restriction when dealing with the Archimedean solids intending to exclude the prisms and antiprisms but, in fact, excluding others as well.) In the course of the proof he finds it more convenient to use another property that both of his rhombic polyhedra possess but which is not included in the definition.

As in much of Kepler's writing, here we can watch thought processes in action. At this point, a modern mathematician would go back and alter the definition so that it contained the properties needed in the proof. In this case, such a proof-generated definition could be phrased as follows: a polyhedron is half-regular if all its faces are equal half-regular polygons and each solid angle is surrounded by a single kind of plane angle. Sometimes the process of juggling definitions

Figure 4.10. A rhombic tiling.

and proofs to achieve the strictest accuracy and widest applicability of theorems can take quite a long time and needs to be repeated several times. At the end, however, we cover over our early attempts at definitions which do not provide the properties we need. This can lead to definitions that appear to be unnecessarily complex until the problems which led to them are explained. The definition of polyhedron developed in Chapter 5 is an example of this kind.

Besides the two rhombic polyhedra that Kepler described, and the rhombohedra, there are two more polyhedra whose faces are all congruent rhombi. They do not, however, satisfy his 'implicit' condition since some of their vertices are surrounded by both acute and obtuse angles. One was discovered by Evgraf Stephanovich Fedorov in 1885. It is an oblate solid with twenty faces—a rhombic icosahedron (Figure 4.11(c)). It can be derived from the rhombic triacontahedron by collapsing a belt of rhombi that run around it. The sequence of polyhedra in Figures 4.11(a)–(c) indicates how such a belt collapses. The other rhombic polyhedron is formed from Fedorov's by collapsing another belt of rhombi. This is illustrated by Figures 4.11 (c)–(e). This results in another rhombic dodecahedron, shown again from a different viewpoint in Figure 4.11(f). It was discovered in 1960 by Stanko Bilinski when he made an exhaustive enumeration of rhombic solids. He called it the rhombic dodecahedron of the second kind to distinguish it from the one described by Kepler.

The Archimedean solids

Pappus, writing in the fourth century AD, attributes to Archimedes a treatise on thirteen polyhedra: 'Figures contained by equilateral and equiangular, but not similar, polygons'. In the fifth book of his *Mathematical Collection*, he describes each of these polyhedra. The revival of interest in solid geometry and the creation of new figures for perspective designs led to the rediscovery of some Archimedean solids. They appear among the drawings in Pacioli's *Divina Proportione*, Dürer's *Unterweysung der Messung*, and Jamnitzer's *Perspectiva Corporum Regularium* (see Figures 3.14–3.16). These examples were produced by truncating the Platonic solids. However, not all the Archimedean solids can be produced in this way and, prior to Kepler, no-one seems to have rediscovered the complete set. It is also possible that Kepler found the pseudo-Archimedean solid known as Miller's solid, (Figure 2.30) for in the passage from *De Nive Sexangula* quoted above (page 152) he makes a passing reference to *fourteen* Archimedean solids.

From his reference to 'Archimedean solids' we can deduce that Kepler was aware of Pappus' work and that, therefore, he knew to search for thirteen polyhedra. However, through his systematic analysis, he found that prisms and antiprisms also satisfy the definition; the latter had not been described before.

He defines these figures as follows:

*A congruence is perfect, but of lower degree, when the plane figures
are regular and all the angles lie on the same spherical surface and
are similar to one another, but the faces are of various kinds, though
the number of each kind must be the same as the number of faces of
one of the most perfect figures, that is, not less than four which is the
minimum number of planes to bound a solid figure.*

*There is an imperfect congruence or figure when other conditions re-
main the same but the larger plane figure does not occur more than
twice.[i]*

The requirement that the number of faces of each kind in one of these figures is
the same as the number of faces in a most perfect polyhedron (that is 4, 6, 8, 12,
or 20 for the Platonic solids, and 12 or 30 if Kepler's rhombic solids are included)
is intended to exclude prisms and antiprisms. Kepler regards these as imperfect
figures being 'discus-shaped' rather than 'globe-shaped' like a sphere—the most
perfect of all solids. Rather than insist that the number of each kind of face is

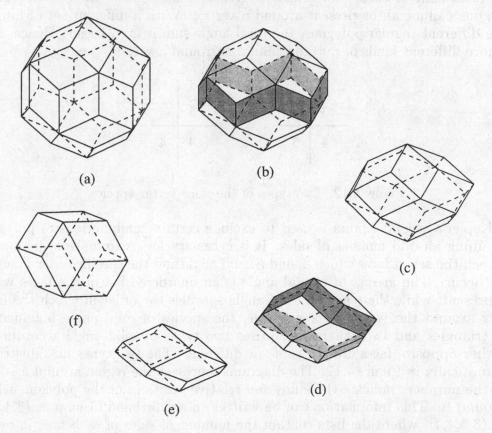

Figure 4.11. Two rhombic polyhedra can be derived from the rhombic
triacontahedron by collapsing 'belts' of rhombi.

at least three, which appears to be a somewhat arbitrary restriction, he links the numbers of faces to the perfect solids which seems more legitimate. However, the condition is far stronger than is needed and excludes more than Kepler intended. There are Archimedean solids that do not possess this property—the snub polyhedra have more than 30 triangular faces.

Kepler's enumeration of these polyhedra proceeds by considering all possible ways that a solid angle can be formed from regular polygons. Two simple observations make the process easier and Kepler states these as propositions preceding the main argument. The first one concerns the number of kinds of face that can surround a vertex.

Lemma. If the faces of a convex polyhedron are all regular polygons then at most three different kinds of face can appear around any solid angle.

PROOF: The four regular polygons having the smallest internal angles are the triangle (60°), the square (90°), the pentagon (108°), and the hexagon (120°). The total sum of these four angles is greater than 360° so these four regular polygons cannot all be present around a vertex. With a different set of four or more different regular polygons, the total angle sum is even larger. Hence, four or more different kinds of polygon cannot surround a vertex. ∎

$$
\begin{array}{c|c} \quad 3 & 4 \quad \\ \hline \quad 4 & 3 \quad \end{array}
\qquad
\begin{array}{c|c} \quad 3 & 3 \quad \\ \hline \quad 4 & 4 \quad \end{array}
$$

Figure 4.12. Two types of the same vertex species.

Kepler's second lemma is used to exclude certain combinations of polygons containing an odd number of sides. It is necessary for us to make a distinction between the set of faces which bound a solid angle and the specific order in which they occur. The *species* of a solid angle is an unordered list of the faces which are present, while the *type* of a solid angle specifies the order in which the faces occur around the vertex. For example, the species of solid angle bounded by two triangles and two squares comprises two types of solid angle according to whether opposite faces are the same or different. The two types are illustrated schematically in Figure 4.12. The diagrams represent the region around a vertex and the numbers indicate the kinds and relative positions of the polygons which surround it. This information can be written in a shorthand fashion as (3,4,3,4) and (3,3,4,4), where the lists contain the number of sides of each face in order. Kepler appears not to have made a distinction between species and types. He enumerates only vertex species, and in the case of the above example, he writes:

Two trigon angles and two tetragon angles are less than four right angles. Thus eight trigons and six tetragons fit together to form a tessareskaedecahedron which I call a cuboctahedron. It is shown here with the number eight [see Figure 4.13].[j]

He does not mention that there are two possible arrangements of the polygons around the vertex. The illustration shows how they are to be arranged; the type (3,3,4,4) is not considered. This case cannot be realised as a perfect polyhedron.

Kepler states that three polygons of different kinds cannot form a solid angle in a perfect figure if any of them has an odd number of sides. An analogue of this result which applies to vertices surrounded by four polygons can be used to exclude the case (3,3,4,4) above. These two results are collected together in the following lemma.

Lemma. A polyhedron in which all the solid angles are surrounded in the same way cannot have solid angles of the following types:

(i) [diagram: a | b over c]   where a is odd and $b \neq c$.

(ii) [diagram: a | b over 3 | c]   where $a \neq c$.

PROOF: In the first case, the fact that all the solid angles have the same type implies that the b-gon faces must alternate with the c-gon faces round the boundary of an a-gon face. But, since a is assumed to be odd, this leads to a contradiction. This is clearly seen in the example shown in Figure 4.14(a) which illustrates the case when $a = 7$.

In case (ii), we consider the way that the faces must be arranged around the 3-gon. At each angle, the face opposite the 3-gon is always a b-gon. Since all the vertices have the same type, the sides of the 3-gon must be attached to a-gons and c-gons, and these must alternate around the 3-gon. This again leads to an inconsistency (see Figure 4.14(b)). ■

Kepler then states the proposition:

There are thirteen solid congruences which are perfect to a lower degree. From these thirteen we obtain the Archimedean solids.[k]

62 DE FIGURARUM HARMON:

Cùm enim misceantur in hoc gradu figuræ diversæ, quare per propos. XXI. miscebuntur aut duarum aut trium specierum figuræ. Quod si duarum, tunc inter eas vel sunt Trigoni vel non sunt.

Igr ex Trigonis & Tetragonis fiunt solida tria, quibus quidem def. IX. competat. Nam illa rejicit formas hasce tres, in quibus solidum angulum claudunt, cum uno Tetragonico plano angulo, tam duo, quàm tres plani Trigonici; aut cum duobus Tetragonicis, unus Trigonicus ; quia in primo casu unus solus Tetragonus est, fit q dimidium Octaëdri, & anguli solidi sunt diversiformes: in secundo duo soli Tetragoni, in tertio duo soli Trigoni: quæ p X, sunt imperfecta congruentia. Restant ergò modi hi, in quibus angulum solidum claudunt 5 lani, Primum, quatuor Trigonici & unus Tetragonicus. Sunt enim minores 4 rectis. Congruunt igitur sex Tetragoni & Triginta duo (id est 20 & 12.) Trigoni, & fit figura Triacontaoctohedrica, quod appello Cubum simum. Hic in schemate sequenti pictus est Numero 12.

Quinq enim Trigonici plani & unus Tetragonicus superant quatuor rectos ; cùm debeant ad solidum claudendum esse minores quatuor rectis, per XVI. Sic etiam quatuor Trigonici & duo Tetragonici, Tres verò Trigonici & duo Tetragonici faciunt quatuor rectos.

Secundò duo Trigonici & duo Tetragonici minus habent quatuor rectis ; Hic igitur congruunt octo Trigoni & sex Tetragoni ad formandum unum Tessareskædecaëdron, quod cuboctaëdron appello. Pictum est hic num: octavo. Duo verò Trigonici cum tribus Tetragonicis superant 4 rectos.

Tertiò unus Trigonicus & tres Tetragonici minus habent 4 rectis. Hic ergò congruunt octo Triangula & octodecim (id est 12 & 6) quadrangula, ad unum Icosihexaëdron, quod appello sectum Rhombū Cuboctaëdricum: vel Rhombicuboctaëdron. Pictus est hic numero 10.

In his igr tribus sunt Tetragoni juxta Trigonos : sequitur ut & Pentagonicos ijs seorsim associemus. Quinq plani Trigonici juxta unum Pentagonicum non stant, quia neq juxta minorem eo, Tetragonicum, stare poterant. Quatuor ergò Trigonici, cum uno Pentagonico, minus efficiunt 4 rectis, & congruunt octoginta (id est 20. & 60) Trigoni, cum duodecim Pentagonis, ad formandum Ennenecontakædyhedron, quod appello Dodecaëdron simum. Pingitur hic numero 13. Et in hoc ordine simorum, Icosaëdron posses esse tertium, quod est quasi Tetraëdron simum.

Tret

Oqi

I. Cubus simus.

M. Rh. Cuboctaedron:

III. Rhombici Cuboctaedron.

IV. Dodecaëdron simū

Figure 4.13. Pages from *Harmonices Mundi* showing Kepler's discussion of the Archimedean solids.

64   DE FIGURARUM HARMON:

que imparilaterarum rejicitur , per XXIII. cum duobus Octogonicis , planum
locum implet: cum majoribus etiam superat 4 rectos; nec assurgit ad solidum angulum formandum.   Ita transactum est cum Tetragono , cum duæ solæ debent
esse planorum species.

Duo Pentagonici cum uno Hexagonico aut quocunq; alio unico reject itium
quid inchoant, per XXIII, quod supra etiam de Trigonico & Tetragonico cum
binis Pentagonicis usurpavimus.   Insuper cum uno Decagonico planit iem ster-
nunt , nec cum illo aut majoribus assurgunt in soliditatem.

X. Trun-   Unus ergò Pentagonicus cum duobus Hexagonicis minus facit 4 rectis; &
cum Icosi- congruunt duodecim Pentagoni cum viginti Hexagonis in unum Triacontakæ-
hedron.

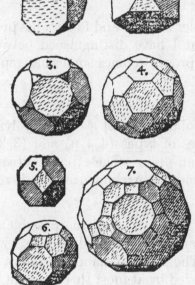

dyhedron , quod appello Truncum
Icosihedron. Formam habes signa-
tam numero 4.   Nec plura expe-
ctanda à Pentagono.   Nam unus
Pentagonus cum duobus Heptagoni-
cis jam superat 4. rectos.

Hexagonicus cum duobus alijs
implet planitiem , cum majoribus
superat 4 rectos. Itaq; hic finis est
mixtorum ex duabus speciebus.

Quod si trium specierum Plana
concurrere possunt ad unum angulum solidum : Primùm anguli duo
plani; unus Tetragoni, alter Pentagoni superant 2 rectos; majores his,
multò magis : tres verò Trigono-
rum trium , æquant 2 rectos : ne-
queunt igr tres Trigonici admitti;
ne summa omniù superet 4 rectos.
Duo verò Trigonici cum uno Tetragonico & uno Pentagonico vel
pro eo Hexagonico , aut quocunque
majori, rejiciuntur, per pr. XXIII.
quia Trigonus imparilatera figura
cingi deberet Tetragono & Pentagono , vel pro eo Hexagono & c.

XI. Rhomb   Unus igitur Trigonicus cum duobus Tetragonicis & uno Pentagonico, mi-
icosidode- nus efficiunt 4 rectis , & congruunt 20 Trigoni cum 30 Tetragonis & 12 Penta-
caedron. gonis, in unum Hexacontadyhedron , quod appello Rhombicosidodecaëdron, seu sectum Rhombum Icosidodecaëdricum. Pingitur num. 11. fol. antecedentis

Unus Trigonicus , duo Tetragonici , cum uno Hexagonico, æquant rectos
quatuor ; cum uno majori ; superant ; nec ad solidum assurgunt. Mutatibus igitur duos Tetragonicos.

Unus Trigonicus, unus Tetragonicus,& duo Pentagonici superant 4 rectos;
multòq; magis si bini majores plani anguli admiscerentur. Desinunt igitur mis-
ceri anguli plani quaterni ad formandum unum solidum; desinit ergò & Trigonus ingredi mixturam triplicem.   Nam unus Trigonicus, unus Tetragonicus &

Figure 4.13 (*continued*).

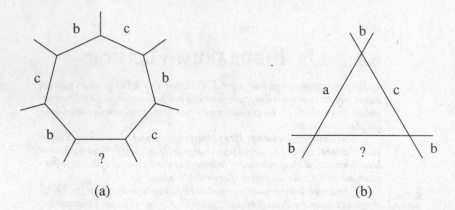

Figure 4.14.

The following proof is based on Kepler's ideas, and follows his presentation fairly closely. The main difference is that I have distinguished between species and types whereas Kepler did not. The proof becomes somewhat repetitive once you have the idea.

Theorem. Suppose that all the solid angles of a convex polyhedron have the same type. Besides the two families of types $(4, 4, n)$ and $(3, 3, 3, n)$ there are thirteen types of solid angle that can occur. These possibilities are realised by the families of prisms, antiprisms, and the Archimedean solids respectively.

PROOF: The theorem is proved by exhausting all the possible combinations of faces which can surround a solid angle and excluding those which cannot be extended in the required manner. The first lemma above shows that the species of solid angle present can be surrounded by at most three sorts of regular polygon, and there must be at least two sorts of polygon (by definition). Following Kepler, these two cases are investigated separately.

First, we consider those species of solid angle where there are two sorts of faces.

(1) species of solid angle bounded by 3-gons and 4-gons only.

Suppose there is only one 4-gon at each solid angle. Then there can be at most four 3-gons since the angle sum of five or more 3-gons and a 4-gon is greater than 360°. Thus there are three possible types of solid angle: (3,3,3,3,4) which is realised in the snub cube; (3,3,3,4) which forms a square antiprism; and (3,3,4) which is excluded because part (*i*) of the previous lemma shows that this type of solid angle cannot be extended.

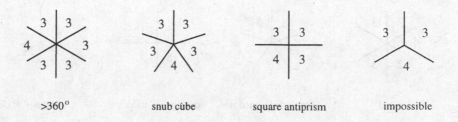

| >360° | snub cube | square antiprism | impossible |

Next, suppose that there are two 4-gons in the species of solid angle. Three 3-gons and two 4-gons fit round a point in a plane so there can be at most two 3-gons at the solid angle. The species of solid angle containing two 4-gons and two 3-gons comes in two types. One of these (3,3,4,4) is excluded by part (*ii*) of the preceding lemma; the other (3,4,3,4) corresponds to the cub-octahedron. The remaining case of two 4-gons and a single 3-gon (3,4,4) forms a triangular prism.

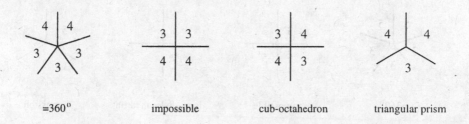

| =360° | impossible | cub-octahedron | triangular prism |

If there are more than two 4-gons at the solid angle then there is a single possibility, namely (3,4,4,4), the rhomb-cub-octahedron. The angle sum of three 4-gons and two or more 3-gons is greater than 360°, and four or more 4-gons suffer the same problem.

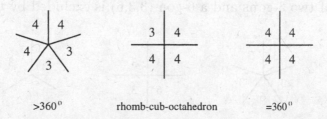

| >360° | rhomb-cub-octahedron | =360° |

(2) species of solid angle bounded by 3-gons and 5-gons only.

The analysis of this case is the same as the preceding one. If there is only one 5-gon at a solid angle then there can be at most four 3-gons. The three possible types give rise to the snub dodecahedron (3,3,3,3,5), and a pentagonal antiprism (3,3,3,5).

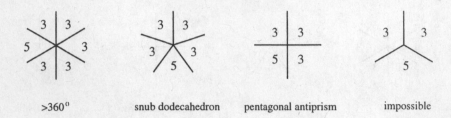

>360°	snub dodecahedron	pentagonal antiprism	impossible

If there are two 5-gons in the species of solid angle then there can be at most two 3-gons. Again, the species containing two 3-gons comes in two types: (3,3,5,5) cannot be extended by part (*ii*) of the lemma; and (3,5,3,5) which is the icosi-dodecahedron. The remaining case is excluded by part (*i*) of the lemma.

There are no other species of solid angle containing 5-gons as even a single 3-gon with three 5-gons has an angle sum larger than 360°.

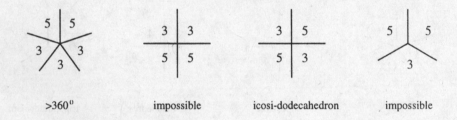

>360°	impossible	icosi-dodecahedron	impossible

(3) species of solid angle bounded by 3-gons and 6-gons only.

In the case when the species contains a single 6-gon there can be at most three 3-gons as four 3-gons and a 6-gon do not form a convex angle but are planar. Three 3-gons and a 6-gon (3,3,3,6) form a hexagonal antiprism, and the case of two 3-gons and a 6-gon (3,3,6) is excluded by the lemma.

=360°	hexagonal antiprism	impossible

If there are two 6-gons at the solid angle then there can be only one 3-gon, otherwise the angle sum is too great. This single case (3,6,6) corresponds to the truncated tetrahedron. More than two 6-gons cannot form a solid angle.

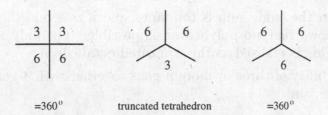

=360°   truncated tetrahedron   =360°

(4) species of solid angle containing 3-gons and n-gons only where $n \geqslant 7$.

None of the species of solid angle containing a single n-gon can form an Archimedean solid. The only possibility is $(3, 3, 3, n)$, the antiprism with an n-gon base.

>360°   n-gonal antiprism   impossible

If there is more than one n-gon at the solid angle then the only possibility is to have two n-gons and one 3-gon. If n is odd then this type of solid angle cannot be extended (by part (i) of the lemma). And if n is even and greater than 10, the sum of the plane angles is greater than 360°. This leaves the cases (3,8,8) which is the truncated cube, and (3,10,10), the truncated dodecahedron.

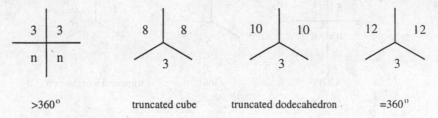

>360°   truncated cube   truncated dodecahedron   =360°

This completes the analysis of all the species of solid angle containing 3-gons and one other type of polygon. Next, the other species which contain only two kinds of polygon are investigated.

(5) species of solid angle containing 4-gons and n-gons only ($n \geqslant 5$).

If there is a single n-gon face then the type of solid angle must be $(4, 4, n)$ as the sum of the plane angles of three or more 4-gons and an n-gon is too large. The allowable case is a prism with an n-gon base.

If there are two n-gon faces then there is only a single 4-gon (otherwise the angle sum would be too large). Thus, the type of solid angle is $(4, n, n)$. If

$n \geqslant 8$ then the angle sum is too large, and if n is odd then part (i) of the lemma shows that no polyhedron is possible. The only remaining case is $(4,6,6)$. This is realised as the truncated octahedron.

The possibility of three or more n-gons together with 4-gons is excluded by their angle sum.

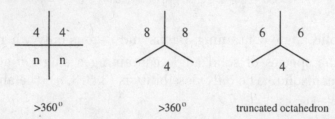

$>360°$ $>360°$ truncated octahedron

(6) species of solid angle containing 5-gons and n-gons only $(n \geqslant 6)$.

A single n-gon cannot form part of a solid angle, for the angle sum of three 5-gons and an n-gon is greater than 360°, and an angle of type $(5,5,n)$ is excluded by part (i) of the lemma.

If there are two n-gons then an angle sum argument shows that the solid angle must be of type $(5,n,n)$. The smallest value of n gives the truncated icosahedron $(5,6,6)$. For any larger value of n the sum of the plane angles is more than 360°.

More than two n-gons leads to an angle sum that is too large.

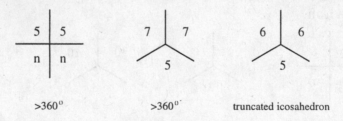

$>360°$ $>360°$ truncated icosahedron

In any other species of solid angle containing only two sorts of polygon the smallest possible angle sum arises from two 6-gons and a 7-gon, and this is larger than 360°. Thus every species of solid angle which contains only two kinds of polygon has now been dealt with. It remains to consider the species involving three kinds of polygon.

(7) species of solid angle containing 3-gons, 4-gons and n-gons $(n \geqslant 5)$.

Assume first of all that there is a single n-gon. If there were one 4-gon there could be at most two 3-gons; the species containing two 3-gons are excluded by part (ii) of the lemma above; and the angle type $(3,4,n)$ is excluded by part (i).

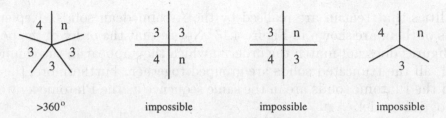

>360°   impossible   impossible   impossible

If there are two 4-gons and a single n-gon at the solid angle then there can
be only one 3-gon, otherwise the sum of the plane angles is at least 360°.
The angle sum is also too large if $n \geqslant 6$. Thus there are two possible types:
(3,4,4,5) which is excluded by part (ii) of the lemma; and (3,4,5,4) which is
the rhomb-icosi-dodecahedron.

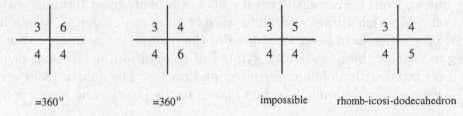

=360°   =360°   impossible   rhomb-icosi-dodecahedron

(8) species of solid angle containing three kinds of face, none of which are 3-gons.

Suppose that four faces form the solid angle. The smallest possible combination is to have two 4-gons, a 5-gon and a 6-gon. The sum of the internal
angles of these polygons is greater than 360° so there must be three polygons
forming the solid angle, all of which are different. Part (i) of the lemma
shows that, in this case, none of the polygons can have an odd number of
sides.

The smallest possible combination of faces is (4,6,8) which corresponds to
the great rhomb-cub-octahedron (or truncated cub-octahedron). The next
smallest combination is (4,6,10) which corresponds to the great rhomb-icosi-
dodecahedron (or truncated icosi-dodecahedron). In all other combinations
of faces the angle sum is too large to produce a solid angle.

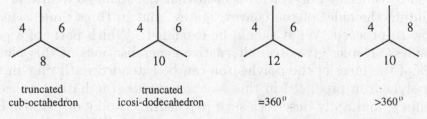

truncated   truncated   =360°   >360°
cub-octahedron   icosi-dodecahedron

All the possibilities of placing regular polygons together to form a solid angle
have now been considered. All the types of solid angle that are not excluded by
the simple conditions in the previous lemmas are candidates for polyhedra which
are 'perfect to a lower degree'. After excluding the prismatic families, the thirteen

possibilities that remain are realised by the Archimedean solids. Kepler's illustrations of them are shown in Figure 4.13. Notice that the order of the polyhedra in the figures does not match the order in which they appear in the enumeration. Instead, all the truncated solids are grouped together. Furthermore, the truncations of the Platonic solids are in the same sequence as the Platonic figures in the cosmological model.   ■

Star polygons and star polyhedra

Star-shaped polygons such as the pentagram have been known since antiquity. A vase dating from the seventh century BC has a pentagram forming part of its decoration. Moorish tilings contain a variety of star polygons. Symbolic and mystical properties have been attributed to the pentagram in many cultures and professions. It has been used as a symbol of recognition, a talisman or charm, and is associated with alchemy, astrology and magic. The golden ratio occurs in several aspects of its geometry—a fact known to the Greeks and also recorded by Pacioli.

Star polygons were first studied mathematically by Thomas Bradwardine (1290–1349) and were later investigated by Charles de Boulles (1470–1533). Kepler also made a study of them and wrote an account of them in book I of *Harmonices Mundi*. He defines a regular polygon as a figure whose sides are all equal in length and whose angles are equal and pointing outwards. He goes on to distinguish two kinds of regular polygon: the fundamental or primary ones whose sides do not cross; and others of a more general type called *stellated* polygons (stellar meaning starlike). These stellated polygons can be derived from the primary ones by extending non-adjacent sides until they intersect. He includes only those stellated polygons that can be traced in a single line. Thus the compound polygons in Figures 4.15(b) and (c) are excluded. Kepler then generalised this procedure so that it could be applied to the regular polyhedra.

When considering polygons it is clear that the stellated figures are produced by extending the sides of the convex figures. But in three dimensions it is less clear how to proceed. What should be extended? Which parts of a polyhedron are its sides? Kepler gives two alternative generalisations of this idea. Firstly, the sides of the faces of the polyhedron can be extended until they intersect. He calls a polyhedron produced in this way an *echinus* (the Latin for 'hedgehog' or 'sea urchin'), and aptly describes such polyhedra as spiky or prickly. His second method involves extending the faces themselves until they intersect, the resulting polyhedron being called an *ostrea* ('oyster'). These two methods are now called *edge-stellation* and *face-stellation*, respectively.

The two star polyhedra which Kepler discovered appear to have been pro-

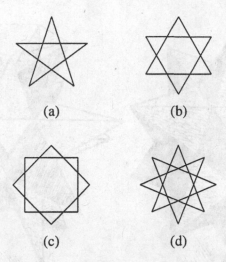

(a) (b)

(c) (d)

Figure 4.15. Star polygons (a), (d) and compound polygons (b), (c).

duced by applying the first method to the Platonic solids. In his account of the resulting polyhedra he gives descriptions only and, unusually, does not pass on any indication of how they were discovered. Extending the edges of a tetrahedron does not produce a new polyhedron since the lines do not intersect apart from at the original vertices. The same is true of the cube and the octahedron. Applying the edge-stellation process to the dodecahedron does produce an example of an echinus; the icosahedron furnishes the other example. These two polyhedra are shown in Figure 4.16 and Plate 8. Kepler's own sketches of them are reproduced in Figure 4.17.

Figure 4.16. Kepler's two star polyhedra.

Figure 4.17. Kepler's sketches of his 'prickly' polyhedra.

In the last few years of his life, Kepler started preparing a treatise on geometry. Among his notes for this unfinished work there is a section entitled *De Auctis* (On Augmented Figures) which deals with the stellation process. In this work he calls the twelve-spiked star polyhedron 'echinus major icosaëdricus' and the twenty-spiked one 'echinus minor icosaëdricus' (larger and smaller icosahedral hedgehogs). Kepler also recognised that the larger hedgehog could be derived from the dodecahedron via the second process of face-stellation. In fact, both of the polyhedra can be produced in this way, and it is the process of face-stellation which is referred to in their current English names: the twelve-pointed one being the *small stellated dodecahedron*, the twenty-pointed one the *great stellated dodecahedron.* These names were introduced by Arthur Cayley in 1859. The stella octangula, or compound of two tetrahedra, can also be produced by face-stellation, this time by extending the faces of an octahedron.

All three stellations can easily be regarded as being formed from convex regular polyhedra with a suitable pyramid erected on each face. From this viewpoint one is led to compare the convex figure with the augmented one. Noticing this, Kepler comments that the stellated figures

> *are so closely related the one to the dodecahedron and the other to the icosahedron that the latter two figures, particularly the dodecahedron, seem somehow truncated or maimed when compared to the figures with spikes.*[l]

The process of augmenting convex polyhedra by adding pyramids of various heights had been used to generate many new polyhedra in the preceding 200 years. It is not surprising, therefore, to find that all of Kepler's stellated forms had already been depicted. Earlier sketches of the stella octangula appear in the works of Pacioli and Jamnitzer. A picture of the small stellated dodecahedron dating from the 1420's can be seen in Saint Mark's Basilica in Venice (see Plate 7). It is inlaid in marble and forms part of the floor decorations in one of the main doorways. The design has been attributed to Uccello. The great stellated dodecahedron was pictured by Jamnitzer as part of one of his polyhedral monuments. The spiky component at the core of the monument shown on page 288 is an example of this polyhedron. However, Kepler's interest in these figures was not merely aesthetic. He considered them in a mathematical context.

The two stellations of the dodecahedron can be interpreted in yet another way. A close examination of the small stellation reveals that the triangular faces are arranged in groups of five so that all the faces in each group lie in a plane. Furthermore, each group surrounds a regular pentagon buried under a pyramid. Together, a pentagon and its five satellite triangles form a pentagram. Therefore, this polyhedron can be regarded as having pentagrams for faces, twelve in total, arranged with five meeting at each vertex. The great stellation can also be regarded as being built up from twelve pentagrams, meeting in threes at the vertices. The models of these polyhedra shown in Plate 8 are coloured so that each pentagram is painted in a single colour.

Under this radically different interpretation the faces of the polyhedra are allowed to pass through each other like the sides of a star polygon. The star pentagon faces are joined side against side as in conventional polyhedra, yet they pass through each other allowing some parts of each face to be seen and confining others to the interior of the polyhedron hidden from sight.

The Dutch artist Maurits Cornelis Escher (1898–1972) used this interpretation in some of his work. Figure 4.18 is a preliminary sketch which he made

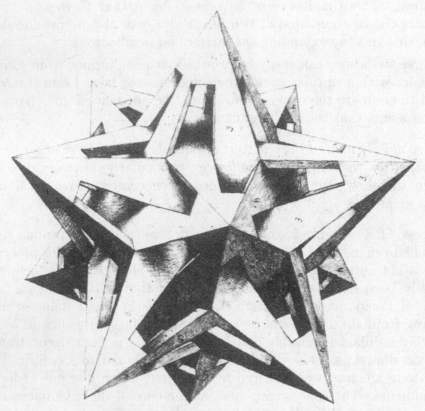

Figure 4.18. A sketch by M. C. Escher showing a cut-away view of the small stellated dodecahedron.

showing a cut away diagram of the small stellation. The sketch clearly shows the faces continuing under each pyramid. In his 1952 print *Gravity*, shown on page 248, Escher adds twelve animals, one standing on each face and poking four limbs and a head tortoise-like through the holes in a pyramid. Escher also made a perspex model of the complete polyhedron and engraved a starfish into each of the twelve pentagram faces.

Regarding the faces of these polyhedra as pentagrams forces us to accept the following surprising conclusion: they are *regular* polyhedra, for their faces are equal regular polygons and the same number surround each solid angle. Including the star polygons among the regular polygons allows a more general type of regular polyhedron to be constructed. Kepler does not recognise that, in this generalised setting, his new regular polyhedra have the same status as the convex regular polyhedra. In the same way that he separated regular polygons into primary and stellated, he regards the convex polyhedra as fundamental and their stellations as secondary: 'nothing but an augmented dodecahedron (but augmented most regularly)'.

Perhaps part of the reason why he was reluctant to raise his star polyhedra to the level of the Platonic polyhedra is that it would ruin his solution of the cosmic mystery. Even after his work on the elliptical nature of planetary orbits, Kepler continued to experiment with his planetary model. In book v of *Harmonices Mundi* he investigates the possibility of replacing the dodecahedron and icosahedron by a single 'echinus major' with its circumsphere bounding the orbit of Mars and its midsphere bounding the orbit of Venus leaving Earth's orbit free and undetermined. But in a later work, *Epitome Astronomiae Copernicanae*, he discards this idea. If such an alteration were made then the reason there are but six planets would evaporate, and the whole harmonious nature of his system would be destroyed.

This unwillingness to abandon his cherished model may go some way to explaining why Kepler did not investigate the properties of his star-faced polyhedra more thoroughly. In the case of the two rhombus-faced polyhedra, he proceeded to analyse them and show that he had found all the possibilities. In contrast, he makes no attempt to establish whether any other polyhedra have properties similar to his star polyhedra. The two stellations are composed of pentagrams meeting in groups of three or five to form the solid angles. Is it possible to construct a polyhedron with four pentagrams surrounding each solid angle? Furthermore, the angle in the corner of a pentagram is 36° so up to nine can be fitted round a vertex. It is conceivable that polyhedra with up to nine pentagrams per solid angle could be constructed. The other star polygons provide further possibilities. Are Kepler's star polyhedra the only ones? Kepler himself is strangely silent on this topic. We shall return to the problem in Chapter 7.

Semisolid polyhedra

Kepler did experiment with other star polygons. Besides the two pentagram-faced polyhedra, he found two semisolid polyhedra: one composed of star octagons, the other of star decagons.

> The sides of the first and fourth points of star octagons and star decagons lie in a line, which passes through two intermediate points, and the stars can be fitted together with such sides joined two by two. The star octagons make a kind of cube, and the star decagons a kind of dodecahedron, figures which have not angles but ears, for when two of the plane angles are fitted together they must leave a gap, which can not be closed.[m]

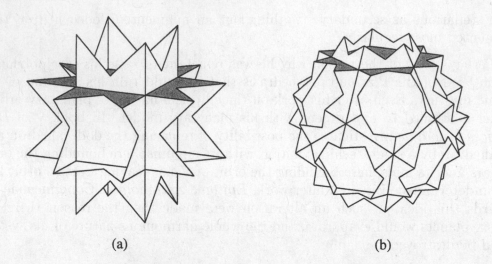

(a)   (b)

Figure 4.19. Kepler's two 'semisolid' polyhedra.

The 'eared cube' and the 'eared dodecahedron' are shown in Figure 4.19. Like the star polyhedra whose faces are pentagrams, the faces of these incomplete polyhedra are star polygons that meet along their edges. As these edges cross each other, the faces pass through each other. In each figure, the shaded parts are the visible parts of a single face.

Recall Kepler's requirement that if a polyhedron does not close up then we must be able to fill in any gaps with regular polygons. The two semisolid polyhedra with 'ears' can be completed by the addition of regular polygons although it is not certain that the way in which it is done is what Kepler had in mind. He made no reference to it.

The 'eared cube' is completed by adding eight triangles to the six octagrams. The result is shown in Figure 4.20(a)—the visible parts of one triangular face are shaded. The sides of the triangles are joined to the unused sides of the octagrams so that two polygons meet at every edge. However, this is not possible without allowing the triangles to intersect each other. The lines where they meet are not counted as edges just as the points where the sides of a star polygon cross are not counted as vertices. The compounds of polyhedra mentioned earlier also exhibit this phenomenon. In those cases, the faces of one polyhedron pass through the faces of another; here the polyhedron intersects itself.

The other 'eared' polyhedron is completed by the addition of twelve pentagons (see Figure 4.20(b)). Curiously, both of these figures arise by extending the edges of an Archimedean solid. They are edge-stellations of the truncated cube and truncated dodecahedron, respectively. This is, perhaps, how Kepler discovered them.

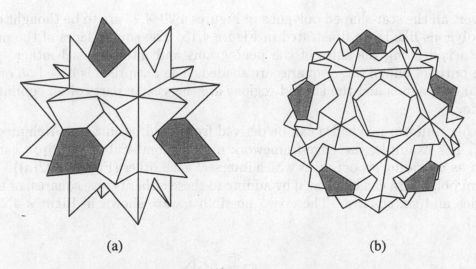

(a) (b)

Figure 4.20. Kepler's semisolid polyhedra can be completed to produce uniform polyhedra.

Uniform polyhedra

The two polyhedra in Figure 4.20 are examples of *uniform* polyhedra. These are to the Archimedean solids what the star polyhedra are to the Platonic solids. All their vertices are surrounded in the same way[2] and their faces are regular polygons (star or convex). Now, however, the faces are allowed to intersect each other. Included under this definition are the Platonic and Archimedean solids, the prisms and antiprisms, and Kepler's two pentagram-faced star polyhedra. There are new possibilities, too.

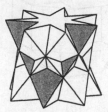

Figure 4.21. Prism and antiprism with star polygon bases.

For a start, we can form prisms and antiprisms with star polygons as bases. An example of each is shown in Figure 4.21. The base for the prism is a pentagram. The sides of the pentagrams are not drawn in full and the top appears as a ten-sided non-convex polygon. (This decagon is sometimes called a *pentacle*.)

[2]We also need to impose some kind of restriction on the symmetry of the resulting figure. To be precise, the polyhedron must be vertex transitive.

However, all the star-shaped polygons in Figures 4.20–4.25 are to be thought of as star polygons like those illustrated in Figure 4.15. The square faces of the prism are attached along the sides of the pentagrams and intersect each other. The visible parts of one of these squares are shaded. The antiprism is based on one of the star heptagons and the shaded regions are the visible parts of an equilateral triangle.

Other uniform polyhedra can be derived from the Platonic and Archimedean solids. For example, the edge-framework of the rhomb-cub-octahedron can be viewed as bounding six octagons which intersect each other (Figure 4.22(a)). Two uniform polyhedra can be formed by adding to these either twelve squares, or eight triangles and six squares. These two possibilities are shown in Figures 4.22(b) and (c).

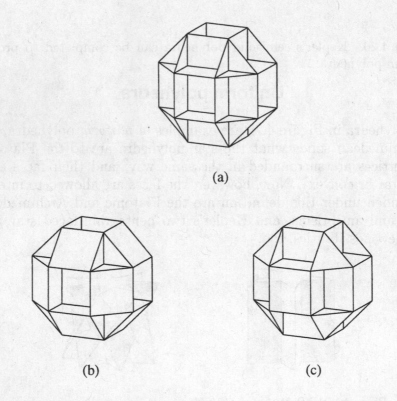

Figure 4.22. Two uniform polyhedra, (b) and (c), with the same edge-skeleton as a rhomb-cub-octahedron.

The edge-framework of an octahedron bounds three squares as shown in Figure 4.23(a). By adding four triangles to these we complete another uniform polyhedron (Figure 4.23(b)). It is commonly called the *heptahedron* since it has seven faces.

The cub-octahedron furnishes two further examples. Its edge-framework

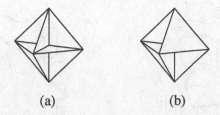

(a)   (b)

Figure 4.23. Four triangles added to three orthogonal squares form the heptahedron.

bounds four hexagons (Figure 4.24(a)), and uniform polyhedra result by adding either the squares or the triangles of the original cub-octahedron (Figures 4.24(b) and (c)).

Apart from the two uniform polyhedra derived from Kepler's semisolid ones, none of these examples contains star polygons. Four uniform polyhedra containing star octagons are shown in Figure 4.25. In the first two examples, the six octagrams occupy equivalent positions. The polyhedra are completed by the addition of (a) six squares and eight triangles, or (b) twelve squares. In the first case, the squares lie in planes parallel to the octagrams. In the second example, the squares are parallel to planes passing through two opposite edges of the surrounding cube. Figure 4.25(c) contains eight hexagons and six octagons besides its six octagrams, and Figure 4.25(d) is completed with twelve squares and eight hexagons. The hexagons in this last example are almost entirely hidden from view. The small triangular regions are, in fact, openings which give access to large cavities inside the polyhedron whose walls are formed by the hexagons.

There are many more uniform polyhedra. As well as those listed above, there are two more composed of a single kind of polygon (the star polyhedra discovered

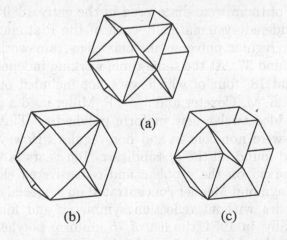

(a)

(b)   (c)

Figure 4.24. Two uniform polyhedra, (b) and (c), with the same edge-skeleton as a cub-octahedron.

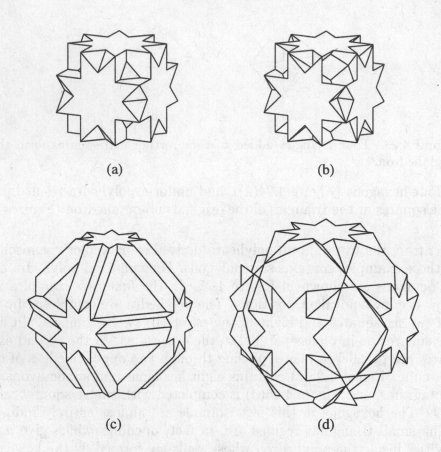

(a)   (b)

(c)   (d)

Figure 4.25. Four uniform polyhedra.

by Louis Poinsot in 1810 discussed in Chapter 7) and 42 having more than one
kind of face. Some of them were discovered in the early 1880's. The Frenchman
A. Badoureau considered systematically each of the Platonic and Archimedean
solids, searching for regular polygons in their edge-frameworks or star polygons
in their faces. He found 37. At the same time, working independently in Austria,
Johann Pitsch found 18, four of which were not included on Badoureau's list.
Fifty years later, H. S. M. Coxeter and J. C. P. Miller used a different method of
enumeration and added twelve new uniform polyhedra. This brought the total
to 75, 53 of which were non-convex and non-regular. However, they could not
prove that they had found all the possibilities. Ten years later M. S. and H. C.
Longuet-Higgins worked on the problem and rediscovered eleven of the twelve.
Jean Lesavre and Raymond Mercier concentrated on a special case: they searched
for uniform polyhedra without reflection symmetry and found five such snub
polyhedra. Eventually, in 1954, the list of 75 uniform polyhedra was published.
It was not until the advent of the computer that J. Skilling was able to show that
the list was complete.

The uniform polyhedra were the main subject for Magnus Wenninger's book *Polyhedron Models*. Starting with the simple Platonic and Archimedean solids, he progresses to some stellated models and then to the 53 other uniform polyhedra. The work contains photos of completed models and instructions on how to make your own. Some of them are extremely intricate. He also names each solid. His names for the completions of Kepler's 'eared' polyhedra (Figure 4.20) are 'quasitruncated hexahedron' and 'quasitruncated small stellated dodecahedron'.

Kepler's influence on the subsequent development of mathematics was minimal. His work on star polyhedra was certainly unknown to Poinsot 200 years on. Later still, Eugène Charles Catalan was unaware that Archimedean polyhedra had been previously investigated. One reason that Kepler's results were overlooked has to do with the fact that much of his work is not published in the form of a scientific exposition, but rather as a saga recording the trials of the process of discovery. He includes in his accounts the motivation, the assaults he has tried, the failures and dead-ends, the moment of inspiration, his feelings when success comes, as well as his solution to the problem at hand. This makes it difficult to isolate the useful results from the discussion in which they are embedded. Consequently, the work of this most creative of people lay neglected for many years.

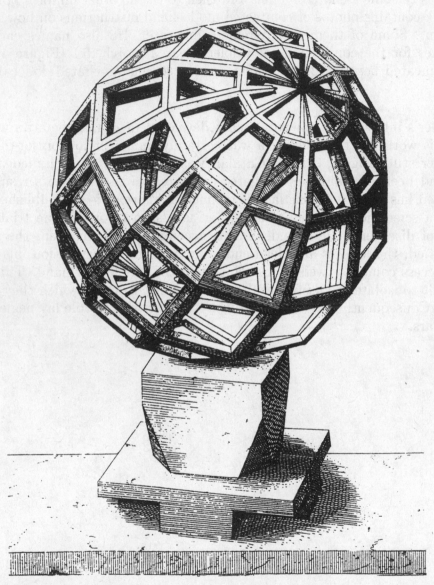

From *Perspectiva Corporum Regularium* by Wenzel Jamnitzer, 1568.

5 Surfaces, Solids and Spheres

> The first important notions in topology were acquired in the course of the study of polyhedra.[a]
>
> Henri Lebesgue

The reader who has made several models may have observed the following phenomenon: as the number of solid angles of a polyhedron increases, the sharpness of its corners decreases. Compare a cube with a dodecahedron, for instance. The eight solid angles in the cube are all acute whereas the twenty solid angles in a dodecahedron are obtuse. This 'sharpness' can be quantified if we cut along one of the edges and open out the faces so they lie flat (Figure 5.1). In any solid angle the sum of its face angles is always less than 360° and it is the size of the deficit which determines the sharpness of the solid angle: the larger the deficit, the more acute the resulting angle.

Examining the deficiencies in the two polyhedra more closely reveals another phenomenon. Each of the eight angles in a cube has a deficit of 90° and each solid angle in a dodecahedron has a deficit of 36°. In both cases, the deficiencies of all the solid angles add up to 720°. These observations illustrate a theorem discovered by René Descartes (1596–1650).

Descartes' theorem appears as part of a work called *Progymnasmata de Solidorum Elementis* (Exercises on the Elements of Solids) along with other propositions in solid geometry and a section on polyhedral numbers—analogues of the familiar triangular and square numbers. He gives formulae for polyhedral numbers corresponding to the five Platonic solids and nine of the Archimedean solids. Descartes did not publish this manuscript and it is a quirk of fortune that its contents have survived.

In the autumn of 1649, Descartes went to Stockholm at the invitation of

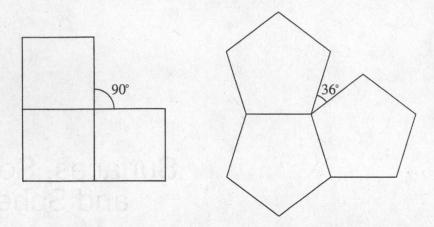

Figure 5.1. Deficiencies of the cube and dodecahedron.

Queen Christina of Sweden, but the severity of the climate was too much for him and he died six months later. His belongings were shipped back to France but suffered accident on route, the box carrying his manuscripts ending up in the Seine at Paris. The papers were rescued from the river, separated and dried. Later, some were published and the remainder were made available for consultation. In 1676 Gottfried Wilhelm Leibniz (one of the founders of the calculus) made copies of several of the latter manuscripts including the work on polyhedra. Descartes' original manuscript has vanished and it is only through the copy that the work is preserved. Even so, it remained unknown until 1860 when the copy was discovered by Comte Foucher de Careil among a collection of uncatalogued Leibniz papers.

Descartes was the first person to study polyhedra in a general context. Prior to this, people had concentrated their attention on specific examples or on particular properties shared by a few polyhedra; the idea of investigating the set of all polyhedra as a whole does not seem to have occurred to anyone. During the two centuries which elapsed while Descartes' work lay forgotten and unknown, another mathematician initiated a general study of polyhedra—a study which had far-reaching effects. Leonhard Euler (1707–1783) revolutionised the theory of polyhedra by introducing new ideas and a new vocabulary. He discovered a formula that related the numbers of the various constituent parts of a polyhedron but he could not explain its origin. Many mathematicians after him struggled to understand the nature of his discovery and to find its underlying causes, the foundations on which it depended. The ensuing debate played a major role in the development of mathematics. Out of the original uncertainty arose a new kind of geometrical discipline—topology.

Plane angles, solid angles, and their measurement

In order to arrive at Descartes' theorem it is necessary to explore the problems of angle measurement. Once this has been done, and some preliminary results have been established, the proof of the theorem becomes almost immediate.

An *angle* is a region of a plane contained between two intersecting lines so that segments of the two lines and their point of intersection form part of its boundary. It is also a measure of the inclination of the lines to each other. Since classical times there have been two units of measurement for angles: the Babylonian system of degrees used in astronomy and trigonometry, and the right angles used in geometry.

Euclid singles out the right angle as the fundamental angle and expresses other angles as multiples of it. The tenth definition of his *Elements* says that when a line stands on another line so that the adjacent angles are equal to each other then each of these angles is a right angle; the fourth postulate is that all right angles are equal. The sizes of plane angles are then expressed as fractions of a right angle. Thus the sum of the interior angles in a triangle is two right angles; each of the interior angles in a regular pentagon is one-and-a-fifth right angles.

Proclus extended Euclid's result on the interior angle sum of a triangle to polygons in general. By dividing an n-sided polygon into triangles he showed that its interior angle sum is $2(n-2)$ right angles. He also proved that the sum of the exterior angles in any polygon was equal to four right angles. The *exterior* angle measures the amount by which one arm of an angle deviates from the straight course of the other (see Figure 5.2(a)). Therefore, when going round a polygon, we would expect that the total sum of all the deviations make a complete turn (5.2(b)). Proclus proves this by noting the fact that exterior and interior angles are *complementary*—they add up to two right angles. Thus the sum of all the interior and exterior angles in an n-sided polygon is $2n$ right angles. Deducting the $2(n-2)$ right angles which comprise the interior angle sum, we see that the exterior angle sum is four right angles—a complete turn.

Another angle that is associated to a plane angle is its *supplementary* angle. This is constructed by erecting a perpendicular on each line forming the angle and extending them until they meet (Figure 5.3(a)). It is a simple exercise to show that an angle and its supplement come to two right angles and thus they are complementary. Therefore, the supplement of an angle and its external angle are equal. Hence, all the supplementary angles in a polygon add up to four right angles. A different proof of this fact is indicated in Figures 5.3(b) and (c): all the 'wedges' of the supplementary angles can be translated to fit together around a point.

Besides the two traditional ways of measuring angles, there is a third, more

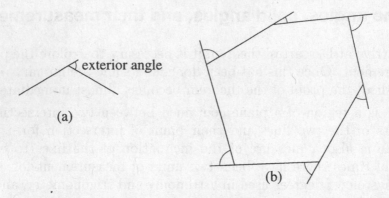

Figure 5.2. Exterior angles.

modern, measure. It uses the length of the arc intercepted by the angle on a circle
of radius one unit centred on the point of intersection of the two lines. This is
called the *radian* measure, and an angle measures one radian when the length of
arc is one unit, that is, when it is equal to the radius of the circle. A whole circle
is 2π radians (or 360° or four right angles). A right angle is $\frac{\pi}{2}$ radians; the angles
in an equilateral triangle are each $\frac{\pi}{3}$ radians. Proclus' theorems can be restated
as: the sum of the interior angles in an n-sided polygon is $(n-2)\pi$, and the sum
of its exterior angles is 2π.

In book XI of the *Elements*, Euclid begins his treatment of solid geometry.
He defines a solid angle as the space which is contained by three or more planes
meeting at a point and not lying in a plane. He then shows that the sum of the
plane angles forming the solid angle must be less than four right angles, but he
makes no attempt to quantify the solid angle itself. There is some difficulty in
this. How should the *size* of a solid angle be measured? How can solid angles be
compared to one another?

If a plane angle is measured by the arc of a circle then perhaps solid angles

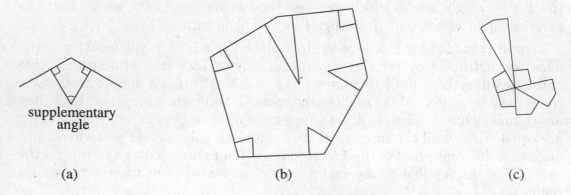

Figure 5.3. Supplementary angles.

can be measured by segments of a sphere. To pursue this analogy, imagine a sphere of radius one unit centred on the apex of a solid angle. It will intersect the plane figures forming the solid angle in arcs of great circles, and these arcs will bound a polygon on the sphere which lies inside the solid angle (see Figure 5.4). This spherical polygon is a measure of the solid angle. The angles between its sides are the dihedral angles between the faces of the solid angle, and the sides are a measure of the plane angles of the faces. In fact, the length of a side is the radian measure of a plane angle. Just as the length of an arc is used as a measure of a plane angle, so the *area* of a spherical polygon can be used to measure a solid angle. The units for this quantity are called *steradians*.

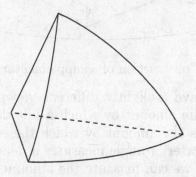

Figure 5.4.

Consider the particular solid angle formed from three mutually perpendicular planes. Every plane angle and every dihedral angle is a right angle. The spherical polygon which measures this angle will cover one eighth of the surface of the sphere. Since the sphere has unit radius, its total surface area is 4π, hence the measure of this solid angle is $\frac{\pi}{2}$. This solid angle is an example of a *solid right angle*. (Recall that plane right angles also measure $\frac{\pi}{2}$.) However, whereas all plane right angles are equal (congruent), there are many solid right angles since the number of spherical polygons whose area is $\frac{\pi}{2}$ is unlimited.

The supplementary angle to a plane angle is constructed by erecting lines perpendicular to those forming the angle. Similarly, a *supplementary solid angle* can be constructed from a given solid angle by erecting planes perpendicular to its faces and edges in such a way that they meet in a point (see Figure 5.5). The dihedral angles of this supplementary solid angle are the supplements of the face angles of the original solid angle. So, for example, in the figure, β is the supplement of α and hence $\alpha + \beta = \pi$ radians. And, as in the case of the supplementary plane angles of a polygon, the supplementary solid angles of a polyhedron can be translated to fit together around a point. Therefore, the sum of the supplementary solid angles is 4π—the whole sphere.

In order to find an analogy to the exterior plane angle in the solid geometry

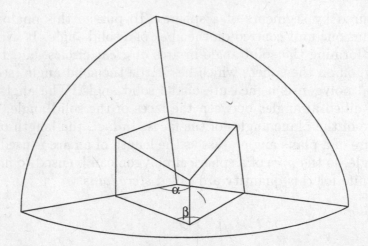

Figure 5.5. Construction of a supplementary solid angle.

setting it is helpful to have a slightly different viewpoint. Rather than think of the exterior angle as the amount by which a line deviates from a straight-on direction, we can see it as the amount by which the interior angle falls short of a straight line. Thus an exterior angle measures a deficit. For an analogous idea applicable to solid angles we can measure the amount by which they 'fall short' of being planar. The angle by which the sum of the plane angles around a solid angle is less than 2π is called its *deficiency*. Solid angles with a large deficiency (like tetrahedral angles, for example) are sharp while those with small deficiencies (dodecahedral angles) are blunt and squat.

Recall that the exterior and supplementary angles of a plane angle are equal. It is a remarkable fact that the deficiency of a solid angle is equal to the size of its supplementary angle. This is a consequence of a certain property of spherical polygons. Unlike plane polygons, whose area depends on their size as well as their shape, the area of a spherical polygon is completely determined by its shape. This result first appeared in print in 1629 in a collection of essays by Albert Girard. In the third essay, entitled *De la Mesure de la Superfice des Triangles & Polygones Sphericques, Nouvellement Inventée*, he showed that the area of a spherical polygon equals the sum of its interior angles minus the sum of the interior angles in a plane polygon having the same number of sides. To express this symbolically, let the angles in an n-sided spherical polygon be $\alpha_1, \alpha_2, \cdots, \alpha_n$. Then its area is given by the formula:

$$\text{area} = \alpha_1 + \alpha_2 + \cdots + \alpha_n - (n-2)\pi.$$

This is called the *Spherical Excess* formula.

We are now in a position where we can prove the crux of Descartes' theorem.

Lemma. The deficiency of a solid angle equals its supplementary angle.

PROOF: Suppose that the solid angle is formed from n faces whose plane angles are $\alpha_1, \alpha_2, \cdots, \alpha_n$. The supplementary solid angle is measured by the area of the segment on a sphere of unit radius which it intersects. This segment is a spherical polygon whose angles are the dihedral angles of the supplementary solid angle. We know that each of these dihedral angles is the complement of one of the plane angles of the faces, that is, the dihedral angles are $(\pi - \alpha_1), (\pi - \alpha_2), \cdots, (\pi - \alpha_n)$. Applying the spherical excess formula gives the area of the spherical polygon to be

$$
\begin{aligned}
\text{area} \quad &= \quad (\pi - \alpha_1) + (\pi - \alpha_2) + \cdots + (\pi - \alpha_n) \; - \; (n - 2)\pi \\
&= \quad 2\pi - (\alpha_1 + \alpha_2 + \cdots + \alpha_n)
\end{aligned}
$$

which is the deficiency of the given solid angle. ∎

Descartes' theorem

> *A very beautiful and general theorem which ought to be placed at the head of the theory of polyhedra.*[b]
>
> E. Prouhet

The most important result in Descartes' *De Solidorum Elementis* is his proposition concerning the total angular deficiency of polyhedra. His other geometric results can be considered as consequences of it. The manuscript begins:

> *A solid right angle is one which embraces the eighth part of the sphere even though it is not formed by three plane right angles.* $\cdots$
> *As in a plane figure all the exterior angles, taken together, equal four right angles, so in a solid body all the exterior angles, taken together, equal eight solid right angles.*[c]

Although he does not define an exterior solid angle, Descartes does state some of their properties. These include the fact that they can be measured by the amount by which the sum of the plane angles bounding the (interior) solid angle is less than four right angles. Thus we have

Theorem. The sum of the deficiencies of the solid angles in a polyhedron is eight right angles. ∎

The manuscript contains no proof of this theorem but the results concerning angle measurement presented in the previous section were all known by 1630, and the proof follows easily from them: the sum of the supplementary solid angles of a polyhedron is known to equal eight right angles (the whole sphere) and the lemma shows that the supplementary angles equal the deficiencies.

Descartes follows his proposition with a simple corollary. Knowing the sum of the deficiencies of all the solid angles enables one to calculate the sum of the plane angles of all the faces. If S denotes the number of solid angles then the sum of the plane angles is $4S - 8$ right angles.

The fact that there can be at most five regular polyhedra can be derived from this result as follows. Suppose that a polyhedron has S solid angles, each of which is surrounded by q faces, and that each face has p sides. Then the sum of the interior angles of each face is $2(p-2)$ right angles and hence every plane angle measures

$$\frac{2(p-2)}{p} \text{ right angles.}$$

Since q plane angles come together at each solid angle, there are qS plane angles in total and their sum is

$$qS\frac{2(p-2)}{p} \text{ right angles.}$$

The above corollary states that the sum of the plane angles is $4S - 8$ right angles so these two expressions can be equated

$$4S - 8 \quad = \quad qS\frac{2(p-2)}{p}$$

and then solved for S to get

$$S \quad = \quad \frac{4p}{2(p+q) - pq}.$$

The denominator of this fraction is equal to $4 - (p-2)(q-2)$ from which it is clear that $(p-2)(q-2)$ must be less than four. As p and q must both be integers greater than two, the five solutions for (p, q) are $(3,3)$, the tetrahedron; $(3,4)$, the octahedron; $(3,5)$, the icosahedron; $(4,3)$, the cube; and $(5,3)$, the dodecahedron.

In contrast to Euclid's geometric proof that there are at most five regular solids, this proof is essentially algebraic in nature. We have calculated p and q. Descartes does not offer a complete proof in his manuscript but notes that if a regular body has F faces and S solid angles then both

$$\frac{2S-4}{F} \quad \text{and} \quad \frac{2F-4}{S}$$

must be integers, and that this is possible only when $S = 4, 6, 8, 12$ or 20 and $F = 4, 8, 6, 20$ or 12 respectively.

This sample of the results contained in *De Solidorum Elementis* illustrates the originality of Descartes' approach to polyhedral geometry. No-one before him had attempted to study polyhedra in general, and some of his results were still new to geometers when Leibniz's copy was discovered two centuries later. Like

Kepler's work, this investigation by Descartes lay dormant for many years and did not influence mathematical progress. In the intervening period the theory of polyhedra developed in a quite different direction.

The announcement of Euler's formula

With the introduction of printing in the fifteenth century access to information was greatly increased. However, the market for mathematics books was limited and they were expensive. Neither were books a suitable medium for the dissemination of new ideas as most results were too short. Even with the advent of journals, supported by the rising academies and learned societies, it could take several years before a paper appeared in print. Thus mathematicians related their discoveries in letters to friends.

Leonhard Euler and Christian Goldbach corresponded for many years. In a letter written in November, 1750 Euler told Goldbach that he had started to investigate polyhedra. He wanted to introduce some sort of order into the diversity of seemingly unrelated solids. He writes

> Recently it occurred to me to determine the general properties of solids bounded by plane faces, because there is no doubt that general theorems should be found for them, just as for plane rectilinear figures, whose properties are:
> (1) that in every plane figure the number of sides is equal to the number of angles, and
> (2) that the sum of all the angles is equal to twice as many right angles as there are sides, less four.
> Whereas for plane figures only sides and angles need to be considered, for the case of solids more parts must be taken into account.[d]

The parts which Euler considered to be important are: the faces and solid angles of the polyhedron, the plane angles and sides of its polygonal faces, and the joints where two faces come together meeting side against side. He could not find a term which had been previously applied to this last kind of part so he chose a name, the Latin word 'acies' meaning ridge or sharp edge. This allowed him to distinguish between the *edges* of the polyhedron and the *sides* of its faces. He proceeded to describe how the numbers of these parts were related to one another. Letting

H be the number of faces   (*hedrae*)
S be the number of solid angles   (*angulorum solidorum*)
A be the number of edges   (*acies*)
L be the number of sides   (*latus*)
P be the number of plane angles   (*angulorum planorum*)

he observed that

> *In each face the number of sides equals the number of plane angles.*
> *Therefore $L = P$.*
> *Two sides meet at every edge so $A = \frac{1}{2}L$. So L (and hence P) is*
> *always an even number.*
> *Every face has at least three sides so $L \geqslant 3H$.*
> *At least three faces surround each solid angle so $L \geqslant 3S$.*

Besides these rather obvious statements, Euler mentioned two further relationships which are far more fundamental. Firstly,

$$S + H = A + 2$$

and also

> *The sum of all the plane angles equals $4S - 8$ right angles.*

The second of these two statements had been previously recorded by Descartes but this was not known at the time. The other relation was totally original and has become known as *Euler's formula* (or his polyhedral formula, to distinguish it from other celebrated relationships he discovered).

It is worth pausing at this point to check some examples. Firstly, we observe that Euler's formula holds for all the Platonic solids. For instance, a cube has 6 faces, 8 solid angles, making a total of 14, and it has 12 edges. An icosahedron has 20 faces, 12 solid angles and 30 edges. As more general examples, there are the families of pyramids and prisms. A pyramid with an n-sided base has $(n+1)$ faces, $(n+1)$ solid angles and $2n$ edges. A prism with an n-sided base has $(n+2)$ faces and $2n$ solid angles, making a total of $3n + 2$, and it has $3n$ edges.

An example which verifies the formula in a particular case is included in Euler's letter. He goes on to derive some consequences of his relations, then remarks:

> *I find it surprising that these general results in solid geometry have*
> *not previously been noticed by anyone, so far as I am aware; and*
> *furthermore, that the important ones $\cdots$ are so difficult that I have*
> *not yet been able to prove them in a satisfactory way.[e]*

A few weeks later Euler presented the first of two papers on his polyhedron formulae to the Saint Petersburg Academy. In it he expanded the summarised account he had sent to Goldbach and verified that the formula holds for several families of solids, but admitted that he could not offer a proof of the general case. His proposed proof appeared in a second paper which he presented the following year.

The naming of parts

The crucial observation that enabled Euler to discover his formula was the isolation and labelling of the elements called edges. Prior to this, people had analysed polyhedra in terms of bases and solid angles in analogy with the sides and angles of polygons. Yet Euler did more than merely find a name for something previously unlabelled. He needed to find a name because of his radically new way of breaking down a polyhedron into its component parts—each of a different dimension. In his first paper he writes:

> Three kinds of bounds are to be considered in any solid body; namely points, lines and surfaces, or with the names specifically used for this purpose: solid angles, edges and faces. These three kinds of bounds completely determine the solid. But a plane figure has only two kinds of bounds which determine it; namely points or angles, and lines or sides.[f]

These new elements are tactile, they are differences in texture. If you hold a model polyhedron you *feel* its flat faces, the ridges where they meet, and the sharp points at the corners.

Although Euler created the term 'edge' to distinguish the lines in the surface of a polyhedron from the lines (sides) bounding its polygonal faces, he retained the term 'solid angle' giving it a new meaning. The term 'apex' had been used for the top of a pyramid but there was no separate term for the tip of a solid angle. Adrien Marie Legendre continued in this vein using 'solid angle' to refer to the points where three or more faces come together. Later French mathematicians introduced the term 'sommet' (meaning summit). Perhaps this choice was influenced by Legendre's translation of 'acies' as 'arête' (meaning ridge of a mountain). The Englishman Arthur Cayley used both 'summit' and 'vertex'; the latter has become the standard English name.

The need to distinguish between different sorts of vertices according to how many faces surrounded them produced more terminology. Cayley wrote of 'trihedral summits' to describe vertices surrounded by three faces. In another context, in order to emphasise the fact that three edges met at a vertex, he called them 'tripleural summits' (from the Greek work for rib). A different terminology arose with the progress of molecular chemistry.

By the middle of the nineteenth century chemists had distinguished elements (containing a single kind of atom) from compounds (composed of atoms of several kinds), and it was clear that the constituent elements of a compound substance were combined in certain fixed proportions. By comparing these ratios for many simple molecules the notion of *valence* developed. This refers to the capacity for the atoms of an element to combine with other atoms. Thus a carbon atom can link to four other atoms, a hydrogen atom to just one. Various kinds of diagrams were used to illustrate these relationships, but the most important of these were those introduced by Alexander Crum Brown (1838–1922). His graphic notation quickly caught on as it neatly explained the phenomenon of chemical *isomerism*— substances having the same composition but different properties. The diagrams showed clearly that the same atoms could be linked in different ways producing molecules with different shapes and structures (see Figure 5.6).

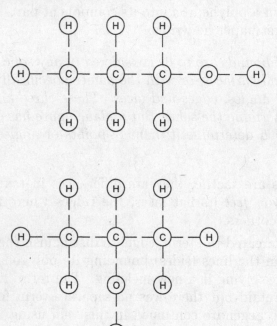

Figure 5.6. Brown's molecular notation explained isomerism.

These diagrams were not only of interest to chemists but were also studied by mathematicians who were interested in finding the number of different isomers with a given composition. These studies were part of the origins of graph theory— the part of mathematics concerned with the study of networks or graphs. The notion of valency is common to both chemistry and mathematics. To a chemist, a carbon atom has valence four since it can make four bonds to other atoms. To a mathematician, a vertex has valence four if it is the endpoint of four edges. In general, a vertex is said to have valence n, or to be n-valent, if it is the meeting point of n edges and, therefore, is surrounded by n faces.

Throughout this book the terms *face*, *edge*, *vertex* and *valence* are used. (The term 'face' was favoured by the French writers and has come to replace 'base'— the term inherited from the Greeks.) When symbols are required to express the quantities of these various elements, the traditional practice of using the initial letters of the words they represent will be followed. This is, of course, language dependent whence Euler's S, A, H for angulorum solidorum, acies and hedrae, and the German E, K, F for Ecken (corners), Kanten (edges) and Flächen (faces). Using symbols corresponding to our chosen English terminology, Euler's formula is expressed

$$V + F = E + 2$$

or alternatively

$$V - E + F = 2$$

where the terms on the left-hand side are arranged in order of dimension.

Consequences of Euler's formula

Before examining some attempts to show that Euler's formula is valid for all polyhedra, some of its many consequences will be derived. Using the Vertices, Edges and Faces terminology, we write Euler's formula as

$$V + F = E + 2. \tag{A}$$

Some of his other observations can be combined and expressed as

$$2E \geqslant 3F \tag{B}$$

$$2E \geqslant 3V. \tag{C}$$

One of the consequences of these formulae which Euler mentioned in his letter to Goldbach is that a polyhedron cannot have 7 edges. For suppose that there exists a polyhedron which has $E = 7$. Then relation (B) implies that $3F \leqslant 14$. Since F is the number of faces, it must be an integer greater than 3, therefore $F = 4$. Similarly, from (C), $3V \leqslant 14$ implies that $V = 4$. Now substituting $V = 4$, $F = 4$ and $E = 7$ into formula (A) produces a contradiction: $4 + 4 \neq 7 + 2$.

The existence of this restriction raises the question of whether there are any other combinations of V, E and F for which there is no corresponding polyhedron. Clearly $F \geqslant 4$, $V \geqslant 4$, $E \geqslant 6$ are restrictions, and now also $E \neq 7$. But can any other choice of V, E and F which satisfies Euler's formula be realised as the number of vertices, edges and faces of some polyhedron? Is there, for example, a polyhedron which has 10 faces and 17 vertices? In fact there is no such polyhedron: if such a polyhedron satisfied Euler's formula it would have 25 edges and these values of V and E do not satisfy relation (C).

The relationships (A), (B) and (C) above can be combined to produce restrictions on F and V. Multiplying (A) by two, and combining the result with (B) gives

$$2V + 2F = 2E + 4 \geqslant 3F + 4.$$

Simplifying produces

$$2V - 4 \geqslant F$$
$$\text{or} \quad V \geqslant \tfrac{1}{2}F + 2.$$

Similarly, combining twice (A) with (C) gives

$$2V + 2F = 2E + 4 \geqslant 3V + 4$$

and thus

$$2F - 4 \geqslant V.$$

Euler knew both of these corollaries. On the left part of Figure 5.7 the two inequalities $V \geqslant \tfrac{1}{2}F + 2$ and $V \leqslant 2F - 4$ are shown graphically. The shaded region indicates combinations of vertices and faces which are impossible. Each open circle indicates a point which could correspond to a polyhedron.

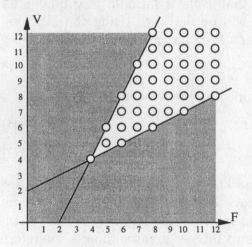 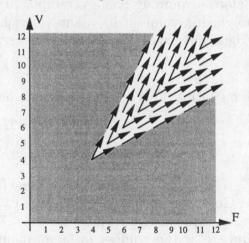

Figure 5.7.

There is, in fact, at least one polyhedron corresponding to each circle. To prove this it is sufficient to construct a polyhedron having the required numbers of faces and vertices for each point indicating a possibility. Firstly, observe that any pyramid with a regular n-sided polygon as a base has $(n + 1)$ faces, $(n + 1)$ vertices and $2n$ edges. Therefore, there is a polyhedron corresponding to each circle on the line $V = F$. Polyhedra corresponding to the other circles can be formed from these pyramids either by truncation or augmentation.

Truncation.

If a 3-valent vertex is truncated then the number of faces is increased by one and the number of vertices by two. Every pyramid contains 3-valent vertices, and the truncation process creates extra 3-valent vertices, so this procedure can be continued indefinitely.

Augmentation.

A low triangular-based pyramid can be erected on any triangular face so that the resulting polyhedron remains convex. This increases the number of faces by two and the number of vertices by one. This process can be applied to pyramids as they contain triangular faces, and the process can be continued since it creates extra triangular faces.

Figure 5.8 shows the results of applying each of these processes to a hexagonal pyramid. On the right of Figure 5.7 the arrows indicate how to get from one polyhedron to another by truncation or augmentation. Examples of convex polyhedra covering all the possibilities can be constructed. ∎

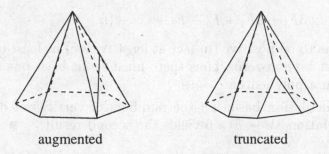

augmented   truncated

Figure 5.8. The (F, V) coordinates can be altered by augmenting or truncating a polyhedron.

Two other consequences that Euler knew are the following:

A polyhedron must contain at least one 3-sided, 4-sided or 5-sided face.

A polyhedron must contain at least one 3-valent, 4-valent or 5-valent vertex.

In order to deduce these results more easily, two further relations will be derived from (A), (B) and (C) above. Multiplying (A) by three gives

$$3V + 3F = 3E + 6$$

and combining this with (C) produces

$$2E + 3F \geqslant 3E + 6.$$

Simplifying gives

$$3F - E \geqslant 6$$

A similar argument using (A) and (B) shows that $3V - E \geqslant 6$.

To show that every polygon contains a 3-sided, 4-sided or 5-sided face let F_n be the number of faces of the polyhedron which have n sides. Thus F_3 is the number of triangular faces. Now, the total number of faces is

$$F = F_3 + F_4 + F_5 + F_6 + \cdots + F_n + \cdots$$

and the total number of sides of all the polygons is

$$2E = 3F_3 + 4F_4 + 5F_5 + 6F_6 + \cdots + nF_n + \cdots.$$

Doubling the relation $3F - E \geqslant 6$ and substituting these values for F and E gives

$$6\left(F_3 + F_4 + F_5 + F_6 + \cdots + F_n + \cdots\right) - \left(3F_3 + 4F_4 + 5F_5 + 6F_6 + \cdots + nF_n + \cdots\right) \geqslant 12.$$

The left-hand side evaluates to

$$3F_3 + 2F_4 + F_5 - F_7 - \cdots - (n-6)F_n \cdots$$

and since this has to be positive (in fact at least twelve) at least one of the terms F_3, F_4 or F_5 must be non-zero. Thus there must be at least one 3-sided, 4-sided or 5-sided face in a polyhedron.

A similar calculation based on the numbers of vertices of different valence and using the relation $3V - E \geqslant 6$ yields the second result. ■

The fact that there are at most five regular polyhedra can also be derived from Euler's formula. To see this, suppose that every face of a polyhedron has p sides. Then there are a total of pF sides joined in pairs to form edges. Thus $pF = 2E$. Suppose also that every vertex is q-valent. Then, since each edge has two ends, $qV = 2E$. The values

$$F = \frac{2E}{p} \qquad \text{and} \qquad V = \frac{2E}{q}$$

can be substituted into Euler's formula to produce

$$\frac{2E}{q} + \frac{2E}{p} = E + 2.$$

Solving for E gives

$$E = \frac{2pq}{2(p+q) - pq}.$$

This fraction has the same denominator as the fraction derived earlier in the chapter when we were looking at regular polyhedra in the context of Descartes' theorem. The same argument shows that $(p-2)(q-2)$ must be less than four, and hence the same solutions for p and q are produced, corresponding to the five Platonic solids.

It is also possible to express F and V in terms of p and q:

$$F = \frac{4q}{2(p+q) - pq} \quad \text{and} \quad V = \frac{4p}{2(p+q) - pq}.$$

By substituting the allowed values for p and q into these formulae, the number of faces and vertices of each solid can be calculated.

Euclid's and Descartes' enumerations of the Platonic solids rely on the geometric properties of angles. This deduction from Euler's formula shows that there is a deeper reason why at most five regular bodies can be constructed. No stage in the proof uses metric information. The faces are not assumed to be equilateral, or equiangular, or even congruent to each other. All that is required is that they have the same number of sides. This, together with the requirement that all the vertices have the same valency, is sufficient.

The fact that there are at most thirteen Archimedean solids can also be derived from Euler's formula. These proofs do not, of necessity, imply that all five regular and all thirteen Archimedean polyhedra exist, they merely exclude any other possibilities. That all the allowable cases can be constructed from equilateral, equiangular polygons seems miraculous.

Euler's proof

A year after announcing the discovery of his formula Euler proposed a proof of its validity. He described a method for removing sections of a polyhedron so that the number of vertices is reduced but so that the sum $V - E + F$ is unchanged. By repeating this process, Euler hoped that the number of vertices could be decreased until only four remained, in which case the polyhedron must be a tetrahedron. Since the sum $V - E + F$ equals 2 for the tetrahedron and is unchanged by the truncation process, the original polyhedron must also have satisfied $V - E + F = 2$.

To elaborate on this strategy, the process of removing a vertex will be described, although, to simplify the situation, it is assumed that all the faces meeting at the vertex are triangular. Let v_0 be the vertex to be deleted and suppose that it is n-valent (Figure 5.9(a)). Let $v_1, v_2, \cdots, v_n$ be the n vertices connected to v_0 by edges of the polyhedron. To form the new polyhedron, remove v_0 and all the edges and faces connected to it. This leaves a (possibly skew) polygonal hole with n sides (see Figure 5.9(b)). To form a closed polyhedron with plane faces, $(n - 3)$ edges are added across this skew polygon, namely those connecting v_1 with $v_3, v_4, \cdots, v_{n-1}$. The covering of the wound is completed with the addition of $(n - 2)$ triangular faces which span the skew polygon and the new edges (see Figure 5.9(c)).

The amputation of the vertex has reduced V (the number of vertices) by one;

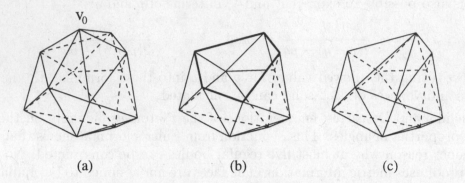

(a)   (b)   (c)

Figure 5.9. Application of Euler's algorithm for removing a vertex.

the number of edges has been changed from E to $E - n + (n - 3)$, or more simply to $E - 3$; the number of faces has changed from F to $F - n + (n - 2)$ or $F - 2$. Although each of V, E and F are altered during the process, the sum $V - E + F$ remains unchanged.

If this procedure were valid in all situations then Euler would have proved that his formula held for all polyhedra. However, in certain circumstances the object that remains after cutting off a vertex does not really qualify as a polyhedron. It is possible to get more than two faces meeting along an edge. A example is shown in Figure 5.10: such a multiple edge is produced if the top vertex is removed. Technically this is described by saying that Euler's algorithm does not stay within its class. This means that the end result of the process is not always the same kind of object that was fed in at the beginning. In this case, various 'degenerate' polyhedra can result.

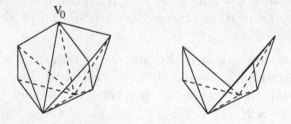

Figure 5.10. Euler's algorithm can produce degenerate results.

Legendre's proof

The first rigorous proof of Euler's formula was given by Adrien Marie Legendre (1752–1833) in his book *Éléments de Géométrie* published in 1794. This book was very popular, went through many editions, and was translated into several languages. Through its large circulation Euler's formula became widely known.

Legendre's proof is based on the geometry of the sphere and, in particular, it makes use of the spherical excess formula for the areas of spherical polygons. Before this can be applied, the polyhedron has to be transformed into a network of polygons on a sphere.

Suppose that a convex polyhedron is made so that the interiors of all the faces are transparent, but so the edges and vertices are opaque forming a kind of framework. Imagine this polyhedron placed inside a sphere so that the centre of the sphere lies inside the polyhedron. A light source at the centre of the sphere will cast shadows of the edges of the polyhedron onto the sphere. The lines on the sphere are shadows of straight line segments and will be arcs of great circles. They divide the sphere up into a network of spherical polygons which fit together in exactly the same way as the plane polygons in the original polyhedron. This method of producing a polyhedral network on a sphere from a given polyhedron is called *radial projection*. A radial projection of the tetrahedron is shown in Figure 2.2.

In order to prove that a polyhedron satisfies Euler's formula, instead of considering the polyhedron itself, Legendre considered its radial projection onto a sphere. This polyhedral network on the sphere will be the same as the original polyhedron in all the essential respects: it will have the same numbers of faces, edges and vertices (though these are now spherical polygons, arcs of great circles, and points where more than two such arcs meet, respectively) and these constituent parts will be connected together in precisely the same way.

Assume that the sphere has radius one unit. Then its surface area is 4π. The surface area is also the sum of the areas of all the faces of the network. As each face is a spherical polygon, its area is equal to the sum of its angles minus $(n-2)\pi$, where n is its number of sides. The sum of all these areas can be separated into three parts:

(*i*) the sum of all the angles of all the spherical polygons. This must be equal to $2\pi V$ since there is a contribution of 2π from each vertex.

(*ii*) the sum of the numbers of sides of all the polygons is $2E$, and each of these contributes π.

(*iii*) there is a contribution of 2π from each face.

Expressed symbolically this becomes

$$\sum_{\text{faces}} (\text{area of spherical polygon})$$

$$= \sum_{\text{faces}} \left((\text{angle sum}) - (\text{number of sides})\pi + 2\pi \right)$$

$$= \sum_{\text{faces}} (\text{angle sum}) \quad - \sum_{\text{faces}} (\text{number of sides})\pi \quad + \sum_{\text{faces}} 2\pi$$

$$= \quad 2\pi V \quad - \quad 2E\pi \quad + \quad F2\pi$$

Thus the total area of the sphere equals 4π and also $2\pi V - 2\pi E + 2\pi F$. Equating these two results and dividing through by 2π gives Euler's formula. ∎

In 1810 Louis Poinsot pointed out that Legendre's proof applies not only to convex polyhedra but also to those non-convex polyhedra that can be radially projected onto a sphere such that each part of the sphere is the image of a unique point in the polyhedron. This last condition ensures that the projections of the faces do not overlap.

Legendre's proof contrasts strongly with Euler's. Euler analyses the way the various parts are connected together and makes combinatorial adjustments that leave $V - E + F$ invariant. Legendre adapts the problem so that he can use the metric properties of the sphere. At first sight this is surprising. The statement of Euler's formula includes only quantities of various elements; it makes no reference to geometric properties and does not prepare the reader for the introduction of spherical geometry in order to establish its validity. Henri Lebesgue commented:

> It is certain that when one has read "Theorem: the numbers F, V, E of faces, vertices and edges of a convex polyhedron satisfy the relation $V + F = E + 2$," in the moments of reflection one takes before passing on to the proof, one does not think of the formula for the area of a spherical triangle.[9]

Although Legendre's proof is ingenious, there is an element of mystery about it. The demonstration is easily seen to be logically correct and, in this sense, it is clear. However, it does not *explain* the conclusion—the reasons for the truth of the theorem remain hidden. Euler's proof, even though it is incomplete, is much more in the spirit of the formula itself.

Cauchy's proof

A proof of Euler's formula that did not rely on metric properties but on arguments concerning the way in which the constituent parts are combined together was given in 1813 by Augustin Louis Cauchy (1789–1857). He also introduced the important notion of deformability. His idea was to choose one face of the polyhedron and then 'transport' the remaining faces so that they formed a tessellation of polygons within this chosen face:

> *Taking one of the faces as a base, and transporting onto this face all*
> *the other vertices without changing their number, one obtains a planar*
> *figure composed of several polygons enclosed in a given contour.*[h]

The number of faces (bounded regions), edges and vertices of this planar network would then satisfy $V + F = E + 1$.

Joseph Diaz Gergonne thought of this in the following way. Imagine that one of the faces of a convex polyhedron is transparent. If your eye were sufficiently close to this face you would be able to see the inside surfaces of all the other faces. You could take a pen and trace over all the edges. The transparent face would then contain a tessellation of polygons fitted together in the same way as the polygons of the original polyhedron. (This is reminiscent of the early explanations of perspective.) Cauchy seems to have thought in terms of continuously deforming (or deflating) the polyhedron so that it lies flat on a plane.

What Cauchy actually proved is that, for any planar tessellation of polygons, $V + F = E + 1$. He then deduced that polyhedra must satisfy Euler's formula. His demonstration proceeds as follows.

By adding diagonals to any non-triangular faces, the planar network can be altered so that the resulting tessellation contains only triangular faces. This modification does not alter the sum $V - E + F$ since if n diagonals are added across an $(n + 3)$-sided polygon, the number of edges increases by n, and n extra faces are created.

The aim now is to remove triangular faces from the boundary of the network, and repeat the process until only a single face remains. An outermost triangle can have either one or two sides forming part of the boundary contour of the network. In the first case, when the triangle is removed, one edge is also removed but the number of vertices is unchanged (see Figure 5.11(a)). In the second case, the removal of the face entails the deletion of two edges and one vertex from the network (see Figure 5.11(b)). In both cases the sum $V - E + F$ is unchanged. Faces can be removed in this way until there is a single face remaining, in which case all of its sides belong to the boundary contour. For this last triangular face $F = 1$, $E = 3$ and $V = 3$ and so $V - E + F = 1$. Since dismembering the network has not altered the left-hand sum, this expression must also hold true of the original tessellation. ∎

This proof was a significant advance for it showed that all polyhedra which can be deflated to lie flat after the removal of one face satisfy Euler's formula. What is less clear is that you must choose the triangles you delete with care. Just as Euler's algorithm can lead to degenerate polyhedra, so Cauchy's algorithm can lead to 'bad'[1] networks. However, the argument can be extended to cover these cases.

---

[1] Graphs with more than one block.

Exceptions which prove the rule

At the same time that Cauchy was developing his proof a Swiss mathematician was compiling a catalogue of polyhedra for which Euler's formula fails. Simon Antoine Jean L'Huilier (1750–1840), professor of mathematics at the Geneva Academy, found three different kinds of exceptional case.

The first kind of exception can occur when two polyhedra are joined to form a larger one as when a small cube is placed on top of a larger one, for example. If the small cube is placed so that none of its edges meet those of the large cube then the new body will have 11 faces—five from the small cube and six from the large one (Figure 5.12). All the vertices and edges of the constituent cubes are also vertices and edges of the composite body. So for this polyhedron $V - E + F = 16 - 24 + 11 = 3$. Solids of this kind can be found in naturally occurring crystals and L'Huilier acknowledges a friend who had allowed him to study his collection of minerals. A similar kind of exception occurs if a depression is gouged out of the centre of a face. A common housebrick is a solid body of this type, and it too satisfies $V - E + F = 3$. In fact, the sum $V - E + F$ can be made as large as desired. L'Huilier gives the example of a tower of prisms piled up like a step pyramid. If there are n prisms in the tower then the solid will have $V - E + F = 1 + n$. The same is true for the column on the right of Figure 5.12.

The source of the discrepancy is the same for all of these exceptional polyhedra: they contain annular faces, that is, faces which have two separate borders. The polyhedra considered previously have not contained faces of this kind.

L'Huilier also described a contrary situation where $V - E + F$ can be made as small as desired, not just zero, but negative. One example he gives is the following. Let a prism be cut by a plane parallel to its base. Draw a polygon in this plane section which lies in the interior of the prism and whose sides are parallel to the sides of the prism. Trapeziums can then be added between the bases of the prism and this internal polygon (see Figure 5.13). The resulting polyhedron can also

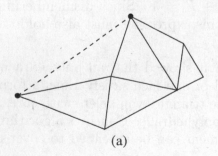   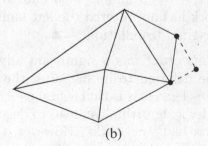

(a)   (b)

Figure 5.11.

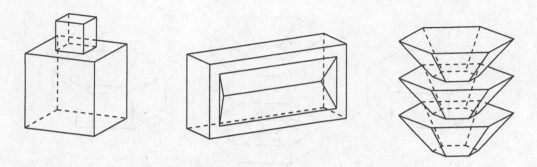

Figure 5.12. Counter-examples to Euler's formula.

be thought of as a solid prism from which two truncated pyramids have been cut out. For this polyhedron $V - E + F = 0$. Other examples which exhibit the same relationship between V, E and F can be constructed by removing a prism-shaped core from a dipyramid—rather like coring an apple. This time the discrepancies arise because there are tunnels through the polyhedron. If n tunnels are bored through a solid body then it will satisfy $V - E + F = 2(1 - n)$.

A third kind of exception occurs when the polyhedron has internal cavities. Such solids arise naturally when one crystal becomes encapsulated within another. For example, opaque lead sulphide crystals are sometimes found in translucent calcium flouride crystals. The translucent perimorphic part of this double crystal is an example of a solid with a cavity. If both its interior and exterior surfaces are regarded as cubic then, for this solid, $V - E + F = 4$. If a solid contains n internal cavities then $V - E + F = 2(1 + n)$.

The different ways that the individual types of exception affect $V - E + F$ can be collected together and incorporated into one formula. If A denotes the number of annular faces, T the number of tunnels, and C the number of cavities in a polyhedron then

$$V - E + F = 2 + A - 2T + 2C.$$

More or less complete versions of this formula were discovered and rediscovered many times over during the course of the nineteenth century.

Besides highlighting these exceptions to Euler's formula, L'Huilier also gave a proof that it holds for convex polyhedra. He did this by adding a vertex in the centre of the polyhedron and then regarding the solid as being composed of pyramids, each with its apex in the centre and a face of the polyhedron as a base (see Figure 5.14). He then showed that the formula holds for pyramids and is still valid when they are glued together to form the original polyhedron.

Another example of a polyhedron which fails to satisfy Euler's formula is Kepler's 'larger icosahedral hedgehog'—the small stellated dodecahedron. For this pentagram-faced polyhedron $F = 12$, $V = 12$ and $E = 30$, and therefore

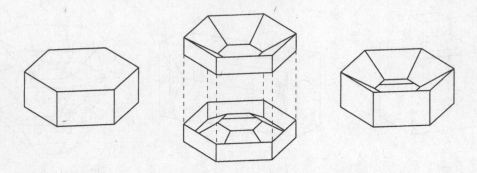

Figure 5.13. L'Huilier's polyhedral torus.

$V - E + F = -6$. This was pointed out (in 1810) by Poinsot, who rediscovered Kepler's pair of star polyhedra along with two more. Of these four self-intersecting polyhedra, two satisfy Euler's formula and two fail. Of course, it is only when Kepler's polyhedron is thought of as having pentagram faces that the discrepancy occurs. If it is regarded as a solid body whose surface consists of 60 triangular faces then there is no disagreement with the formula.

In the 1830's further counter-examples to Euler's formula were recorded by the mineralogist Johann Friedrich Christian Hessel. He had observed double crystals and so reproduced L'Huilier's example of polyhedra with internal cavities. (It was Hessel who gave the specific example of a compound mineral cited above.) He also found two other kinds of exception which have no tunnels, no cavities and no annular faces, and yet still fail to satisfy the formula. Examples can be formed by joining two pyramids together either edge to edge, or vertex to vertex (see Figure 5.15). The first of these counter-examples contains an edge which is a side of more than two faces. It is the possible production of this kind of solid

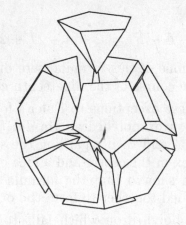

Figure 5.14. L'Huilier proved Euler's formula by building up polyhedra from pyramids.

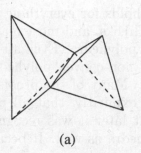
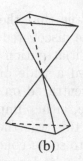

(a)   (b)

Figure 5.15. Hessel's counter-examples to Euler's formula.

that invalidates Euler's proof and it seems sensible to exclude such things from consideration. So, for a solid to be classed as a polyhedron we could stipulate that exactly two faces must meet at every edge.

What about the second of Hessel's counter-examples—should it be classed as a polyhedron? And if not, why not? Several of the beautiful polyhedral forms displayed in the stellations of the icosahedron (Chapter 7) are connected in this way. Should all the exceptions be ruled out as 'improper' polyhedra? Or does the fact that Euler's formula can be generalised to cope with objects having cavities and tunnels mean that these more general objects are examples of polyhedra? If so, then our notion of polyhedron has certainly stretched a bit. It is in this state of uncertainty that we realise we have been using 'polyhedron' as an undefined, or at best ill-defined, term!

What is a polyhedron?

> *"When I use a word" Humpty Dumpty said, in a rather scornful tone, "it means just what I choose it to mean—neither more nor less".*
>
> *"The question is", said Alice, "whether you can make words mean so many different things".[i]*
>
> Lewis Carroll

At different times, to different people, the word 'polyhedron' has conjured up a wide variety of images, some of which are incompatible with each other. It is not unknown for the same person to use different interpretations on different occasions. It is the lack of a precise definition which led to the misunderstandings about the domain of validity of Euler's formula. The list of exceptional cases which fail to satisfy it highlighted the problem. Some people regarded such examples as 'pathological'—deliberate attempts at sabotage, created with malicious intent to discredit a theorem. The exceptions were ignored, hushed up, or dismissed as being weird creations to which the theorem was obviously never intended to apply.

It is probably true that the formula holds for everything that Euler had in mind at the time he discovered it. The Greeks, and most mathematicians after them, restricted their attention to convex polyhedra. Even after the Renaissance artists had produced a wide diversity of non-convex polyhedral forms, mathematicians still concentrated on convexity. Many of the proofs of the validity of the formula assume, either implicitly or explicitly, that all polyhedra are convex. Traditionally this was the case, but later it was recognised that some of the proofs applied to some concave polyhedra as well. It became apparent that Euler's formula did not depend on convexity, but corresponded to something more fundamental. Trying to discern the essential ingredient upon which the all-encompassing proof would rely was not easy as the appropriate concepts had yet to be developed. Writing in 1858 Poinsot noted:

> What makes the theory of polyhedra very difficult is that it requires an essentially new science, which may be called 'geometry of position' because its principal concern is not the size or proportion of figures, but the order and [relative] position of the elements composing them.[j]

And indeed a new discipline was born out of the struggle to find the foundations on which the formula rested—a discipline related to geometry as algebra is related to arithmetic. It concentrates on the relationships and connections between the various constituent elements; specific details such as size, area, angles, and in fact all metric properties are ignored, just as algebraic equations express general relationships between numbers but do not deal with particular cases. Originally called *analysis situs* (*geometrie de situation* in French, *Geometrie der Lage* in German), Johann Benedict Listing coined the name *topology*—the name by which this new geometry is now known.

An important step in the creation of topology, and one which greatly contributed to a resolution of the Euler formula puzzle, was the paradigm transition from solids to hollow shells. For many centuries, 'polyhedron' was synonymous with 'convex solid'. Euclid occasionally used the term 'polyhedron'. For example, in book XII, proposition 17 he constructs a polyhedral solid whose surface lies between two concentric spheres. We met an example as Campanus' sphere in Chapter 3. But he does not define the term, leaving its meaning to be abstracted from the particular cases he does describe.

More often, he used the term 'solid figure'. Later authors also refer to 'solid bodies' when writing about polyhedra. We still talk of Platonic and Archimedean solids even though our models are usually hollow. Descartes and L'Huilier viewed the study of polyhedra as part of solid geometry. Legendre defines a polyhedron to be a solid whose surface consists of polygonal faces. In contrast, others have thought a polyhedron to be the surface itself.

During the fifteenth and sixteenth centuries artists began to concentrate on the surfaces containing objects, and broke them down into planar pieces to assist with their perspective constructions. Albrecht Dürer carried this further and created nets which close up to form polyhedral surfaces. Kepler considered the ways that polygons can be fitted together to form tessellations of the plane and 'congruences in space'. He probably made models in this way reinforcing his mental image of a hollow surface. His star-faced polyhedra can only be understood as surfaces that pass through themselves. Cauchy certainly regarded polyhedra as surfaces composed of polygons. Deforming, or deflating, a polyhedron to lie flat in a plane cannot be performed on a solid. In a later paper, Cauchy analysed collections of rigid polygons joined by hinged edges—an idea abstracted from paper models, perhaps. Cauchy viewed polyhedra as more than just a surface, he saw a potentially flexible, deformable surface (though not yet the 'rubber sheet' made up of undulating faces with curved edges—this did not appear until the 1860's).

Euler, too, recognised that it was the surface of the solid that was important:

> The consideration of solid bodies therefore must be directed to their boundary; for when the boundary which encloses a solid body on all sides is known, that solid is known.[k]

After Cauchy's proof appeared, attention shifted from solids to the consideration of surfaces. Consequently, the notion of cavity no longer applied. In the example of the double crystal, although the outer part is a single solid, its boundary consists of two distinct pieces. Each part of its disconnected surface satisfies Euler's formula. By concentrating on connected surfaces (those in one piece), cavities ceased to be of interest and were forgotten.

Solids with tunnels presented more of a problem. It does not make sense to speak of tunnels running through a surface, and the notion of tunnels through a solid had to be translated into an analogous concept applicable to surfaces. Yet even the notion of tunnels is unclear, for if tunnels are branched, how should they be counted? Before trying to count tunnels, it would be useful to be able to detect them—to know whether there are any tunnels at all. Reinhold Hoppe described the essential characteristic of a tunnel:

> Let the polyhedron be of some stuff that is easy to cut like soft clay,
> let a thread be pulled through the tunnel and then through the clay.
> It will not fall apart.[l]

If there are no tunnels through a solid then every cut separates it into two pieces.

The idea of cutting up a polyhedron applies equally to solids and surfaces: you cut through a solid, or along a curve in a surface. A closed curve in a surface is called *non-separating* if the surface remains in one piece when the curve is cut. Some surfaces possess non-separating curves, in others every curve separates the surface into two pieces. Non-separating curves on a surface correspond to tunnels through a solid, and Hoppe defined the number of tunnels as the maximum number of cuts which leave the surface connected.

Although annular faces were incorporated into formulae for many years, in the end they, like cavities, were discarded. Faces of the conventional kind could be identified in several ways. Firstly, there are curves on an annular face, both of whose ends are on the boundary of the face, which do not separate it. If the curve is cut, the face remains in one piece. In a conventional face all such curves separate it. Alternatively, conventional faces have a single boundary contour whereas annular faces have two disconnected boundaries. From this viewpoint excluding annuli is analogous to excluding cavities: the boundary of the solid and the boundaries of all its faces must be connected.

Some of the properties of polyhedral solids have now been recast and identified with equivalent properties of polyhedral surfaces: tunnels correspond to non-separating curves, cavities to disconnected surfaces. Yet the original problem remains unresolved—What is a polyhedron?

August Ferdinand Möbius' answer to this question appears in an 1865 paper, the same one in which he described his famous one-sided strip.[2] He attempted to preclude the problematic polyhedra with the following definition: A polyhedron is a system of polygons arranged in such a way that

(i) The sides of exactly two polygons meet at every edge.

(ii) It is possible to travel from the interior of one polygon to the interior of any other without passing through a vertex.

To Möbius, a polyhedron is a surface composed of polygons; the second condition ensures that the surface is connected. Moreover, Möbius has tried to exclude *singularities* (places where the surface pinches together) of the kind seen in Hessel's counter-examples. This marks a considerable change in emphasis from the 'flat-faced solid' sort of definition in common use 50 years before.

If one were introduced to polyhedra for the first time through this definition it is probable that the second of Möbius two conditions would appear contrived or irrelevant. It is difficult to imagine what purpose such a restriction could serve unless one can think of objects that fail to satisfy it. The motivation behind it is to exclude singular vertices—those of the form shown in Figure 5.15(b). Despite good intentions, however, this kind of behaviour can still slip through the net: the 'croissant'-shaped polyhedron shown in Figure 5.16 satisfies Möbius definition. This illustrates the difficulty involved in trying to achieve precision in definitions.

---

[2]From entries in his diary we know that he had discovered the 'Möbius strip' as early as 1858. J. B. Listing also discovered it independently around the same time.

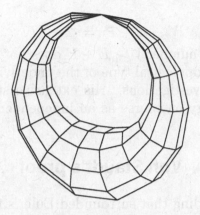

Figure 5.16. A polyhedral croissant.

Actually, Möbius' definition is almost complete. The 'croissant' has crept in because condition (*ii*) refers to the complete polyhedron. The problem is cured by modifying the definition so that it applies *locally* to each vertex, not *globally* to the polyhedron as a whole.

We can now give a full answer to the question at hand.

Definition.   A *polyhedron* is the union of a finite set of polygons such that

(*i*) Any pair of polygons meet only at their sides or corners.

(*ii*) Each side of each polygon meets exactly one other polygon along an edge.

(*iii*) It is possible to travel from the interior of any polygon to the interior of any other.

(*iv*) Let V be any vertex and let $F_1, F_2, \cdots, F_n$ be the n polygons which meet at V. It is possible to travel over the polygons F_i from one to any other without passing through V.

In this context, polygon means a planar figure bounded by straight lines that is topologically equivalent to a disc. The restriction excludes the star polygons used by Kepler and polygons such as annuli whose boundaries are not connected. The first condition excludes star polyhedra of the kind described by Poinsot (see Chapter 7) and other self-intersecting polyhedra. Conditions (*ii*) and (*iv*) exclude singular edges and vertices, and condition (*iii*) ensures that the poly-hedron is connected. Of the five original sources of counter-examples to Euler's formula—cavities, annular faces, tunnels, self-intersections, and singularities—only one remains: a polyhedral surface can have tunnels.

With 'polyhedron' now well defined, Euler's formula can be seen to depend only on the type (or genus) of the polyhedral surface. L'Huilier's version of the formula can be revised to give the following. If a polyhedral surface has T tunnels then

$$V - E + F \ = \ 2 - 2T.$$

The observation that the number $V - E + F$ does not depend on the particular polyhedron, only on the topological type of the underlying surface, was the starting point for Listing's investigations. His extensive study was very influential and helped to set topology on course as an independent and valuable branch of mathematics.

Von Staudt's proof

The search for understanding that surrounded Euler's formula focused attention on several concepts, notably convexity and connectedness (the number of tunnels). The former is a geometric property and is peripheral to the point of being a diversion; the latter is a topological notion and lies at the heart of the problem.

The importance of connectedness and its relevance to the Euler formula was first explained by Karl Georg Christian von Staudt (1798–1867) in his book *Geometrie der Lage* published in 1847. His proof is very illuminating—an example of what mathematicians consider a *good* proof. It does not merely demonstrate that a theorem is true, but shows *why* it is true without introducing any extraneous ideas.

The equation $V + F = E + 2$ describes how the quantities of various constituent parts of a polyhedral surface are related. Since metric properties are not mentioned, they should be avoided in a proof. Legendre's use of spherical geometry and the areas of spherical polygons introduces ideas which are completely foreign to those used to state the theorem. Furthermore, the scope of his proof is limited by the need to radially project the polyhedron on to a sphere.

Various kinds of limitation feature in the many proofs of Euler's formula. These ought to be reflected in restrictions on the type of objects which satisfy the hypothesis of the theorem, but the ambiguity in the term 'polyhedron' allowed each writer to choose a convenient interpretation in his individual circumstances, often subconsciously. Even though the theorem statement usually read 'the number of faces plus the number of vertices of a polyhedron exceeds by two the number of edges' the proofs that followed were not all attempts to demonstrate the same theorem.

The following proof does not suffer from these deficiencies. The hypotheses of the theorem state necessary and sufficient conditions on the polyhedron for the formula to be satisfied, and as a consequence of the proof, the foundations of the theorem become clear: the polyhedron must be 'spherical'.

Theorem. Let P be a polyhedron (as defined above) such that

(*i*) any two vertices are connected by a path of edges, and

PLATE 1 John Robinson's sculpture *Prometheus' Hearth*. (Courtesy of the artist.)

PLATE 2 A group of striated pyrite crystals. (Liverpool Museum collection.)

The original colour versions of all plates in this section can now be found
at www.cambridge.org/9780521664059

PLATE 3 Archimedean solids: cub-octahedron, rhomb-cub-octahedron and great
rhomb-cub-octahedron.

PLATE 4 Archimedean solids: great rhomb-icosi-dodecahedron, rhomb-icosi-dodecahedron
and snub dodecahedron.

PLATE 7 A marble tarsia in Saint Mark's Basilica, Venice.

PLATE 8 The small stellated dodecahedron and great stellated dodecahedron.

PLATE 9 The great dodecahedron and great icosahedron.

PLATE 10 The complete icosahedron.

PLATE 11 Compounds of two tetrahedra and four tetrahedra.

PLATE 12 Compounds of five tetrahedra and ten tetrahedra.

PLATE 13   Compounds of five octahedra and five cubes.

PLATE 14   Compounds of three octahedra and three cubes.

PLATE 15 Compounds of four octahedra and four cubes.

PLATE 16 Compounds of two dodecahedra and five dodecahedra.

(*ii*) any closed curve on the surface separates P into two pieces.

Then P satisfies Euler's formula: $V + F = E + 2$.

PROOF: The argument proceeds by dividing the edges of the polyhedron into two kinds and counting the number of each sort. To keep a record of which edges have been counted we can imagine colouring them in two colours. (In fact, the reader may find it helpful to carry out the steps involved in the proof on a model of a dodecahedron.)

Suppose we colour an edge of P red, say. (In actual fact, since edges are lines which have no thickness, we colour a small sausage-like region around the edge—see Figure 5.17.) The two vertices at its ends will also have been coloured red. Choose another edge which has a red vertex at one end and an uncoloured vertex at the other end, and colour this edge red as well. Proceed in this fashion until no more edges can be added.

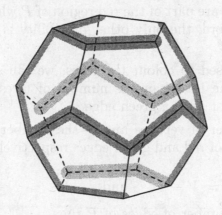

Figure 5.17.

When this colouring process has stopped, all the vertices will have been coloured red. For, if some vertex remains uncoloured then, by hypothesis (*i*), there is a path of edges which connects this vertex to a red vertex and the number of red edges can be increased by colouring this path. Therefore, when the process terminates, every uncoloured edge has a red vertex at both ends.

We now determine the number of edges that have been coloured so far. When the first edge was coloured, two vertices were also coloured. In colouring every subsequent edge only one extra vertex was coloured. Since all the vertices are coloured, the number of coloured edges must be one less than the number of vertices.

The uncoloured portion of the polyhedron is formed from the interiors of all the faces and the uncoloured edges. Suppose that this uncoloured part is not connected. Then there must be a loop of red edges separating it into at least two distinct pieces. But in order to complete a loop of red edges, the edge between two previously coloured vertices must be coloured red, and this was not allowed. Hence the uncoloured piece must be connected.

To examine the structure of the uncoloured part of the polyhedron we can colour each face green (say) in turn. To do this, we choose any face and colour its interior and any of its sides which are uncoloured edges of P. Then choose an uncoloured face which has exactly one green side and colour it green. Proceeding in this way, we continue to enlarge the area of green until no more faces can be coloured—either because all faces are green, or because the uncoloured faces have more than one green side. Suppose that the second case occurs, that is, that an uncoloured face meets two or more green faces. Then there would be a closed curve running through this face and through the green region that did not cross any red edges, and which had vertices lying on both sides of it. By hypothesis (ii), this means that the curve separates the vertices into two disconnected sets. However, all the vertices are part of the red region of P which is connected, giving a contradiction. Therefore, the only other possibility is that all the faces have been coloured.

From the method used to colour the faces, we can see that the number of green edges must be one less than the number of faces since the colouring of every face except the first adds a green edge.

Let V be the number of vertices and F the number of faces of P. Let E_R and E_G be the numbers of red and green edges respectively. Then we have shown that

$$E_R = V - 1 \quad \text{and} \quad E_G = F - 1.$$

If E denotes the total number of edges of P then

$$E = E_R + E_G = (V-1) + (F-1) = V + F - 2$$

and hence

$$E + 2 = V + F. \qquad \blacksquare$$

In the course of the preceding proof, we have divided the polyhedron into two parts by colouring it red and green. Both of these parts are connected and have a single boundary contour—the line where red and green meet. This means that both parts can be deformed into discs. The surface formed by gluing two discs together along their perimeters is a sphere, so the polyhedron can be deformed into a sphere. On the polyhedron itself the two 'discs' will be interleaved in a complex pattern. A tennis ball provides a simpler example of a sphere formed from two interleaved (topological) discs.

If a polyhedron satisfies Euler's formula it can be deformed into a sphere. The converse is also true: if a polyhedron can be deformed into a sphere then it satisfies Euler's formula. One way to prove this is to show that such polyhedra satisfy conditions (i) and (ii) of the previous theorem. That this is so is a consequence of the famous Jordan Curve Theorem, one of the first results in topology, which states that a closed curve in a plane separates it into two pieces. In light of this, the first condition also holds since the definition of 'polyhedron' being used excludes annular faces.

Alternatively, the attempted proof given by Euler can be reinterpreted in the context of polyhedral surfaces in which case it is seen to be valid.

Problem. Construct your own proof of Euler's formula by applying his ideas to networks on a sphere. An outline of the stages involved in such a proof is as follows.

(1) Assume the polyhedral surface can be deformed into a network on a sphere.

(2) Show the network can be triangulated without altering $V - E + F$.

(3) Show how to reduce the number of vertices in the network so that the result is still triangulated and $V - E + F$ is unchanged.

(4) Reduce the number of vertices to four and show that the formula holds in this case.

(5) Deduce that the formula holds for the original polyhedron.

(†) Euler's formula shows no preference for either V or F, they are both of an equal status and the formula is symmetric in these two quantities. However, in many attempts to prove the validity of the formula this intrinsic symmetry is neglected and therefore is unexplained. Euler concentrated on reducing the number of vertices, while Cauchy chose to remove faces. When viewed as a formula about networks on smooth surfaces, von Staudt's proof can be phrased so that the origin of this symmetry becomes clear. In this setting the red region is seen to be a neighbourhood of a spanning tree of the edge-network, and the green region is a neighbourhood of the complementary tree spanning the (combinatorial) dual network.

Complementary viewpoints

Although Descartes' theorem and Euler's formula appear to be concerned with different aspects of polyhedral geometry and topology, they are in fact interrelated in a very strong way: they are logically equivalent and can be derived from one another.

Many people have observed that Euler's formula arises as a simple consequence of combining two statements in Descartes' manuscript:

> There are always twice as many plane angles as sides [edges] on the
> surface of a solid body for one side [edge] is always common to two
> faces.
>   ⋮
> I always take α for the number of solid angles and ϕ for the number
> of faces. $\cdots$ The actual number of plane angles is $2\phi + 2\alpha - 4$.[m]

This has led some to conclude that Descartes was aware of Euler's formula. However, Descartes himself does not seem to have connected the two statements. He
did not possess the requisite concepts to see the benefit of doing so. Leibniz also
failed to see any connection when he copied the manuscript.

Conversely, Euler did not notice Descartes' theorem, nor did any of his suc-
cessors so that it was still unknown to geometers when the Leibniz' manuscript
was unearthed. It is as though the same coin were being viewed from opposite
sides: Descartes, approaching polyhedra with the classical concepts of faces and
solid angles, derived metrical properties by analogy with polygons; Euler, gain-
ing information by scientific induction (observing many particular instances and
extrapolating to polyhedra in general), produced topological results.

People sometimes state that they find it surprising that Euler's formula was
unknown to the Greeks. (Some even assert that Archimedes was aware of it,
but any evidence would have been destroyed in the fire at Alexandria.) I find
this extremely improbable since they, like Descartes, were not in possession of
topological concepts. A more likely person to have discovered the formula before
Euler is Kepler. He thought of polyhedra as surfaces, he produced stellations by
extending their edges, and he had an obsession for finding relationships between
things. However, there is no reason to suppose that he was familiar with Euler's
formula.

Neither Descartes' theorem nor Euler's formula is easy to establish directly.
However, to show that they are logically equivalent is straightforward. The sum
of all the plane angles of a polyhedron can be found by adding up all the interior
angles of the faces. The sum of the interior angles of a face with n sides is $(n-2)\pi$
radians. So

$$\text{sum of the plane angles} \quad = \quad \sum_{\text{faces}} (\text{interior angle sum})$$

$$= \quad \sum_{\text{faces}} \big((\text{number of sides})\pi - 2\pi\big).$$

Since the sum of the numbers of sides of all the faces is twice the number of edges, this equals

$$2E\pi - F\,2\pi.$$

The sum of the plane angles can also be expressed in terms of the deficiencies of all the solid angles:

$$\text{sum of the plane angles} \quad = \quad \sum_{\text{vertices}} \Big(2\pi - (\text{deficiency})\Big)$$

$$= \quad V\,2\pi - \sum_{\text{vertices}} (\text{deficiency}).$$

Equating these two expressions shows that the sum of all the deficiencies of the solid angles equals $2\pi(V - E + F)$:

$$\sum_{\text{vertices}} (\text{deficiency}) \quad = \quad 2\pi\,(V - E + F).$$

Now, if it is known that a polyhedron has $V - E + F = 2$ (so that it satisfies Euler's formula) then, by applying the above equality, it must also have the sum of its deficiencies equal to $2\pi \cdot 2$ which is 4π, and hence the polyhedron satisfies Descartes' theorem. Conversely, if it is known that a given polyhedron obeys Descartes' theorem then it must also satisfy Euler's formula. The two results are completely interdependent.

Furthermore, the argument which shows this equivalence does not require any restriction on the topological type of the polyhedron—it can be convex, concave, or have tunnels. If the concept of the deficiency of a vertex is interpreted in a more general context which allows negative deficiencies when the angle sum is greater than 2π radians then the relationship between the Euler number $V - E + F$ and the total deficiency still holds. This is closely related to a celebrated result in differential geometry—a branch of mathematics in which smooth surfaces (rather than polyhedral ones) are studied.

The Gauss–Bonnet theorem

In 1827 Carl Friedrich Gauss (1777–1855) introduced the idea that a smooth (differentiable) surface has an intrinsic geometry. This geometry is based on measuring the lengths of arcs in the surface, and from arclength other geometric notions such as angles and area can be defined. Having the notion of arclength also means that shortest curves, or *geodesics*, can be considered. In the plane the geodesics are straight line segments; on a sphere the geodesics are arcs of great circles.

The shape and local properties of a surface affect its geometry. We encountered an example of this earlier in the chapter: the interior angle sum of a triangle depends on the surface it sits in. That the sum of the interior angles of a spherical polygon should be related to its area may have seemed puzzling. After all, basic dimension analysis suggests that something is amiss since area is a two-dimensional quantity whereas angles are one-dimensional. In fact, the angles of a polygon are related to its area in almost all geometric surfaces. It is only because the most familiar geometry is done on a *flat* plane that the idea seems strange.

Gauss introduced a precise measure of the *curvature*, or non-planarity, of a surface. Since a surface can be flatter at some points than at others, this curvature is defined for each point on the surface, and can vary from place to place on the surface. For a plane it takes the value zero at every point; for the sphere, every point has positive curvature.

There is also another kind of curvature known as *geodesic curvature*, which applies to lines in a surface. It measures how much the line deviates from being a geodesic. When the line is a geodesic, its geodesic curvature is zero.

Now suppose that we have a polygon P sitting in a smooth geometric surface, each of whose sides is a smooth curve. The exterior angles of this polygon can be defined at each corner. For a conventional planar polygon, the sum of the exterior angles is 2π. This is a measure of the total angle one turns through when traversing the boundary of the polygon. For polygons in non-planar surfaces, it is not only the exterior angles which contribute to this total angle—the bending or turning of the sides of the polygon has to be taken into account, as does the curvature of the surface itself since it affects the amount of turn on each side of P. All this information is collected together in the Gauss–Bonnet Formula:

$$\sum_{\text{corners of } P} (\text{exterior angle}) \ + \ \int_{\partial P} k_g \, ds \ + \ \int_P k \, dA \ = \ 2\pi.$$

The first term in this expression is the sum of the exterior angles. The second term measures the curvature on the sides of the polygon P: k_g stands for geodesic curvature, and since this is a continuous rather than a discrete quantity, summation is replaced by integration with respect to arclength, ds. The third term takes into account the curvature of the surface itself: k is the Gaussian curvature and it is integrated over the area (dA) of the polygon.

Two examples will help to clarify this. If the polygon is bounded by geodesics then $k_g = 0$ and the middle term is zero. If the polygon is planar then the Gaussian curvature is also zero and we have the statement that, for planar polygons bounded by straight lines,

$$\sum_{\text{corners of } P} (\text{exterior angle}) \ \ + \ 0 \ + \ 0 \ = \ 2\pi.$$

If the polygon is on a sphere of unit radius then the Gaussian curvature $k = 1$

and the third term equals the area of the polygon:

$$\int_P k \, dA \; = \; \text{area of } P.$$

If the polygon is bounded by arcs of great circles then, again, k_g is zero. Rearranging the order of the terms in the Gauss–Bonnet formula gives

$$\text{area of } P \; = \; 2\pi \; - \sum_{\text{corners of } P} (\text{exterior angle}).$$

Rewriting this in terms of the interior angles of the polygon rather than its exterior angles produces

$$\text{area of } P \; = \; 2\pi - \sum_{\text{corners of } P} (\pi - \text{interior angle}).$$

If P is a polygon with n sides, and therefore n corners, then this equals

$$\text{area of } P \; = \sum_{\text{corners of } P} (\text{interior angle}) \quad - (n-2)\pi$$

which is the spherical excess formula for the area of a spherical polygon with n sides.

By covering a geometric surface with a network of polygons, and adding up their individual contributions to get the total area (rather like Legendre did) it is possible to derive the

Gauss–Bonnet Theorem. Suppose that a connected network on a smooth surface S has V vertices, E edges and F faces. Then

$$\int_S k \, dA \quad = \quad 2\pi \, (V - E + F)$$

This theorem is truly amazing. It defies our intuition. For on the left-hand side is a quantity defined completely in terms of the intrinsic geometry of the surface: curvature is a metric property. Yet on the right-hand side is an expression which is completely independent of any metrical information: the network is purely topological. This paradoxical situation reflects the interrelation between Descartes' and Euler's theorems. Choosing a fixed network and altering the geometry of the surface shows that the total curvature is independent of the geometry. Conversely, fixing the surface and varying the network shows that $V - E + F$ does not depend on the network chosen—yet another derivation of Euler's formula.

Augustin Louis Cauchy (1821)

Equality, Rigidity and Flexibility

Euclid's "principle of superposition" ⋯ raises the question of whether a figure can be moved without changing its internal structure.[a]

H. S. M. Coxeter

When are two polyhedra the same? At first glance this question appears straightforward enough. Two things are the same when no observation can distinguish between them. To show that two polyhedra are different it is sufficient to find a property of one not possessed by the other. This may be something easily determined such as the number of square faces, for example, or it may involve the way that the faces are arranged. However, this is not a general solution to the problem. Even though it may be easy to find differences, what kinds of properties should we allow ourselves to use in order to distinguish between two polyhedra? Is it only the combinatorial properties of polyhedra which need to be taken into account, or do metric properties matter as well? Are size, or position, or orientation in space relevant?

A definition that tells you whether or not two objects are thought of as the same is called an *equivalence relation*. The definition of equivalence lists the properties which are to be regarded as important. These can be anything. For example, polyhedra which have the same number of faces can be defined to be equivalent. When you have an equivalence relation defined on a collection of objects, you can classify them into sets so that all the items belonging to one class are equivalent, and things from separate classes are different. In some contexts we might want a fairly coarse equivalence relation which regards only a few properties as relevant and which leads to large equivalence classes. For example, a topologist interested only in combinatorial properties would like any prism on a quadrilateral base to be considered equivalent to a cube. A crystallographer, on the other hand, would need to retain the geometric information.

The question 'When are two polyhedra the same?' is a request for an equivalence relation to be defined on the set of polyhedra. It is the goal of this chapter to find one. However, we are not merely searching for a relation which is convenient for some people some of the time, but one which matches our intuitive notion of when polyhedra are 'the same'.

Disputed foundations

The problem of how to define equality for polyhedra has a long history. Early references occur in Euclid's definitions in the *Elements*. The first appears in his list of basic premises: axiom 4 states

> *Things which coincide are equal.*

This axiom is the basis of proofs which use superposition arguments. These are derived from the experience (direct or imagined) of drawing two figures on a piece of paper, cutting round one of them and placing it directly over the other.

This method of proof is used in the very first theorem in the *Elements*. After three propositions which describe constructions, proposition 4 of book I states a general result about triangles: If the two sides and the included angle of a triangle are equal respectively to two sides and the included angle of a second triangle then the two triangles are equal in all particulars. In his *Commentary on the First Book of Euclid's Elements* Proclus describes the ideas that lie behind the proof:

> *The proof of this theorem, as anyone can see, depends entirely on the common notions [axioms] and grows naturally out of the very clarity of the hypotheses. Because two sides are equal respectively to two sides, they coincide with one another; and because the angles contained by these sides are equal, they also coincide. Since the angle coincides with the angle and the sides with the sides, the lower extremities of the sides also coincide; and if they coincide, the base coincides with the base; and if three sides coincide with three sides, so does the triangle with the triangle and everything with everything. Visible equality, therefore, in things of the same form is manifestly the ground of the entire proof.[b]*

Note that besides using the axiom that things which coincide are equal, the discussion also supposes the converse: things which are equal can be made to coincide.

Even in Euclid's time, there was an unease about the fourth axiom and its connotations. Euclid himself rarely made use of superposition arguments if he could avoid them. The unease was still felt 2000 years later. In 1844, the philosopher Arthur Schopenhauer remarked that he was surprised mathematicians attacked the parallel postulate rather than the axiom that figures which coincide are equal. For, either coincident figures are automatically identical or equal, and hence no axiom is required; or else coincidence is something empirical which belongs to external sensuous experience, not to pure intuition. Furthermore, the axiom presupposes the mobility of figures, but that which is movable is matter and therefore is outside geometry.

The objection to the definition is fundamental: superposition arguments rely on the implicit assumption that a figure retains all its properties (such as size and shape) when moved from one place to another. To assume that the properties of a figure remain unchanged as it is moved around is to make some strong assumptions about the nature of space.

When Euclid starts to deal with three-dimensional geometry, he introduces two further definitions of equality which relate to solid bodies. Definitions 9 and 10 at the beginning of book XI read:

> 9. Similar solid figures are those contained by similar planes equal in multitude.
>
> 10. Equal and similar solid figures are those contained by similar planes equal in multitude and magnitude.

The first of these definitions implies that two figures are *similar* if the faces of one can be matched to the faces of the other so that corresponding faces have the same shape and are similarly surrounded. The second definition says that if, in addition, corresponding faces have the same size, then the figures are identical or *equal*.

These definitions, too, have received much criticism. The problem is that the equality of the figures is defined in terms of the equality of their constituent parts. While there is nothing intrinsically wrong in this (we can define an equivalence relation however we like), the usefulness of a definition is determined by the extent to which it agrees with our intuition. A problem arises because there are some polyhedra which definition 10 asserts are equal but which we intuitively regard as different.

One such pair is shown in Figure 6.1. Each of the polyhedra has twenty triangular faces: the one on the left is the regular icosahedron, the other has a depression where the top has been pushed in. In this case, the confusion can be avoided if, besides giving information on the size, shape and arrangement of the faces, we specify the convexity characteristic of each edge—whether the edges are mountain folds or valley folds. However, even such a detailed description is not

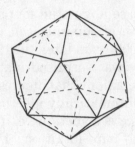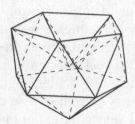

Figure 6.1. The Platonic icosahedron and a non-convex isomer.

always sufficient to uniquely determine a polyhedron. The net shown at the top of Figure 6.2 can be folded up to give two distinct models. Both are derived from a triangular prism by twisting the top relative to the base. In fact, if you make a model out of thin card, you will find that the two positions are interchangeable—under light pressure the polyhedron will jump from one state to the other. This 'jumping octahedron' was discovered by Walter Wunderlich. Its two positions are illustrated in Figure 6.2, both from above and from a more general viewpoint.

Two more examples of nets which can be folded up in more than one way were described by Michael Goldberg. They have similar architecture and properties. One is built from 20 equilateral triangles and looks like two dipyramids stuck together (Figure 6.3(a)). Its three stable positions can be interchanged by applying light pressure. The second example has twelve isosceles triangular faces whose apex angle is slightly more than 103°. This tristable dodecahedron is shown in Figure 6.3(b).

An example of a 'flexible' deltahedron was discovered by Paul Mason. It is formed by erecting a pyramid on each face of a cube and separating one of them by a square antiprism (Figure 6.4). (In fact, the model remains flexible if the isolated pyramid is replaced by a square face.) A model of this polyhedron can be easily manipulated and we cannot tell by playing with it how many genuinely stable positions it has. But by applying powerful mathematical methods, we can prove that some part of the model must be distorted very slightly as it is flexed. Examples of this kind are sometimes called *shaky* polyhedra (or are said to be infinitesimally flexible).

It is probable that Euclid was thinking only of convex solids when he made his definitions. However, even by restricting attention to convex figures, it is not obvious that two figures comprised of similar faces similarly arranged are necessarily 'equal'. How can we be certain that a situation analogous to those described above is impossible after convexity has been imposed? Our experience with models does suggest that convex figures cannot be assembled in more than one way. Even so, definitions 9 and 10 are not as self-evident as other definitions or axioms. The ideology underpinning mathematics demands that any assertion

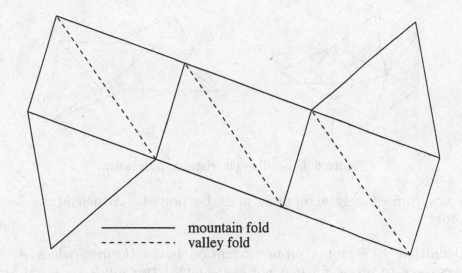

mountain fold
valley fold

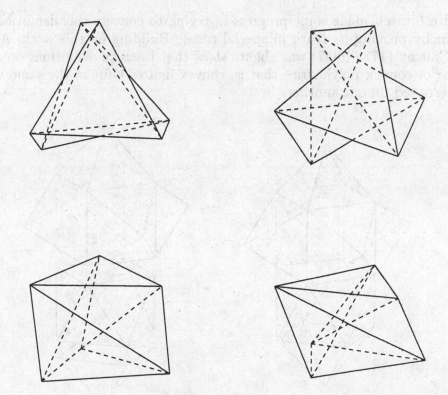

Figure 6.2. Wunderlich's bistable octahedron.

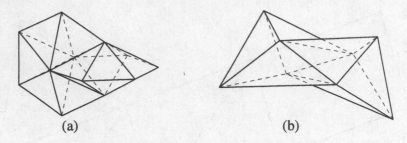

(a)   (b)

Figure 6.3. Goldberg's tristable polyhedra.

that is not immediately acceptable must be proved. Adrien Marie Legendre
remarked:

> *Definition 10 is not a proper definition but a theorem which it is
> necessary to prove, for it is not evident that two solids are equal for
> the sole reason that they have an equal number of equal faces, and if
> true should be proved by superposition or otherwise.*[c]

Legendre himself made some progress in trying to convert this definition into a
theorem by proving its truth in special cases. Building on this work, Augustin
Louis Cauchy (1789–1857) was able to show that Euclid's definitions do agree in
the case of convex polyhedra—that is, convex figures built in the same way can
be superposed on one another.

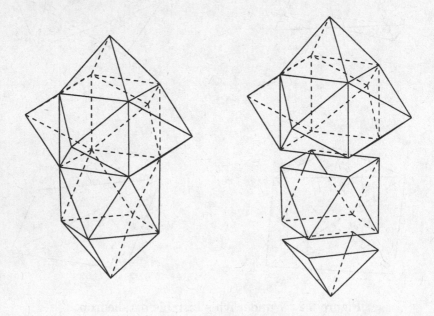

Figure 6.4. Mason's 'flexible' deltahedron.

Stereo-isomerism and congruence

Before discussing Cauchy's work, it is necessary to introduce a few definitions of our own which allow more precise expressions of the content of Euclid's definitions to be formulated.

In Chapter 2 the notion of *isomerism* was borrowed from chemistry to describe the relationship between two polyhedra which have the same constituent faces. Miller's solid and the rhomb-cub-octahedron are one example of a pair of isomers, and some of the other Archimedean solids also have various isomeric forms. In the two isomeric forms of Wunderlich's 'jumping octahedron' the faces are even assembled in the same way. Adopting further chemical nomenclature, two polyhedra which are related in this way are called *stereo-isomers*: their faces differ only in their positions in space, not in their relative positions in the polyhedra. Stereo-isomers can be unfolded to produce the same net.

The pair of icosahedra in Figure 6.1 are examples of stereo-isomers. The various forms of Goldberg's tristable polyhedra are also stereo-isomers. Miller's solid and the rhomb-cub-octahedron are not stereo-isomers: even though they are comprised of the same faces they cannot be made from the same net. If we are allowed to turn a net over and convert valley folds into mountain folds then mirror-image polyhedra such as the two snub cubes become stereo-isomers.

The two icosahedra in Figure 6.1 cannot be interchanged without being dismantled then reassembled. Isomers of this type are called *configurational* isomers. The two forms of the jumping octahedron can be interchanged: by adding energy to the system, it can be forced to flip over into the opposite state—all the intermediate positions are unstable. This behaviour occurs because the material from which the model is made is easily deformed. As the model jumps, the faces buckle. If the model were made from a stronger material so that the faces could not be distorted then the two forms of the polyhedron would both be rigid—the 'excited' intermediate states would no longer be accessible. So, strictly speaking, these polyhedra must be considered as configurational isomers.

This terminology is again borrowed from chemistry. In that context, stereo-isomers are categorised into two types according to the ease with which they can be interconverted. Those which can only be interchanged by breaking and remaking bonds are called configurational isomers. This requires the input of a substantial amount of energy (sufficient to break bonds) and so each form of the isomer tends to be stable.

The interconversion of two isomeric molecules can occur in ways that do not require bonds to be broken. Simple rotation about a bond produces molecules with different shapes. This kind of continuous deformation requires very little energy and isomers of this kind are relatively unstable and interconvert freely. This raises the question of whether some kind of continuous deformation can occur in polyhedra. Is there something like Mason's deltahedron which is truly flexible?

The reader who has made models will probably have noticed that a partly made model is often flexible. It can be deformed with very little effort, the edges between adjacent faces acting as hinges. This situation may continue until the addition of the final face, after which the model becomes more rigid and stable. This is especially true of small convex models when no flexing at all can be detected in the completed model. If the model has a large number of faces, or if it is 'angular' and non-convex with re-entrant angles some movement may be seen. But the deformations of these polyhedra are not continuous motions of the polyhedral surface. They can usually be traced to a distortion of the model— often a separation of the faces at a re-entrant vertex, or the bending of edges and faces. This is the case with the jumping octahedron. We shall insist that the faces remain rigid throughout the deformation, only the dihedral angles may vary with the edges acting as hinges.

Experience of polyhedral models (convex ones at least) suggests that poly-hedra are rigid—they cannot be deformed without being broken apart. Suppose for a moment that a flexible polyhedron did exist. Then it could be continuously deformed from one position into another. Its initial and final forms would not be coincident so, by Euclid's fourth axiom, these two forms are not equal. But the two forms are derived from the same polyhedron by flexing so they must have equal faces similarly arranged. Hence by definition 10 of book XI, they are equal. The two polyhedra are simultaneously equal and not equal; the same yet different.

This paradox may lead one to believe that flexible polyhedra do not exist. However, the contradiction does not arise from the assumption that flexible poly-hedra exist, but from the assumption that the two definitions of the word 'equal' are equivalent and can be used interchangeably. We have already seen from the examples above that the two definitions do not always agree. However, in this case it is not clear which definition is intuitively correct.

In this situation, it is essential to understand precisely what is meant by 'equal' each time it is used. This will depend on the definition being invoked at the time. In definition 10 of book XI of the *Elements*, two polyhedra are defined to be equal if they are stereo-isomers. The other definition, Euclid's fourth axiom, corresponds to our modern notion of congruence.

The term 'congruence' can be defined in various ways and these differ in their choice of primitive concepts. Congruence itself can be used as an undefined term in the same way as 'point' or 'line'. In this case a list of statements must be given which describes the way in which congruence is assumed to behave. These axioms of congruence include:

A thing is congruent to itself.

Things congruent to the same thing are congruent to each other.

These two together replace the fourth axiom since they imply that things which coincide are congruent. The axioms also list some basic objects that are taken to be congruent. One such axiom is

If two sides and the included angle of a triangle are congruent respectively to the two sides and their included angle of a second triangle then the two triangles are congruent.

This last statement was given in the form of a theorem by Euclid which he proved using superposition but, as we have seen, such an argument relies on unstated assumptions. David Hilbert (1862–1943), whose research into the foundations of mathematics led to formal systems of axioms for several areas of mathematics including geometry, realised that it had to be treated as an axiom if congruence were taken to be an undefined term.

If *motion* is regarded as an undefined term then superposition arguments can be made rigorous and Euclid's fourth axiom can be taken as a definition of congruence. Of course, some axioms of motion are needed to specify the properties that motion is assumed to possess: figures must retain their shape and metric properties during motion, for example. Alternatively, *distance* can be taken as an undefined term. A motion of space can then be defined as a continuous one-to-one transformation which associates to each point of space a unique image point. The motions which preserve the distances between points are called rigid motions (or isometries) and these can be used to model superposition arguments. Whichever of these formal definitions of congruence is used, the intuitive idea of placing one cut-out figure over another so that they coincide is made explicit, and this provides a solid basis for investigation.

If two polyhedra are congruent then clearly they are also stereo-isomers. However, the converse is not necessarily true. When discussing equality, it is essential

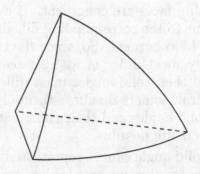

Figure 6.5.

to know which equivalence relation is being used: what does 'equal' mean in the current context. Euclid's two definitions give rise to two different equivalence relations, and the conflict concerning a possible flexible polyhedron arose because the two were mixed. All that can be said about two forms of a flexible polyhedron is that they would be stereo-isomers but would not be congruent.

Cauchy's rigidity theorem

> *Cauchy's most important contribution to geometry is his proof of the statement that up to congruency a convex polyhedron is determined by its faces.[d]*
>
> H. Freudenthal

The proof that Euclid's two different definitions of equality are consistent when restricted to convex polyhedra came in 1813 when Cauchy proved what is known as the Rigidity Theorem: a closed convex polyhedron is rigid. Although it is often expressed in these terms, the proof actually shows the stronger result that convex stereo-isomers are congruent. From this it is easy to derive the rigidity of convex bodies: a flexible convex polyhedron would still be convex after a slight deformation. Thus there would be two convex stereo-isomers which were not congruent and this violates the theorem. Hence, a flexible convex polyhedron does not exist.

Theorem.   Closed, convex polyhedra which are stereo-isomers are congruent.

PROOF: The theorem is proved by assuming that a pair of convex stereo-isomers exists which are not congruent, and then deriving a contradiction. This is achieved by analysing the relationship between local geometrical properties and the global topological nature of the polyhedra.

So suppose there exist two convex polyhedra which are stereo-isomers but which are not congruent. They are bounded by the same kinds of faces similarly arranged and corresponding faces are congruent. If every dihedral angle in one polyhedron were congruent to the corresponding dihedral angle in the other then the two polyhedra would be congruent. So, since the two polyhedra are assumed not to be congruent, they must differ in at least one of their dihedral angles, and consequently some of their solid angles must differ as well. The differences between the two polyhedral isomers manifest themselves at the solid angles, so studying the variety of solid angles and their combinations provides information about what sorts of isomers are possible.

The properties of a solid angle can be represented by a polygon on a sphere. As in Chapter 5, a sphere of radius one unit is placed with its centre at the vertex of the solid angle. The faces of the polyhedron surrounding this vertex intersect the sphere in arcs of great circles forming a spherical polygon (Figure 6.5). The lengths of the sides of this spherical polygon are a measure of the plane angles of the faces which form the solid angle, and the angles of the polygon are equal to the dihedral angles between these faces.

Corresponding faces of the two stereo-isomers being considered are congruent, so corresponding plane angles must also be congruent. Therefore, the spherical polygons formed from corresponding solid angles must have sides of the same length. Any differences between the two polygons must be in their angles, which vary as the dihedral angles of the polyhedron vary.

As an example to show the power of this method of description, consider a convex polyhedron in which every vertex is 3-valent. In this case, there are three faces meeting at every solid angle, so all of the spherical polygons which describe these solid angles are triangles. As with planar triangles, a spherical triangle is completely determined by the lengths of its three sides. This means the spherical polygon associated to a solid angle of any convex stereo-isomer of this polyhedron must be congruent to the spherical polygon of the corresponding solid angle in the original polyhedron because corresponding sides must have the same length. This means that their corresponding angles are also congruent, and therefore the dihedral angles of the two isomers are congruent, implying the congruence of the two polyhedra.

To apply this idea to polyhedra in general, it is necessary to know how other kinds of spherical polygon with sides of fixed length can be deformed. In particular, it is useful to know how the angles may change as this gives information on the way in which the dihedral angles of the polyhedron can vary. This motivates the following discussion and lemma.

The spherical polygon can be thought of as a chain of rigid rods connected by hinged vertices. We are interested in the different ways in which this chain can be placed so that it forms a convex polygon, and also in the relationships between corresponding angles of the various positions. An initial reference position can be chosen for the polygon and then other positions can be compared to it. In this way, the angles of the polygon in any of its positions can be labelled '+' or '−' according to whether they are larger or smaller than their corresponding angles in the reference polygon.

Lemma. When reading round the labels on the angles of the polygon in cyclic order, at least four changes of sign must be encountered.

PROOF: There must be an even number of changes of sign when reading round the polygon since the sets of '+' labels and '−' labels must alternate. So to show that there are at least four it is sufficient to show that there cannot be zero or two changes of sign.

Firstly, consider a polygon in which all the sides except one have fixed length.

(In the 'closed chain of rods' image, one of the rods has been replaced by an elasticated cord.) If the polygon is triangular then it is clear that as the angle between the two rigid sides varies, the length of the variable side also changes, increasing or decreasing as the angle increases or decreases. If the polygon has n sides and all the angles between the rigid sides are varied together then the length of the variable side also changes—increasing when all the angles increase, and decreasing when they all decrease.[1] So, if the variable length side is now made rigid, all the angles between the rods cannot simultaneously all increase nor all decrease.

Therefore, if the angles do change, some must increase and others decrease. This means that both '+' labels and '−' labels must appear at the vertices. Therefore, there are at least two changes of sign around the polygon. Suppose that there are exactly two. Then the polygon can be divided by a line into two parts so that all the angles in one part are labelled '+' and all the angles in the other part are labelled '−' (see Figure 6.6). If all the angles labelled '+' were to increase then the length of this diagonal line would also increase; and if all the angles labelled '−' were to decrease then the length of the line would decrease in response. But the line cannot grow both longer and shorter at the same time, therefore the angles of the polygon cannot be separated into the two sets and there must be at least four changes of sign around the polygon.

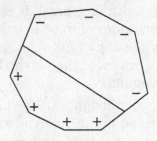

Figure 6.6.

Until now, it has been assumed that all the angles are labelled so that none of the corresponding angles are equal. But suppose that only some of the angles alter while the others remain unchanged. Another polygon can be inscribed in the first by connecting the labelled vertices with lines (see Figure 6.7). The shaded regions in the figure are rigid because their sides in common with the original polygon are rigid, and the angles between these sides do not change. Hence the sides in common with the inscribed polygon are also rigid. Thus all the angles of this inscribed polygon are labelled and all of its sides are rigid. The same argument as above shows that there must be at least four changes of sign around

---

[1] Although this is easily believable, it is tricky to establish rigorously as we shall see in a later section.

this inscribed polygon, and hence at least four changes of sign around the original polygon. ∎

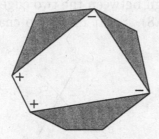

Figure 6.7.

This lemma provides sufficient information to complete the proof of the theorem.

Suppose that one of the two non-congruent stereo-isomers is taken as the reference point and the other is compared to it. Then its dihedral angles can be labelled '+' or '−' according to whether they are larger or smaller than the corresponding dihedral angles in the reference polyhedron. These labels can also be attached to the angles in the spherical polygons since they are equal to the dihedral angles. We now proceed to count the total number of changes in sign (denoted by T) which are encountered when reading round all of the spherical polygons in turn. This will be done in two ways one of which produces a lower bound and the other an upper bound. Further analysis leads to the following contradiction

$$T \;\leqslant\; \text{(upper bound)} \;<\; \text{(lower bound)} \;\leqslant\; T.$$

STEP 1. The lower bound.
The lemma shows that there must be at least four changes of sign when reading round each spherical polygon, so the total number of changes of sign must be at least four times the number of vertices. Letting V denote the number of vertices, this gives the lower bound for T as

$$T \;\geqslant\; 4V. \tag{A}$$

STEP 2. The upper bound.
To calculate an upper bound for T we look at the faces of the polyhedron. Suppose, initially, that every dihedral angle of the polyhedron is labelled. Two edges of the polyhedron which are adjacent sides of a face are also adjacent edges at a vertex of the polyhedron. Suppose that a face is triangular and has edges e_1, e_2 and e_3. The labels on the dihedral angles associated with the edges can be

used to label the edges themselves. Suppose that e_1 is labelled '+'. In order that the maximum number of changes of sign occur around the vertices, the edges adjacent to e_1 must be labelled '−'. This forces both e_2 and e_3 to be labelled '−'. There is now no change of sign between the two edges meeting at the third corner of the triangle (see Figure 6.8). So at most two changes of sign are contributed to T by triangular faces.

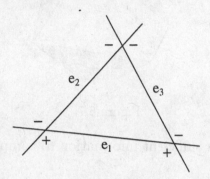

Figure 6.8.

A four-sided face or a pentagonal face can contribute at most four changes of sign to T (see Figure 6.9), and similarly a hexagonal or heptagonal face can contribute at most six changes of sign, and so on.

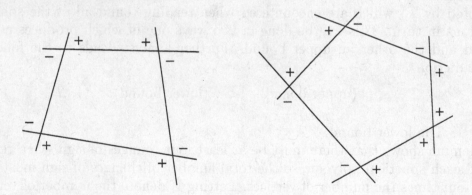

Figure 6.9.

Adding all these contributions together gives an upper bound for T. Let F_n denote the number of faces of the polyhedron which have n sides. Then

$$T \;\leqslant\; 2F_3 + 4F_4 + 4F_5 + 6F_6 + 6F_7 + \cdots. \tag{B}$$

STEP 3. The contradiction.
To show the contradiction contained in statements (A) and (B), Cauchy used

Euler's formula (of which he had recently given a proof). Let the number of edges and faces of the polyhedron be denoted by E and F respectively. Then

$$F = F_3 + F_4 + F_5 + F_6 + F_7 + \cdots$$
$$E = \tfrac{1}{2}(3F_3 + 4F_4 + 5F_5 + 6F_6 + 7F_7 + \cdots).$$

Euler's formula can be written $V = 2 + E - F$. Multiplying this by four and substituting in the above values for E and F gives

$$4V = 8 + 2F_3 + 4F_4 + 6F_5 + 8F_6 + 10F_7 + \cdots. \tag{C}$$

A straightforward comparison shows that the expression on the right-hand side of this equation is strictly greater than the upper bound for T derived in B. But $4V$ is a lower bound for T, so the contradiction has been established.

The extension of this proof to the case where only some of the dihedral angles differ (and hence not all of the edges are labelled) is analogous to the extension of the polygonal case in the previous lemma. The edges which are associated to labelled dihedral angles can be labelled themselves. There are at least four changes of sign round a vertex so if one edge incident to a vertex is labelled then at least three other edges incident to that vertex must also be labelled. Hence, the labelled edges form a network which partitions the faces of the polyhedron into groups, each of which is rigid. The set of edges of this network may not be connected, in which case Euler's formula needs to be adapted, becoming $V \geqslant 2 + E - F$, and (C) becomes an inequality. Using the same argument as before, we can reach the same contradiction. ∎

Cauchy's early career

In 1805 Cauchy entered the École Polytechnique aged 16. Two years later he graduated to the École des Ponts et Chaussées and was trained in civil engineering. After finishing his education he was employed on some major building projects: the construction of the Ourcq canal, and the harbour at Cherbourg which was to be Napoleon's naval base for his assault on England. It was the influence of Joseph Louis Lagrange which set Cauchy on a mathematical path.

Cauchy's work on polyhedra was the first research he undertook. Lagrange suggested that he investigate how to enumerate the regular star polyhedra— a problem posed by Louis Poinsot in 1810. His solution (which is discussed in Chapter 7), together with his derivation of Euler's formula, formed his first paper. He submitted it to the Institut de France (which took over the functions of the Académie des Sciences during the Napoleonic era). The commission that evaluated his work comprised Legendre and Etienne-Louis Malus (1775–1812). They gave Cauchy a very favourable report and suggested that he continue his studies of polyhedra. A year later he submitted his proof that Euclid's definitions of equality agree for convex polyhedra. This time Malus was harder to convince. Cauchy wrote to his father who was acting as intermediary:

> *If M. Malus seemed unsatisfied with the proof I sent you it probably has to do with the fact that you did not advise him of what I had taken care to tell you; namely, that my proof rests on several lemmas that are easy to prove. It does not, therefore, surprise me in the least that M. Malus has concluded that I assumed what could not be assumed. But, that is not the question: if I had the time, I would have sent you the proofs of the lemmas I used. Today, I will reduce the question down to the matter of knowing whether or not my proof is acceptable, assuming the lemmas are established. As to the form of the proof that I used, I think it would be not only difficult to change it, but downright impossible. The reason is that until now a geometric argument has not been given, except in terms of reductio ad absurdum, of the theorem that in 2-dimensional geometry is analogous to the one in 3-dimensional geometry that I dealt with: I mean the theorem by which it is proved that two triangles are equal if their three sides are equal. If one should establish this latter theorem without using either trigonometry or reductio ad absurdum, I would agree that my proof ought not to be admitted. It thus seems impossible to banish the reductio ad absurdum proof from geometry; and this is particularly true in the present case. In fact, in order to prove that under certain conditions only one polyhedron can be constructed, it is necessary to see that after the first figure has been constructed subject to the given conditions then one cannot construct a second figure without encountering a contradiction. I insist on this argument because the type of proof I gave seems to me to be inherent in the nature of the theorem in question. Moreover, it is precisely what M. Legendre used in establishing several particular cases of the same theorem.[e]*

Four weeks after Cauchy submitted his paper, the evaluating commission (composed of Legendre, Carnot and Biot) gave it a glowing report.

Once Cauchy's proof had been accepted and appeared in print, it became highly regarded and was reproduced in several nineteenth-century geometry books. It displayed originality and ingenuity and did not require a mastery of difficult techniques to be able to follow. Even so, developing the brilliant insights which make it work required the mind of a fine mathematician.

Cauchy went on to make numerous contributions to many branches of mathematics. The sheer volume of his output, second only to Euler, is made up of several books and around 800 papers which occupy the two dozen volumes of his collected works. His huge productivity led to a rule still enforced by the Académie des Sciences limiting papers in its journal *Comptes Rendus* to a maximum length of four pages. Cauchy submitted articles so frequently that the Académie could not keep up with the rising cost of printing.

Cauchy's two landmark papers on polyhedra were the only work he did in the field. His later works were mainly in analysis (the study of convergence or divergence of infinite series), functions of real and complex variables, differential equations and mathematical physics. But he is probably best remembered by the mathematical community for his demands for rigour in analysis and calculus. His inspiration helped to convince many of the need to banish the intuitive arguments which often led to false conclusions.

Steinitz' lemma

In spite of Cauchy's great achievement in proving the rigidity theorem, as time went by, small cracks appeared in his argument. The most serious flaw went unnoticed for more than a century. In the 1930's Ernst Steinitz noticed a defect in the very foundation of the theorem—the lemma concerned with the effects of varying the angles in a polygon. The entire proof rests on this one result:

> *If, in a plane or spherical convex polygon $ABCDEFG$ all of whose sides AB, BC, CD, $\cdots$, FG with the exception of AG have fixed length, one increases (or decreases) simultaneously the angles B, C, D, E, F, G between these same sides then the length of the variable side increases in the first case, and decreases in the second.*

Cauchy thought that changing the varying angles one at a time would produce the same effect as making all the changes simultaneously. In this way he could repeatedly use the analogous result for triangles (proved by Legendre) to show that altering each angle in turn increases the length of the extendable side. However, because the lemma deals with convex polygons, it is necessary to ensure that all the intermediate stages are also convex. The example shown in Figure 6.10 illustrates this point: when only one angle is changed and the polygon becomes non-convex. Steinitz, who was first to notice this deficiency, provided a correct proof. His proof, however, is quite long and this has prompted others to search for shorter ones. The version given below is due to Isaac J. Schoenberg.

Lemma. Let $A_1, A_2, \cdots, A_n$ and $B_1, B_2, \cdots, B_n$ be the corners of two convex n-sided polygons whose sides are such that $A_1A_2 = B_1B_2$, $A_2A_3 = B_2B_3$, $\cdots$, $A_{n-1}A_n = B_{n-1}B_n$ and whose angles are such that $\angle A_2 \leqslant \angle B_2$, $\angle A_3 \leqslant \angle B_3$, $\angle A_{n-1} \leqslant \angle B_{n-1}$, where at least one inequality is strict. Then $A_1A_n < B_1B_n$.

PROOF: The proof uses induction on the number of sides of the polygon. As Cauchy observed, the lemma is true for triangles—the case $n = 3$. We now assume

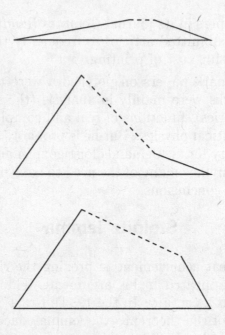

Figure 6.10.

that the lemma holds for all polygons with fewer than n sides and establish the result for n-gons.

STEP 1. Suppose that at least one of the angle relations is an equality so that $\angle A_i = \angle B_i$. The triangles with vertices A_{i-1}, A_i, A_{i+1} and B_{i-1}, B_i, B_{i+1} are congruent so the lengths $A_{i-1}A_{i+1}$ and $B_{i-1}B_{i+1}$ are equal. Therefore, the polygons

$$A_1, \cdots, A_{i-1}, A_{i+1}, \cdots, A_n \text{ and } B_1, \cdots, B_{i-1}, B_{i+1}, \cdots, B_n$$

are both convex and satisfy the conditions of the lemma. They have $(n-1)$ sides and so, by the inductive hypothesis, we have $A_1A_n < B_1B_n$.

STEP 2. Assume now that all the angle inequalities are strict so that $\angle A_i < \angle B_i$ for all $i = 2, 3, \cdots, (n-1)$. Let θ be an angle between $\angle A_{n-1}$ and $\angle B_{n-1}$. Consider the polygon whose vertices are $A_1, A_2, \cdots, A_{n-1}, A_\theta$, where the angle $A_{n-2}A_{n-1}A_\theta = \theta$ and the side $A_{n-1}A_\theta = A_{n-1}A_n$ (see Figure 6.11). While this polygon remains convex we have that $A_1A_n < A_1A_\theta$ (by step 1).

If the polygon $A_1, A_2, \cdots, A_{n-1}, A_\theta$ is convex for all values of θ then when $\theta = \angle B_{n-1}$ it has one angle in common with the polygon $B_1, \cdots, B_n$. Applying step 1 again we see that $A_1A_\theta < B_1B_n$. Hence

$$A_1A_n \; < \; A_1A_\theta \; < \; B_1B_n.$$

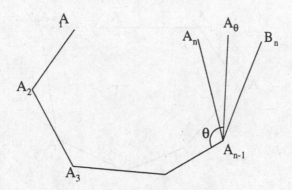

Figure 6.11.

STEP 3. We have still to deal with the case when the polygon $A_1, A_2, \cdots, A_{n-1}, A_\theta$ is not convex for all values of θ between $\angle A_{n-1}$ and $\angle B_{n-1}$. Let $\widehat{\theta}$ be the last value of θ for which the polygon is convex (Figure 6.12). By step 1 we have

$$A_1 A_n \; < \; A_1 A_{\widehat{\theta}}. \tag{A}$$

The angle at A_1 is now $180°$ so

$$A_1 A_{\widehat{\theta}} \; = \; A_2 A_{\widehat{\theta}} - A_2 A_1. \tag{B}$$

The two polygons whose vertices are $A_2, A_3, \cdots, A_{n-1}, A_{\widehat{\theta}}$ and $B_2, \cdots, B_{n-1}, B_n$ are convex and both have $(n-1)$ sides. Applying the induction hypothesis to these polygons gives

$$A_2 A_{\widehat{\theta}} \; < \; B_2 B_n. \tag{C}$$

Since $B_1 B_2 B_n$ is a triangle, its sides satisfy

$$B_1 B_2 + B_1 B_n \; \geqslant \; B_2 B_n. \tag{D}$$

Putting all this together gives the following:

$$
\begin{aligned}
A_1 A_n \; &< \; A_1 A_{\widehat{\theta}} && \text{from (A)}\\
&= \; A_2 A_{\widehat{\theta}} - A_1 A_2 && \text{from (B)}\\
&< \; B_2 B_n - B_1 B_2 && \text{from (C)}\\
&\leqslant \; B_1 B_n. && \text{from (D)}
\end{aligned}
$$

The proof is now complete. ■

Rotating rings and flexible frameworks

There are some kinds of structures of a polyhedral nature which are not rigid but which move freely in some way. One kind of these flexible structures seems to have been discovered several times over. An example was described by Max Brückner

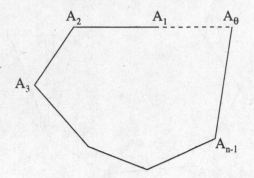

Figure 6.12.

in his 1900 compendium on polyhedra *Vielecke und Vielflache*. These structures are formed by joining identical tetrahedra together in a chain. A tetrahedron is attached to two others along opposite edges so that the connecting edges act as hinges. When the chain of tetrahedra is sufficiently long the two ends can be brought together to form a closed ring. The flexibility of the hinged joints allows the ring to move and it can be rotated continuously through its centre. A ring of as few as six tetrahedra has this remarkable property. You can verify this by making a model from the net shown in Figure 6.13. (Each column of triangles folds up into a tetrahedron. The net will naturally form a chain and you will need to join the two ends together.) Examples can also be constructed from irregular tetrahedra or sphenoids. Rings of scalene sphenoids have a twisted appearance and when they are rotated the tetrahedra tumble into the centre in sequence.

Doris Schattschneider and Wallace Walker have covered the faces of various rings of tetrahedra (along with the Platonic solids and the cub-octahedron) with

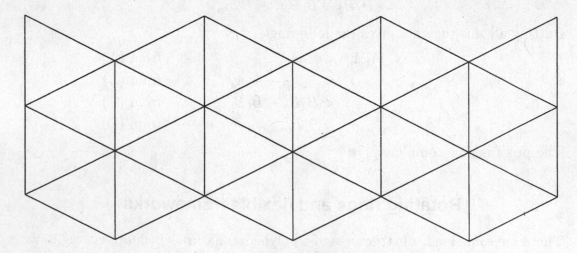

Figure 6.13. A net for a rotating ring of six tetrahedra.

some of the repeating patterns drawn by M. C. Escher. These retail under the name 'Kaleidocycles'. I also have an educational ring of tetrahedra marketed as a 'Tectonocycle'. This is a ring of six tetrahedra which have a hexagonal appearance like the model made from Figure 6.13. Each of the four apparent hexagons contains a map of the Earth at a particular era in history. By rotating the ring so that different faces become visible, the movement of landmasses under the influence of plate tectonics can be traced.

An attempt to construct ordinary polyhedra which flex was made by a French engineer named Raoul Bricard. In 1897 he discovered some flexible 'octahedra'. However, his examples of flexible polyhedra all possess self-intersections: models cannot be made without some face passing through another. The flexibility of these polyhedra can be demonstrated by models in which the interiors of the faces are omitted. The resulting skeletal framework of edges and vertices can be thought of as a set of rods joined at the vertices so that they can rotate freely. Each face of the polyhedron is a triangle so the faces are rigid even though their interiors are missing. Two of Bricard's three flexible octahedra are described below.

Bricard's first flexible octahedron.

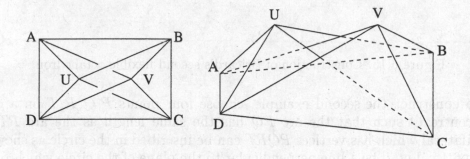

Figure 6.14. Construction of Bricard's first flexible octahedron.

To construct the first type of octahedron start with a quadrilateral $ABCD$ which is non-planar and such that the pair of opposite edges AD and BC have the same length, and similarly for the pair AB and CD. Even though the quadrilateral is not planar, it still has 2-fold rotational symmetry. The vertices U and V lie above the quadrilateral and are chosen so that the rotational symmetry is preserved. Four of the faces of this octahedron are the triangles of the pyramid with base $ABCD$ and apex U. The other four faces are the triangles of the pyramid on the same base with apex V. This construction is illustrated in Figure 6.14. Notice that the sides of the triangular faces ADV and BCU are linked so these faces must intersect. (These faces have been omitted in the figure.) A model of the edge-and-vertex framework can easily be made from drinking straws and string. The straws form the bars of the framework and a single piece of string can be threaded through all the straws (because the octahedron is 4-valent) and knots placed at the joints.

Bricard's second flexible octahedron.

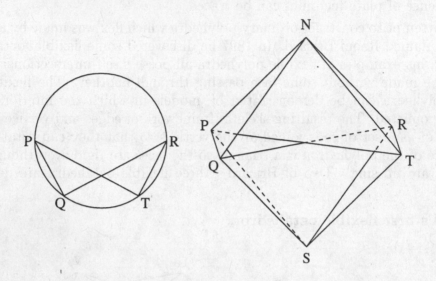

Figure 6.15. Construction of Bricard's second flexible octahedron.

To construct the second example, choose four points P, Q, R, T on a circle with centre O such that the arc PQ has the same length as the arc RT. A quadrilateral which has vertices $PQRT$ can be inscribed in the circle as shown in Figure 6.15. Let L be a line perpendicular to the plane of the circle which passes through O. Choose two points N and S on L which lie on opposite sides of O and which are equidistant from O. Again, the faces of the octahedron are formed from the triangular faces of two pyramids—one with apex N and base $PQRT$ and the other on the same base with apex S. In the figure faces QRN and PST are omitted.

Are all polyhedra rigid?

The flexible structures described in the preceding section are not genuine polyhedra. Cauchy's proof of the equivalence of Euclid's two definitions implies that convex polyhedra must be rigid. But what about polyhedra in general—are they all rigid? People had speculated that this was the case long before Cauchy provided the first mathematical evidence in favour of the conjecture. In 1766

Leonhard Euler wrote

> *A closed spatial figure allows no changes as long as it is not ripped apart.[f]*

The conviction that all polyhedra are rigid became known as the Rigidity Conjecture.

Research on rigidity developed in many directions as people sought to extend Cauchy's theorem. Some people adapted Cauchy's ideas to differential geometry and proved analogous results about the rigidity of smooth surfaces. This led to the study of *infinitesimal* rigidity. Something which is infinitesimally rigid is also rigid in the normal sense. Polyhedra which are not infinitesimally rigid are called infinitesimally flexible or *shaky*. Mason's deltahedron shows that the two kinds of rigidity are not identical for it is both shaky and rigid.

People have also studied the rigidity of the edge-skeletons of polyhedra. A convex polyhedron with triangular faces has a rigid skeleton. Other skeletons, like that of a cube for example, are flexible. Since any polygon can be cut up into triangles by adding diagonals, any convex skeleton can be triangulated by adding extra edges. Alexander Danilovich Alexandrov modified Cauchy's theorem and showed that a convex skeleton which is triangulated by adding extra edges becomes rigid. The additional edges must end at the original vertices or at new vertices on the original edges. If new vertices are added in the interior of a face then an infinitesimal flexing can occur.

Herman Gluck studied Alexandrov's work and in the early 1970's he proved that, for non-convex polyhedra, almost all triangular skeletons are infinitesimally rigid. This curious result implies that almost all polyhedra are rigid. The Rigidity Conjecture was now known to be true in nearly every case. Gluck remarked

> *Euler's conjecture in this case is therefore "statistically" true.[9]*

The reader may feel that expressions such as 'nearly every' and 'almost all' are too vague to appear in the statement of a theorem. However, they can be given precise interpretations. As an example to illustrate this usage, imagine part of a plane is coloured white and that a mathematical figure composed of points and lines lies in this plane and is coloured black. We can think of this as a diagram on a page. The expression 'almost all' can be used in the same way it is used above to describe the relative numbers of black and white points: almost all the points on the page are white. The probability that a point of the page chosen at random is coloured black is given by the ratio of the area of the diagram to the area of the page. But since mathematical lines are idealised and have no thickness, the diagram has zero area. So 'statistically' all the points are white.

The technical statement of Gluck's theorem requires a knowledge of the topological notion of an open dense set. (In the above example, the white points form an open dense subset of the plane.) The theorem states that the infinitesimally rigid polyhedra are open and dense in a space of all polyhedra. Gluck's proof uses methods from algebraic geometry. Ethan Bolker used the same machinery to derive properties which a flexible polyhedron would be forced to possess. For example, the flexing would only stop when one face collided with another. Mason's deltahedron does not have this property. Its apparent motion cannot be continued this far without some edges tearing and the polyhedron breaking apart. Therefore, it must be rigid.

Just as the observation that almost all points on a page are white does not allow one to conclude that *every* point is white, so Gluck's proof that almost all polyhedra are rigid does not imply that flexible polyhedra cannot exist. It does mean, however, that if such polyhedra do exist then they are extremely rare.

This tantalisingly small window of opportunity attracted the attention of a young American Ph.D student: Robert Connelly resolved to find the elusive missing piece of the jigsaw. The fact that almost all polyhedra are rigid means that it is extremely unlikely that a flexible polyhedron can be found by accident. Such a thing would need to be explicitly constructed to break the conjecture. Connelly started investigating skeletons or frameworks. In particular, he studied frameworks constructed from planar polygons by adding two extra vertices, one on either side of the plane—a process called *suspension*. He wanted to discover which polygons have flexible suspensions.

Using suspension and other constructions, Connelly found several classes of flexible frameworks including Bricard's flexible octahedral skeletons. Unfortunately, all of them suffered from the problem that the faces could not be filled in. There were always places where faces had to pass through each other for the mechanism to flex. Connelly asked himself how bad these self-intersections needed to be. Was there anything in his catalogue of flexible frameworks which could be considered as having the fewest self-intersections? He found that one of Bricard's octahedra can be adapted to produce an immersed polyhedral surface with only two self-intersections, and these were particularly nice: places where edges crossed through each other; there were no intersections in the interiors of the faces. All he needed was some way to remove these last two obstructions.

At a 1975 topology conference in Cornell University, Connelly heard that another mathematician, working in different area of mathematics, had been working on the same problem and had found a flexible polyhedron. Disheartened by the news, but curious to find out more, Connelly traced the rumour back to its source. Eventually he discovered that it referred to himself!

Two years after this meeting, Connelly was back at Cornell after taking visiting fellowships at Paris and Syracuse. He was puzzling over his lists of flexible frameworks once more. Then, in one of those rare moments when fortune smiles on the prepared mind, things clicked into place. He recalls vividly this moment of illumination. It was a warm June day and he glanced up at the clock: it was 3pm. He spent the next few weeks working with renewed enthusiasm checking every detail to make sure nothing had been overlooked, to ensure that his quest was really over. It was true. He had constructed a genuinely flexible polyhedron— one without self-intersections. He had created a counter-example to the Rigidity Conjecture; a black point in the vastness of the white page.

The Connelly sphere

Connelly constructed his flexible polyhedral sphere as follows.

Take a planar framework of twelve edges and six vertices like that illustrated in Figure 6.16(a)—the two crossover points in the interior are not vertices. This framework is flexible and is one of Bricard's first type of octahedra. In this case some of the edges intersect other edges, and the whole of every face meets parts of some other faces. The polyhedron does not remain planar when it is flexed. This polyhedron can be converted into one with fewer self-intersections and which is still flexible. To achieve this, each triangular face is replaced by a group of three triangular faces forming a pyramid without a base. The four faces which surround vertex V are replaced by four pyramids lying above the plane of the framework like mountains, and the four faces surrounding U are replaced by pyramids beneath the framework forming pits (Figures 6.16(b) and (c)). The resulting surface has two points of self-intersection which occur where one edge passes through another. The regions of the polyhedral surface around each of these points is illustrated in Figure 6.16(d).

The next stage in the process is to find a way of removing these two remaining intersection points. This is achieved by the insertion of what Connelly calls *crinkles*. A crinkle is derived from Bricard's second type of octahedron by removing two faces. This creates a flexible surface with no self-intersections, and which has a boundary that can be attached to another surface.

A crinkle can be inserted into the edge of a polyhedron as follows. Let E be the edge and let F_1 and F_2 be the faces which are joined along E. Let C be a circle which meets E in two points, P and R, such that the plane containing the circle bisects the angle between the two faces F_1 and F_2. Let N be a point on F_1 and S be a point on F_2 such that the line NS passes through the centre of the circle C and is perpendicular to the plane of C. Remove the triangles PRN from F_1 and PRS from F_2 as shown in Figure 6.17(a). Choose two points Q and T on C with Q under E and T above E so that the distance from P to Q is the same as the distance from R to T. A crinkle can then be constructed from the points P, Q, R, T, N and S which plugs the hole in F_1 and F_2. The crinkle is isolated in Figure 6.17(b) and shown inserted into the fold in 6.17(c). As the crinkle is constructed from a flexible octahedron, the edge E still acts as a hinge between the two faces attached to it.

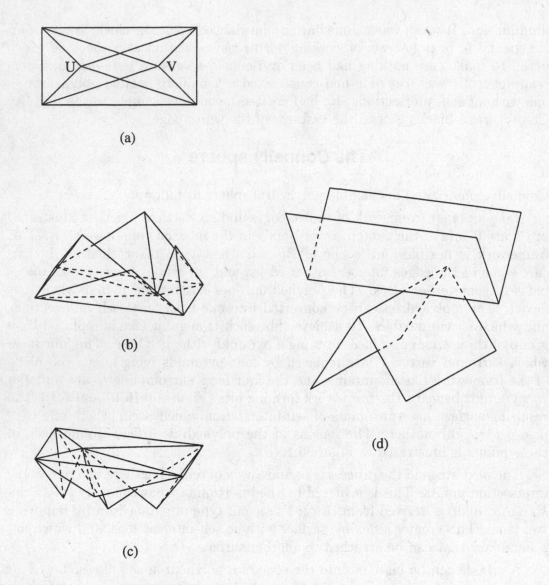

Figure 6.16. Construction of the Connelly sphere.

The construction of the flexible polyhedron is completed by inserting a crinkle into one of the two edges at each intersection point in the surface.

Further developments

Connelly sent a detailed description of the flexible sphere to his colleagues at the Institut des Hautes Études Scientifiques in Paris. They held a contest to see whether simpler examples could be found. Connelly had paid little attention to the complexity of his polyhedron. He was satisfied just to have found just one

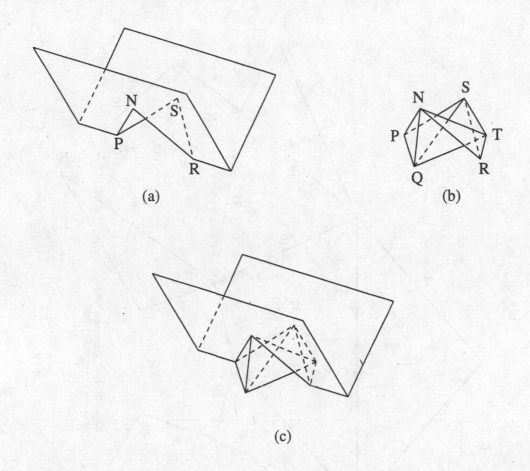

(a)

(b)

(c)

Figure 6.17. Inserting a crinkle into a fold.

example. Nicolaas H. Kuiper and Pierre Deligne modified Connelly's example by expanding the crinkles and amalgamating some of the vertices. They obtained a flexible polyhedron with 18 faces and 11 vertices. Klaus Steffen took a more independent approach and found an example with only 14 faces and 9 vertices. A net for this polyhedron is shown in Figure 6.18. The reader is encouraged to make a model and work out how it flexes. Observe how the flexing is limited by the collision of several faces. Earlier, we noted that this must be the case for a genuinely flexible polyhedron. A Russian mathematician, I. G. Maksimov, has recently shown that all polyhedral spheres with triangular faces and fewer than nine vertices are rigid. This implies that Steffen's flexible polyhedron has as few vertices as possible.

An intriguing property of all the known flexible polyhedra is that their volume does not change as they are flexed. This raises the question of whether or not it is possible to construct polyhedral bellows—does there exist a flexible polyhedron whose volume varies as it is flexed? Dennis Sullivan conjectured that no such

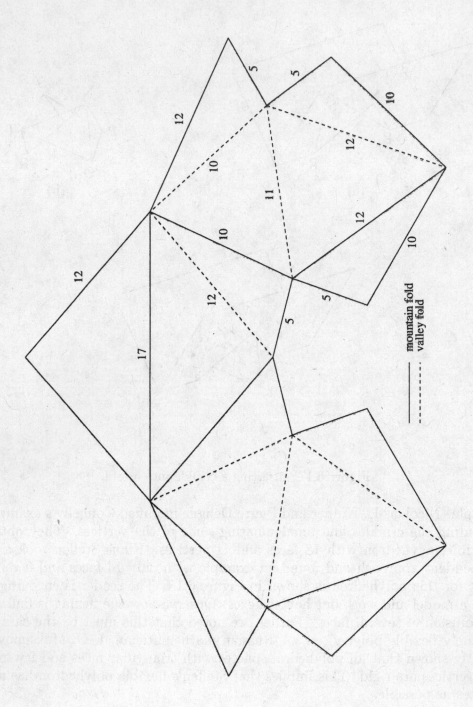

Figure 6.18. A net for Steffen's flexible polyhedron.

bellows exist and that all flexible polyhedra flex with constant volume. Connelly has nicknamed this the Bellows Conjecture.[2]

When are polyhedra equal?

At the beginning of this chapter we started on a quest to search for an equivalence relation defined on polyhedra which captures our intuitive idea of when two polyhedra are the same. The superposition arguments used in Greek times convey a great deal of what is involved perceptually and, until the discovery of flexible polyhedra, congruence was probably the best choice of equivalence relation.

However, the existence of flexible polyhedra leads us to reconsider whether congruence still matches our experience. For surely, one of these polyhedra remains the same polyhedron while it is being flexed, even though it passes through a continuum of non-congruent positions during the process. Our equivalence relation needs to be modified to take this into account. We define two polyhedra to be equal if one is congruent to the other, or if one can be deformed into a polyhedron congruent to the other by flexing.

Definition. Two polyhedra P_1 and P_2 are equal if either

(*i*) P_1 is congruent to P_2, or

(*ii*) there is a continuous deformation of P_1 which preserves the metric properties of the faces and which results in a polyhedron congruent to P_2.

With this definition, a rigid polyhedron is equal only to itself and is the unique member of its equivalence class. The equivalence class of a flexible polyhedron contains all of the polyhedra into which it can be deformed and these are all regarded as equivalent forms of one polyhedron. We could remove the dependence on scale by replacing congruence with similarity. Then polyhedra of different sizes would also be equivalent. An unfortunate aspect of this equivalence relation is the complexity which can be involved in deciding whether two stereo-isomers can actually be interconverted by a continuous deformation. To prove that Mason's deltahedron is rigid takes a lot of high powered mathematics.

We can again adapt some chemical terminology to describe the way in which the different positions of a flexible polyhedron are related. Isomers of a molecule which differ only by a rotation of one part of the molecule about a bond are called rotational isomers, or rotamers for short. By analogy, we can call isomers of a polyhedron which are related by a continuous deformation *fleximers*. With this terminology our equivalence relation becomes easy to state: two polyhedra are equal if, and only if, they are fleximers.

---

[2]This has recently been established. See R. Connelly, I. Sabitov and A. Walz, 'The Bellows Conjecture', *Contributions to Algebra and Geometry* **38** no 1 (1997) pp1–10.

7

Stars, Stellations and Skeletons

In Chapter 4 we saw how Kepler produced two starlike polyhedra by extending the edges of some of the Platonic solids (see Figure 4.17). These figures can be regarded as regular polyhedra but, to give this interpretation, the figures must be seen as being composed of star pentagons rather than naively from triangles. The faces then seem to pass through one another even though they still meet edge to edge. We have also seen examples of a generalised kind of Archimedean polyhedron that are constructed from regular polygons (convex or star like) but whose faces intersect in places that are not edges. These are the uniform polyhedra (also discussed in Chapter 4). The Russian Evgraf Stephanovich Fedorov called such self-intersecting, or re-entrant, polyhedra *koilohedra* (from the Greek for concave).

In these examples, the term 'polyhedron' is interpreted in a broader sense than the topologist's polyhedral surface defined in Chapter 5. This more general type of polyhedron is the subject of this chapter. As a polyhedron will always be some kind of collection of polygons, whatever other conditions are imposed, we shall begin by studying the kinds of building blocks available.

Generalised polygons

A polygon is often visualised as a fragment of a plane bounded by segments of straight lines. However, a polygon can also be viewed as a collection of distinct points (called vertices) connected by straight lines with the constraint that exactly two lines meet at every vertex. The lines are the sides of the polygon. This transition from a filled-in polygon to one which is a circuit of edges is analogous to the transition from a solid polyhedron to a surface discussed in Chapter 5.

For the moment, we shall restrict attention to planar polygons, those whose vertices lie in a plane. Some examples are illustrated in Figure 7.1. At first sight, it appears that some of the polygons in the figure do not satisfy the requirement that exactly two sides meet at every vertex. However, the points where four line segments meet are not vertices; they are points where two sides pass through each other.

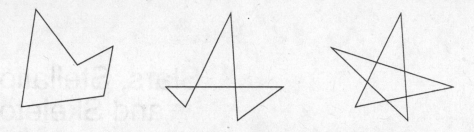

Figure 7.1. Some general polygons.

In the special case of a regular convex polygon, all of the vertices lie on a circle and are equally spaced around it. The sides of the polygon join adjacent vertices on the circle. If these vertices are connected in any other way then the resulting polygon must have some sides which pass through each other. If all the sides are the same length, so that they connect vertices which are the same distance apart on the circle, then a regular star polygon is formed. For example, in a pentagram the five vertices are equally spaced around a circle and the sides connect every second vertex.

Regular star polygons share many of the properties of the traditional (convex) regular polygons: equal angles at the vertices, equal length sides, the vertices all lie on a circle, as do the midpoints of all the sides. One feature which distinguishes a convex polygon from a star polygon is the number of times it winds around a point in its interior. If you trace along the sides of a regular convex polygon then, when you return to the starting point, you will have encircled the centre point of the polygon exactly once. However, if you repeat this procedure on a regular star polygon, you will wind around the centre more than once before retracing your path. In the case of the star pentagon, you go round twice. The star pentagon is said to *cover* its centre point twice.

Another way to find this covering number is to count the number of times a line from the centre of the polygon to a point outside it crosses the sides of the polygon. The line should avoid vertices and places where sides pass through one another. For a convex polygon, the line has to cross one side to reach the exterior. For the star pentagon it must cross two. This method works for any polygon so long as all the interior angles are less than 180°.[1]

---

[1]It can be made to work for general polygons by careful counting. The polygon has to be oriented and the direction taken into account each time a side is crossed.

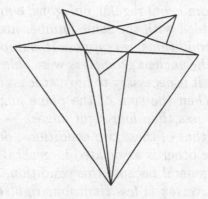

Figure 7.2. A pyramid with a pentagram base.

The notion of covering number can be extended to polyhedra. Here, we will restrict attention to polyhedra whose dihedral angles are all less than 180°. We say that a polyhedron covers a point in its interior n times if a line from the centre of the polyhedron to the exterior intersects n faces of the polyhedron. As before, the line should avoid the vertices, edges and other places where the faces intersect each other. For example, M. C. Escher's cut-away diagram of the small stellated dodecahedron (Figure 4.18) helps us to see that it covers an interior point twice.

Poinsot's star polyhedra

Early in the nineteenth century, unaware of the previous work by Kepler, Louis Poinsot (1777–1859) investigated the possibility that polyhedra could be built from these generalised polygons. He allowed star polygons to be used as faces but, unlike Kepler, he also investigated the possibility that vertex figures could be star polygons. Recall that a vertex figure is a (spherical) polygon which describes how the faces are arranged around a vertex. It is formed as the intersection of the faces that meet at a vertex with a small sphere centred on that vertex. Figure 7.2 shows an inverted pyramid with a pentagram base. The vertex figures at the base are all isosceles triangles, and the vertex figure of the apex is a star pentagon. Since the pentagram covers its centre point twice, we say that the triangular faces cycle around the vertex twice. In general, if a vertex figure covers its centre point n times then the faces meeting there will cycle around the vertex n times.

Poinsot begins his paper by discussing the properties of generalised polygons. When he turns his attention to polyhedra, he notes that the proofs which show that there can be at most five regular polyhedra depend on an implicit assumption:

> Until now, we have known only five perfectly regular bodies, that is to

say those formed from equal regular polygons, equally inclined to one another, and assembled with the same number around each vertex.

By imposing these conditions, we suppose it is impossible to construct any more and that the ancient geometers were able to make a complete enumeration. First, it is necessary to have at least three planes to form a solid angle, and then the sum of the plane angles which form the solid angle must be less than four right angles. ···

However, I observe that of these two conditions, only the first is absolutely necessary, the other is associated in general with what is called convexity. I say in general because this condition, that the sum of the angles around each vertex is less than four right angles, does not always lead to a convex surface: one which has the property that no line cuts the surface in more than two points. But one tacitly supposes this third condition, so much so that we only make the five combinations which give rise to the five regular bodies that we know.

But if, while conserving the general definition of regular solid, we extend the idea of convexity, we see the possibility to construct new regular polyhedra, not only with the new polygons I have just considered but even with the ordinary regular polygons.[a]

For Poinsot, there is the hidden assumption that regular polyhedra are convex. In the above passage he states the modern very restrictive definition of convexity: no straight line meets the polyhedron in more than two points. This definition can be applied to any geometric object. However, in his discussion of polygons, he suggests that an alternative definition of convexity would be to require that all the interior angles of a polygon be less than 180°. Under this notion of convexity, the regular star polygons would then be considered convex.

He proposes a similar alternative for polyhedra: a polyhedron is convex if all its dihedral angles are less than 180°. Under this definition, polyhedra like Kepler's pair made up of pentagrams are convex. Poinsot describes another example. Recall that the Platonic icosahedron can be decomposed into a pentagonal antiprism and two pentagonal pyramids. This shows that a pentagon can be inscribed in the icosahedron (see Figure 7.3(a)). Do this in as many ways as possible and you end up with twelve pentagons which pass through each other (Figure 7.3(b)). If we define an edge to be the line where the sides of two pentagons are joined, and ignore the other intersections, then we get a polyhedron. Its edges and vertices happen to be the same as those of the original icosahedron. Furthermore, all its dihedral angles are equal and are less than 180°. So, according to Poinsot, this is another regular convex polyhedron. Curiously, a very similar looking figure can be found among Wenzel Jamnitzer's sketches in his *Perspectiva Corporum Regularium*: see the centre–left of his plate reproduced in Figure 3.16.

In Legendre's proof of Euler's formula (see Chapter 5), the polyhedron is projected onto an enclosed sphere to produce a network of spherical polygons. The areas of these polygons can be expressed in terms of the numbers of vertices, edges, and faces of the original polyhedron. Equating their total area to the area of the sphere produces the proof. Poinsot used a similar idea to derive a formula which must be satisfied by his generalised regular polyhedra. This then allowed him to deduce some possible structures for them. His argument runs as follows.

Suppose that the faces of a star polyhedron are equal regular polygons of the conventional kind, and that the vertex figures are allowed to be any regular polygon—simple or star. Suppose that all the faces of such a polyhedron are p-gons, and that all the vertices are surrounded in the same manner with q faces coming together at each vertex and cycling around it n times. Suppose also that the whole polyhedron covers an interior point N times. This polyhedron can be projected onto an inscribed sphere to give a network of vertices, edges, and spherical polygons. The spherical polygons are all equal and regular so all the interior angles of these polygons are equal. Let α denote this common interior angle. At the vertices of the network, q spherical polygons come together and the total angle formed by all the polygons at the vertex is $q\alpha$. These faces make n complete cycles around the vertex so the total angle is also $2\pi n$. Equating these gives

$$2\pi n = q\alpha \qquad \text{which implies} \qquad \alpha = 2\pi\frac{n}{q}.$$

The area of each of the spherical p-gons is given by

$$p\alpha - (p-2)\pi = p \cdot 2\pi\frac{n}{q} - p\pi + 2\pi.$$

Suppose that the polyhedron has F faces. Then the sum of the areas of these

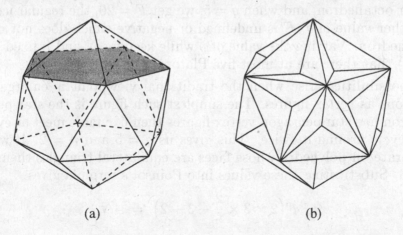

(a)   (b)

Figure 7.3. The great dodecahedron (right).

faces must equal the area of the sphere multiplied by the number of times it is covered by the polyhedron. Thus

$$F\left(p \cdot 2\pi \frac{n}{q} - p\pi + 2\pi\right) = 4\pi N$$

and dividing through by π gives

$$F\left(2p\frac{n}{q} - p + 2\right) = 4N.$$

Poinsot used this last formula to deduce possible forms of star polyhedra. The formula has five variables so, in order to explore the possibilities, some of them must be fixed and the consequences of the choice followed through. We can choose to concentrate on a particular kind of face, or covering, or any other combination of the variables.

Firstly, we can check that the formula holds for conventional polyhedra. In this case, a polyhedron covers its insphere once and the faces cycle around each vertex once so we substitute $N = 1$ and $n = 1$. If we further suppose that all the faces are equilateral triangles then p is set to 3. Substituting all these values into the formula gives

$$F\left(2 \times 3 \times \frac{1}{q} - 3 + 2\right) = 4 \times 1.$$

Rearranging this, we get a formula relating the number of faces of the polyhedron to the valency of its vertices.

$$F = \frac{4q}{6 - q}.$$

There must be at least three faces around any vertex so $q \geqslant 3$. For $q = 3$ we have $F = 4$ giving the regular tetrahedron. When $q = 4$ we get $F = 8$ giving the regular octahedron, and when $q = 5$, we get $F = 20$, the regular icosahedron. For any other value of q, F is undefined or negative which does not make sense for a polyhedron. Varying the value of p while keeping N and n fixed at 1 shows (yet again) that there are at most five Platonic polyhedra.

New possibilities arise when the traditional viewpoint is enlarged to allow star polygons as vertex figures. The simplest such figure is the star pentagon. If a polyhedron has star pentagon vertex figures then five faces meet at every vertex and they cycle around it twice. This gives us $q = 5$ and $n = 2$. If we continue to concentrate on polyhedra whose faces are equilateral triangles then, again, we have $p = 3$. Substituting these values into Poinsot's formula gives

$$F\left(2 \times 3 \times \frac{2}{5} - 3 + 2\right) = 4N$$

which reduces to

$$7F = 20N.$$

The smallest values of F and N which satisfy this equation are $F = 20$ and $N = 7$. This suggests the possible existence of a polyhedron having twenty triangular faces which come together at the vertices in sets of five arranged to give star pentagon vertex figures. This polyhedron would cover an enclosed sphere seven times. Remarkably, such a polyhedron does exist! It is shown in Figure 7.4 and Plate 9. The shaded regions in the figure are the visible parts of one of the triangular faces.

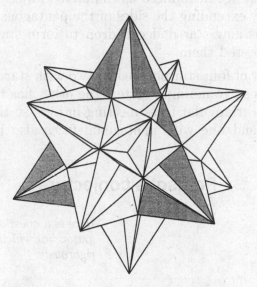

Figure 7.4. The great icosahedron.

Continuing this investigation of polyhedra with triangular faces, the next simplest star polygon which could appear as a vertex figure is a star heptagon. There are two star heptagons, one which covers an inscribed circle twice, the other three times. Substituting the values $p = 3$, $q = 7$, $n = 2$ into the formula leads to the equation

$$5F = 28N$$

and substituting the values $p = 3$, $q = 7$, $n = 3$ gives

$$11F = 28N.$$

The problem now arises as to whether or not polyhedra can be constructed which have the required properties. Poinsot's formula gives conditions which must be satisfied by a generalised regular polyhedron, but does every set of values for p, q, n, N, and F which satisfy the formula correspond to an actual polyhedron? In particular, can either of these potential polyhedra with star heptagon vertex figures be realised? We shall return to this problem later.

The investigation can be continued using other star polygons as vertex figures and different polygons for the faces. In the case of pentagonal faces meeting with star pentagon vertex figures we get $p = 5$, $q = 5$ and $n = 2$. These values substituted into the formula give $F = 4N$. The solution $F = 12$ and $N = 3$ of this equation corresponds to the polyhedron described earlier (see Figure 7.3).

Poinsot also considered whether the star polygons can be used as faces of polyhedra and rediscovered the two polyhedra described by Kepler 200 years earlier. However, they are not produced through a careful analysis like the other two star polyhedra but are introduced as completed objects. He notes that they can be constructed by extending the sides of the pentagons in both the Platonic dodecahedron and his new star dodecahedron to form star pentagons. This is perhaps how he discovered them.

He therefore knew of four star polyhedra: two with star pentagon faces one of which has three faces surrounding a vertex, the other has five faces surrounding each vertex; one with triangular faces meeting in fives so that the vertex figures are star pentagons; and one with pentagonal faces also having star pentagon vertex figures.

Poinsot's conjecture

> *Here is a question which deserves investigation but which seems difficult to resolve rigorously.*[b]
>
> L. Poinsot

Having found four star polyhedra, Poinsot naturally desired to know whether he had found all the possibilities. He observed that the face planes of each of these four polyhedra coincide with those of a Platonic solid (either the dodecahedron or the icosahedron) and he wondered whether this would be true of all regular star polyhedra. He put forward evidence for and against this hypothesis but was unable to decide one way or the other.

First he argued that one can start to build a model of any regular star polyhedron. Imagine constructing a model of a triangular-faced polyhedron where seven faces meet at each vertex and so that they cycle around a vertex twice. Start with seven equilateral triangles and arrange them around a vertex so that they are equally inclined to each other, and proceed to build up the model by attaching triangles, completing vertices or forming new ones. If this assembly of triangles closes up then the number of faces will be a multiple of 28 and the resulting polyhedron will cover an inscribed sphere the same multiple of 5 times. The crucial point is whether the polyhedron ever closes up or whether an unlimited number of triangles can be assembled without this happening.

If such a polyhedron does exist then its inscribed sphere will touch the centre of every face, and these touching points will be evenly and uniformly distributed

over the sphere. These points can be regarded as the vertices of a convex[2] polyhedron sitting inside the sphere. Poinsot is now caught in a dilemma. On one
hand he cannot see any reason why this convex polyhedron, whose vertices are
uniformly and regularly distributed over the sphere, should be anything other
than a regular polyhedron. This would resolve the problem of whether or not the
polyhedron imagined above can be constructed. If such a polyhedron did exist,
the number of its faces, and hence the number of vertices of the convex polyhedron inside the sphere, would be a multiple of 28. But none of the convex regular
polyhedra has a number of vertices which is a multiple of 28, so the polyhedron
cannot exist.

On the other hand, what does it mean to say that a set of points is regularly
distributed on a sphere? Is a convex polyhedron which has such a set of points
for vertices necessarily a regular solid? After all, the 30 midpoints of the edges of
a regular icosahedron all lie on a sphere, and this is surely an evenly distributed
arrangement of points. However, these points are not the vertices of a Platonic
figure but of an Archimedean solid—the icosi-dodecahedron composed of triangles and pentagons. Another Archimedean solid results from the midpoints of
the edges of a cube: the cub-octahedron. If the face centres of a regular star
polyhedron were distributed over a sphere like the 30 midpoints of the edges of
an icosahedron then his previous argument would not apply.

In this discussion, Poinsot displays two competing thought processes which
mathematicians go through when trying to determine the truth of a conjecture.
While considering a possible method for proving his conjecture, Poinsot is also
looking for obstacles to his strategy. A knowledge of the situations where a
method fails is often very useful. Sometimes the obstacles can be overcome;
sometimes a new strategy of attack is required; in other situations they can
lead to new examples which require the original conjecture to be modified or
abandoned. As Poinsot remarks, one of the things which makes the theory of
polyhedra so difficult is the ease with which the study of a few isolated examples
leads to speculation and conjecture about polyhedra in general.

Cayley's formula

An appendix to Poinsot's paper shows how Euler's formula can be generalised so
that it applies to polyhedra with star polygon vertex figures. Suppose a polyhedron covers an enclosed sphere N times and that the faces which meet at a vertex
cycle around it n times. Using Legendre's proof as a basis, Poinsot showed that

$$nV + F = E + 2N.$$

---

[2]From here onwards, convex has its modern restrictive meaning: an object is convex if it
meets any straight line in at most two points.

where V, E and F denote the number of vertices, edges and faces (as usual). When N and n equal one this reduces to Euler's original formula. The formula fails to work when the faces are star polygons.

Some 50 years later, Arthur Cayley presented a different way of measuring the number of times a star polyhedron covers an enclosed sphere. This number has come to be called the *density* of a polyhedron. Density is calculated in a similar way to the covering number: we count the number of times a line from the centre of the polyhedron to its exterior cuts the faces. The difference is that we take into account how many 'layers' of each face the line goes through. Convex polygons have only a single layer so the density of a polyhedron with convex faces is the same as its covering number. For star polygon faces, the number of layers depends upon which point of the polygon the line passes through.

The sides of a planar polygon divide up the plane into regions, and the number of layers is constant in each region. To work out how many layers of a polygon cover a region, trace a path around the polygon and count how many times you wind around a point inside the region before retracing your steps. For example, the five corner regions of a pentagram have only one layer, but the central region has two layers (see Figure 7.5). We also define the density of a polygon to be its greatest thickness. For star polygons this is the same as its covering number.

Figure 7.5. The layers of a pentagram.

With this new way of counting the intersections of the faces with a line, we can find the density of the star polyhedron in Escher's sketch (Figure 4.18). A line from the interior of the polyhedron will pass through the central region of one face, and a corner region of another. The first intersection passes through two layers, and the second intersection contributes only one layer. Therefore, the density of the polyhedron is 3.

Using the idea of density, Cayley was able to obtain a generalised version of Euler's formula which is satisfied by all four of Poinsot's star polyhedra and in which the symmetry between V and F is restored. If d_V and d_F are the densities of the vertex figures and of the faces respectively, and D is the density of the polyhedron then

$$d_V V + d_F F = E + 2D.$$

Again, when all the densities are 1, this reduces to Euler's original formula.

The names which Cayley used for Poinsot's four star polyhedra have become their accepted English names. The one with 20 triangular faces is called the *great icosahedron*, the one with 12 pentagonal faces is called the *great dodecahedron*. The two polyhedra which have 12 star pentagonal faces are called the *small stellated dodecahedron* and the *great stellated dodecahedron*, the small one having 12 vertices, the great having 20 vertices.

Cauchy's enumeration of star polyhedra

The problem of how many regular star polyhedra can be constructed was resolved by Augustin Louis Cauchy in 1812. His solution, together with his proof of Euler's formula, formed the contents of his first paper. Using concepts that we would now call symmetry and transitivity, he interpreted what is meant by regularity in a new way. However, it is important to remember that the fundamentals of symmetry were not fully developed until later in the nineteenth century when crystallographers used symmetry to explain the shape and structure of crystals.

Cauchy applied the principles of rotational symmetry in an intuitive way. Rather than use the standard definition of regularity, which requires the congruence of the faces, solid angles and dihedral angles, he observed that all the Platonic solids can be superposed on themselves in more than one way. For example, let P and Q be two cubes of equal size. We can choose a face of each cube and then superpose P on Q so that the chosen faces are also superposed. This is possible because all the faces of a cube play equivalent roles in its structure. Furthermore, having aligned the chosen faces, we can choose a side of each face and superpose the cubes in such a way that both the chosen faces and their chosen edges coincide.

The classical definition of regularity forces a regular polyhedron to possess Cauchy's 'multiple coincidence' property. The defining properties are sufficiently explicit to be used as a set of instructions for making a regular polyhedron. All you need to know is what kind of polygon to use for the faces and how many to place around each vertex. You start by taking a regular polygon to form the first face. Then take another polygon and attach it to the first. Since both polygons are the same, and both are regular, it makes no difference which two sides are joined to make the first edge. A vertex can be built up by surrounding it with the appropriate number of polygons. Since all the dihedral angles must be the same, each vertex figure must be a regular polygon. Thus, there is no choice in the shape of the solid angle. Continue to add faces building up the model until it closes up. At no stage in the process is there a choice: the faces are all the same, the solid angles are all the same, the faces are rigid, the vertex figures are rigid.

When you examine the completed model, you cannot tell how the polyhedron was constructed. All the faces and vertices are equivalent. Knowing the instructions does not help because everything looks the same. You cannot tell which was the first face. It is because the construction could have started with any face, and the rules completely determine the rest of the shape, that the polyhedron can be superposed on itself in the way Cauchy observed.

With a slight modification to the rules, this description of building up a regular polyhedron applies to star polyhedra as well as to the Platonic solids. To describe Poinsot's polyhedra, it is no longer sufficient to give only the valence of the vertices: one must give the vertex figure.

This property of regular polyhedra can also be used in reverse as a definition of regularity. The fact that any face can be superposed on any other means that all the faces are congruent. Furthermore, since each face can be rotated to superpose one side on any other side, the faces must be regular polygons. By combining these operations we can superpose an edge of the polyhedron on any other so all the dihedral angles must be the same. In this way, we can recover the classical definition from Cauchy's observation—the two viewpoints are equivalent.

With this new way of thinking about regularity we can now turn our attention to Cauchy's solution of Poinsot's question.

Lemma.   The face-planes of a regular star polyhedron coincide with those of a Platonic solid.

PROOF: Suppose you have two copies of a regular polyhedron. One can be brought into coincidence with the other in several ways. Firstly, a face of one can be superposed on any face in the other. Furthermore, any side of the face in the first polyhedron can be paired with any side of the chosen face in the second copy of the polyhedron. This is possible for both convex and star regular polyhedra.

Now imagine yourself transported to the centre of a regular star polyhedron. If the faces are opaque then your view of the polyhedron will be limited to a convex kernel whose faces lie in the face-planes of the star polyhedron.

What happens to this kernel when the star polyhedron is superposed on a second star polyhedron? The face-planes of the second star polyhedron are the same as those of the first so the kernels of the two polyhedra coincide. Since a face of the first star polyhedron can be superposed on any face of the second star polyhedron, a face of the first kernel can be superposed on any face of the second kernel. Therefore, the faces of the convex kernel are all congruent. This implies (via a corollary of Euler's formula) that its faces have three, four or five sides.

Because we can align a given side of a face in the first star polyhedron with any side of the chosen face in the second, a face of the kernel can be superposed on itself in more than one way. If the faces of the star polyhedron have n sides then each face of the kernel can be rotated onto itself in at least n ways. If the faces of the kernel have p sides then p must be a multiple of n. But $p = 3$, 4 or

5 and n is at least 3. So, in fact, $p = n$. This means that a side of a face of the first kernel can be superposed on any side of the chosen face in the second kernel. Thus, the kernel is a convex regular solid. ■

This lemma answers Poinsot's question 'Do the face-planes of a regular star polyhedron have to coincide with those of a Platonic solid?' in the affirmative and so excludes his two hypothetical examples having triangular faces and star heptagon vertex figures. However, we still need to show that he found all the possibilities.

To enumerate the star polyhedra, Cauchy analysed the faces of the Platonic solids to see which could be extended to bound regular polygons. In the case of the cube it is easy to see that its face-planes never meet again when they are extended. They bound only the cube. The same is true of the tetrahedron. The other cases need more careful consideration.

First, we observe that the face-planes of the octahedron, dodecahedron, and the icosahedron come in parallel pairs. If we place a model of one of them on a table, there is a natural choice of a base face (in contact with the table surface) and a top face (opposite the base). To construct regular star polyhedra we shall examine the possible ways that groups of faces can be extended to bound a regular polygon in the top face-plane. For a set of face-planes to intersect the top face-plane in lines that bound a regular polygon, they must be arranged symmetrically around the axis passing through the centres of the top and the base. We shall refer to such a symmetrically placed set of faces as a group.

The faces of the octahedron can be divided into four groups: the top, the base, the three faces adjacent to the top, and the three adjacent to the base. The base and the top are parallel and do not meet. The faces adjacent to the top bound only the top face and do not give anything new. The three face-planes adjacent to the base do bound a new regular polygon in the top face-plane—a triangle whose sides are twice as long as those of the octahedron (Figure 7.6). Four of these triangles form a Platonic tetrahedron. When all eight of the faces of the octahedron are extended to form these triangles, Kepler's stella octangula is formed. This is a compound polyhedron and not what we are searching for.

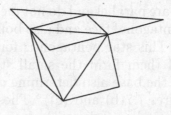

Figure 7.6.

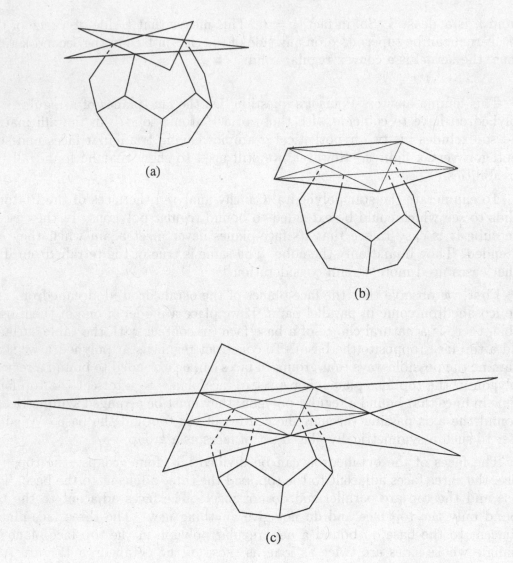

(a)

(b)

(c)

Figure 7.7.

The faces of the dodecahedron can also be divided into the same four groups: top, base, the five faces adjacent to the top, and the five adjacent to the base. Again, the base and the top are parallel and do not meet. The face-planes adjacent to the top bound the top pentagonal face and also bound a pentagram in the same face-plane (Figure 7.7(a)). This star pentagon is formed by extending the sides of the top face. Twelve of them form the small stellated dodecahedron. The five face-planes adjacent to the base also determine a pentagon and a pentagram in the top face-plane (Figures 7.7(b) and (c)). These are the faces of the great dodecahedron and the great stellated dodecahedron, respectively. This exhausts the dodecahedral family of regular polyhedra.

Figure 7.8.

The faces of the icosahedron divide into eight groups. As usual, the top, the base, and the three faces adjacent to the top are three of them. To see the others, label each of the faces adjacent to the base 'A', and apply successive letters to the faces around its top vertex as shown in Figure 7.8. Each set of three face-planes labelled with a given letter forms a group. When searching for regular polygons in the top face-plane we need only consider these five lettered groups of face-planes—the other groups clearly do not bound any new polygons.

Figure 7.9 shows the polygons formed by groups 'A', 'C' and 'E'. The shading indicates which faces belong to the group. The three face-planes labelled 'A' bound a large triangle which is the face of the great icosahedron (see Figure 7.9(c)). The three face-planes of group 'C' or of group 'D' determine a smaller triangle (Figure 7.9(a)), eight of which fit together to form a Platonic octahedron. When all the faces are extended in this manner a compound of five octahedra is produced (see Plate 13). The three face-planes of group 'B' or of group 'E' determine another triangle, one of intermediate size (Figure 7.9(b)). Four of them fit together to form a Platonic tetrahedron. Extending all the faces in this way produces a compound of five tetrahedra (see Plate 12).

This completes Cauchy's exhaustive search for regular star polyhedra. Only the four described by Poinsot are produced.

Face-stellation

The process of producing new polyhedra by extending the faces of a given one is called *face-stellation*. It was first described by Kepler. Although he probably discovered his star polyhedra by extending edges rather than faces, he knew that the small stellated dodecahedron could be obtained from the Platonic dodecahedron by face-stellation.

The stellation process may seem clear enough but there is some ambiguity about how we should interpret the result. For example, is the great dodecahedron composed of twelve regular pentagons, or 60 isosceles triangles. Is a stellated polyhedron a solid whose face-planes coincide with those of the original (solid)

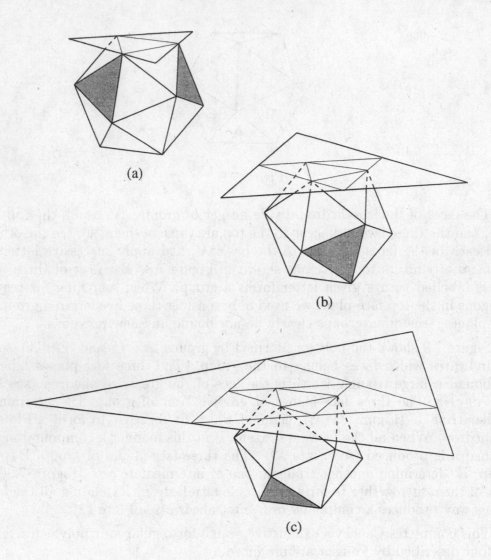

Figure 7.9.

polyhedron, is it the bounding surface of such a solid, or is it a set of intersecting polygons.

This freedom of interpretation means that there are complementary ways to think about the process of face-stellation. Cauchy's approach is essentially two-dimensional: choose one face-plane and see how the others intersect it. From the information in this one plane, we then deduce possible faces for stellated forms. Another approach is three-dimensional: the stellations are thought of as being built up from layers of solid cells. To show how these different means of expression are related we will examine the stellations of the dodecahedron.

The three-dimensional approach.

Each face of a convex polyhedron lies in a unique plane, and the set of all these face-planes partitions space into a collection of cells. The original polyhedron must be one of the cells, and there will always be some unbounded cells. The situation becomes interesting when there is more than one bounded cell. This happens when the dihedral angles of the polyhedron are greater than 90°.

The bounded cells come in layers surrounding the central, core polyhedron. They can be stuck together to form new polyhedra whose faces will lie in the same planes as those of the original polyhedron.

A regular dodecahedron is surrounded by three kinds of finite cell. The first layer consists of twelve pentagonal-based pyramids (see Figure 7.10). The second layer consists of 30 wedges that sit between the pyramids. The wedge on the left of Figure 7.11 sits behind and to the right of the pyramid. Notice how each successive layer completely covers the faces of the previous layer. The final layer of cells is a set of 20 spikes each of which is an asymmetric triangular dipyramid. These fit into the hollows between the wedges. On the left of Figure 7.12 the sharp end of the spike points backwards, and the spike fits in a depression hidden from view on the upper right of the second stellation. The figures are all drawn to the same scale.

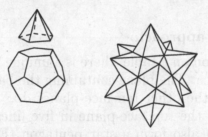

Figure 7.10. The first stellation of the dodecahedron is built from a layer of 12 pyramids.

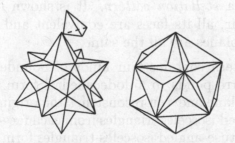

Figure 7.11. The second stellation of the dodecahedron is built from a layer of 30 wedges.

Figure 7.12. The third stellation of the dodecahedron is built from a layer of 20 spikes.

The two-dimensional approach.

If we place a dodecahedron on a table, there is a unique plane parallel to the table top containing its top face. The plane containing the base does not meet this top face-plane, but each of the other ten face-planes do. The five planes adjacent to the top face intersect the top face-plane in five lines. These lines bound the top pentagonal face and also form a star pentagon (Figure 7.7(a)). When the five planes containing the faces adjacent to the base are extended to meet the top face-plane, five more lines of intersection are produced which pass through the vertices of this pentagram (Figure 7.7(c)). These ten lines of intersection form a pattern called a *stellation pattern*. It is shown in Figure 7.13. As the dodecahedron is regular, all its faces are equivalent and the stellation patterns formed in all the face-planes are all the same.

The stellation pattern contains four kinds of bounded region. In this case, each kind of region corresponds to a dodecahedral form. Take the central pentagon from each face-plane and you reconstruct the original dodecahedron. Take all the small acute-angled isosceles triangles from each face-plane and the first stellation appears. The obtuse-angled isosceles triangles form the second stellation—the great dodecahedron. The remaining triangles form the great stellated dodecahedron. The lines in the stellation pattern do not bound any more finite regions and so this is the last of the dodecahedron stellations.

Figure 7.13. The stellation pattern of the dodecahedron.

In a more complex stellation pattern, such as that of the icosahedron, the regions have to be chosen carefully. Just selecting corresponding regions in each face-plane may result in a set of disconnected polygons with free edges not attached to other faces. To enumerate the stellated forms, it is necessary to work out which sets of regions fit together to form closed polyhedra.

Stellations of the icosahedron

All the stellations of the dodecahedron happen to be regular star polyhedra. Stellating the icosahedron, on the other hand, is far more challenging. Its stellation pattern is shown in Figure 7.14. It consists of the 18 lines where a face-plane is met by the face-planes of all the faces except the opposite one (which is parallel to it). These lines divide the face-plane into 66 finite regions, in addition to the central triangle which is a face of the icosahedron. Try to imagine how many face patterns can be built by piecing these together in different ways.

We can also consider this abundance of possibilities from the three-dimensional viewpoint. The 20 face-planes of the icosahedron partition space into 473 bounded cells of some 11 or 12 kinds (depending on whether you class mirror images as the same or different). With such a large number of blocks from which to start building up stellations, some care is needed in deciding which sets of cells are to be counted as proper stellated forms. Although we can construct $2^{12} - 1$ combinations of cells from the icosahedral decomposition of space, this does not mean we can build over 4000 stellations of the icosahedron. We need to decide which combinations should be included. What should be the acceptance criteria by which we judge their validity? What properties should a stellation possess?

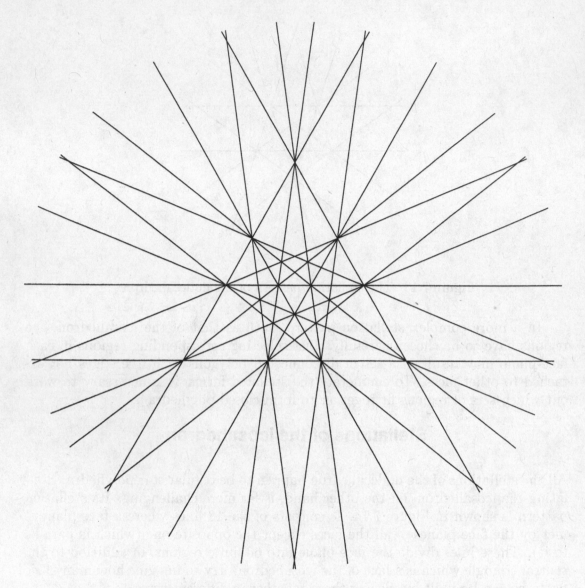

Figure 7.14. The stellation pattern of the icosahedron.

A few examples of stellations of the icosahedron were known, and recognised as such, at the turn of the twentieth century. Some are identified in Max Brückner's book *Vielecke und Vielflache* published in 1900. Besides the great icosahedron (Plate 9) and the compounds of five octahedra (Plate 13) and five and ten tetrahedra (Plate 12) known to Cauchy, he described six others. They are shown among the Figures 7.15–7.33 and Plate 10. All the stellations in the figures are drawn on the same scale and from the same viewpoint.

Nine more stellations were discovered by Albert Harry Wheeler bringing the total to 19 stellated forms plus the icosahedron itself. Most of his stellations are shown in the figures. One of them (Figure 7.27) is cheiral like the compound of five tetrahedra.

Inspired by Wheeler, an exhaustive search was made for more stellations. As mentioned above, in order to perform such an enumeration, a definition of what is meant by 'stellation' is required. The following criteria were proposed by J. C. P. Miller.

(i) The faces of the stellated form must lie in the face-planes of the original polyhedron.

(ii) All the regions composing the faces must be the same in each plane; these regions need not be connected.

(iii) The regions included in a plane must have the same rotational symmetry as a face of the original polyhedron. Together with (ii), this implies that the stellation process preserves the rotational symmetry of the original polyhedron.

(iv) The regions included in a plane must be accessible in the completed stellation.

(v) Compounds of simpler stellations are excluded. More specifically, we disallow unions of two stellations with no face-to-face contact except combinations of mirror images.

Applying these rules to the manifold possibilities reduces their number substantially. There remain 31 stellations with mirror symmetry, and 27 cheiral stellations (or pairs of enantiomorphs).

The complete list was published in 1938 in a delightful little booklet called *The Fifty-nine Icosahedra*. Of the four authors J. F. Petrie produced the fine illustrations, H. T. Flather made a complete set of models which are now housed in the Mathematics Department at Cambridge[3], and H. S. M. Coxeter and Patrick Du Val wrote the text each supplying a different method of enumeration: Coxeter used the two-dimensional approach and Du Val worked with cells.

Another problem to be overcome is how to describe each stellation in a systematic and economical way. The method Du Val used to solve this problem was to classify the cells into their various kinds and then list the combination of cell types which forms each stellation. As an example, we shall apply the method to the dodecahedron to see how it works.

A number can be associated with each cell which is a measure of its distance from the central cell—the dodecahedron itself. The index of a cell is the number of

---

[3] They can be viewed by appointment. Write to the Head of the Department of Mathematics, University of Cambridge, 16 Mill Lane, Cambridge. CB2 1SB. England.

face-planes a straight line from the cell to the centre passes through (compare this with the definition of covering number or density). So, for example, dodecahedron cells have following indices:

0 : the dodecahedron

1 : the 12 pyramids

2 : the 30 wedges

3 : the 20 spikes.

A *layer* of cells is defined to be all the cells that have a given index. These layers form a series of concentric shells which envelop the core. Each layer is denoted by a lower-case letter running in sequence from the centre outwards. Thus the dodecahedron is layer **a**, the pyramids are layer **b**, the wedges are layer **c**, and the spikes are layer **d**. When describing stellations it is also helpful to have symbols for the set of all cells with index less than or equal to a given number. These are denoted by upper-case letters. The set of all cells of index at most 0 (which is just the central dodecahedron) is denoted **A**. The set of cells with index at most 1 (the small stellated dodecahedron) is denoted **B**. Similarly, the great dodecahedron, denoted **C**, is formed by adding layer **c** to **B**. The third stellation, **D**, is the set of all (bounded) cells taken together.

For any initial polyhedron, the stellations **A**, **B**, **C**, **D**, **E**, ... are called the *main sequence*. They form a natural progression and are the only stellations that can be sensibly ordered: **B** is the first stellation, **C** is the second, and so on. Notice that Wheeler did not find the entire main sequence for the icosahedron—he missed **E** (Figure 7.18).

All the stellations in the main sequence of the icosahedron are illustrated here. Figure 7.15(b) shows stellations **A**, **B** and **C**. The *triakis icosahedron*, **B**, is formed by erecting a pyramid on each face of the icosahedron. (The term 'triakis' is borrowed from crystallography.) **C** is the compound of five octahedra. Figure 7.17 shows the result of extending the faces of this compound until they meet again. Figure 7.20 shows the sixth stellation, **G**, familiar as Poinsot's great icosahedron. The final stellation, **H**, composed of all the cells, is shown in Plate 10. It is known as the *complete icosahedron*.

This system of nomenclature can be used to label the cells and their combinations which arise by stellating any convex polyhedron. In the case of the icosahedron the situation becomes more complicated, and more interesting, because three of the eight layers contain different kinds of cell. One layer even contains cheiral cells. Layers **e** and **g** can both be separated into two kinds of cells. These are denoted e_1 (shown in Figure 7.21) and e_2, g_1 (shown in Figure 7.30) and g_2. All four of these stellations exhibit an unusual feature which has not appeared in stellations seen earlier. They are not polyhedra in the sense of the polyhedral surfaces described in Chapter 5 but have singular points where they are vertex-connected. Other stellations of the icosahedron, such as f_1, are edge-connected having more than two faces meeting at an edge. These are naturally occurring instances of Hessel's counter-examples to Euler's formula (see Figure 5.15).

Layer f is the most unusual of all. The cells in one set, f_2, are completely disconnected (see Figure 7.33). The remaining cells, f_1, can be split into two mirror-image forms. One cheiral set is shown in Figure 7.32. It is denoted f_1 — note the change in typeface. It is the inclusion of just one of these two cheiral sets that produces the cheiral stellations such as the compound of five tetrahedra, Ef_1.

By comparing the figures with one another you may begin to be able to visualise the different building blocks and the ways they combine. For example, by adding cells g_1 (7.30) to Fg_2 (7.31) we get the great icosahedron G (7.20). Adding g_1 to the compound of ten tetrahedra (7.22) produces a non-convex deltahedron (7.24). The inclusion of the twelve spikes of f_2 is especially easy to identify. Wheeler's cheiral stellation (7.27) is formed by adding them to a compound of five tetrahedra. Figures 7.24 and 7.25 show a similar relationship. The last polyhedron in Table 7.16, $e_1f_1g_1$, has tunnels through it. It is not shown here but is easily described: it is the figure formed by removing De_2f_2 (Figure 7.29) from Fg_1 (Figure 7.25).

These examples must suffice though there are many more. Some are polyhedral surfaces which are not spherical but have tunnels through them. Some are very intricate and simple line drawings of them are difficult to interpret. As usual, the best way to understand is to play with models.

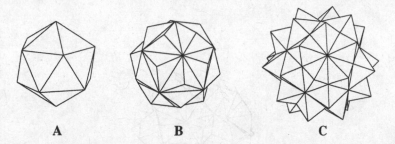

A   B   C

Figure 7.15. The icosahedron and the first two stellations in its main sequence. In the Du Val notation these are labelled **A**, **B** and **C**. The first stellation is sometimes called the triakis icosahedron, and the second is familiar as the compound of five octahedra.

Label	Other names	Cauchy	Brückner	Wheeler	Figure
A	icosahedron	✓	✓	✓	7.15
B	triakis icosahedron		✓	✓	7.15
C	5 octahedra	(✓)	✓	✓	7.15
D			✓	✓	7.17
E					7.18
F				✓	7.19
G	great icosahedron	✓	✓	✓	7.20
H	complete icosahedron		✓	✓	Plate 10
Ef_1	10 tetrahedra	(✓)	✓	✓	7.22
Ef_2			✓	✓	7.23
Ef_1g_1			✓	✓	7.24
Fg_1			✓	✓	7.25
Ef_1	5 tetrahedra	(✓)	✓	✓	7.26
Ef_1f_2			✓	✓	7.27
De_1			✓	✓	7.28
De_2f_2			✓	✓	7.29
g_1				✓	7.30
Fg_2				✓	7.31
e_1					7.21
f_1					7.32
f_2				✓	7.33
De_2				✓	
$e_1f_1g_1$				✓	

Table 7.16.

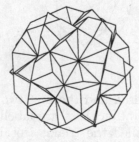

Figure 7.17. Icosahedron stellation **D**.

Figure 7.18. Icosahedron stellation **E**.

Figure 7.19. Icosahedron stellation **F**.

Figure 7.20. Icosahedron stellation **G**. This is also the great icosahedron discovered by Poinsot, one of the four regular star polyhedra.

Figure 7.21. Icosahedron stellation e_1.

Figure 7.22. Icosahedron stellation **Ef₁**. This is also familiar as the compound of ten tetrahedra.

Figure 7.23. Icosahedron stellation **Ef₂**.

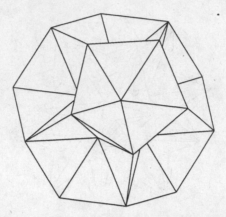

Figure 7.24. Icosahedron stellation $\mathbf{Ef_1g_1}$. This is also an example of a non-convex deltahedron: all its faces are equilateral triangles.

Figure 7.25. Icosahedron stellation $\mathbf{Fg_1}$.

Figure 7.26. Icosahedron stellation **Ef₁** (*dextro*). This is the compound of five tetrahedra.

Figure 7.27. Icosahedron stellation **Ef₁f₂** (*dextro*).

Figure 7.28. Icosahedron stellation **De$_1$**.

Figure 7.29. Icosahedron stellation **De$_2$f$_2$**.

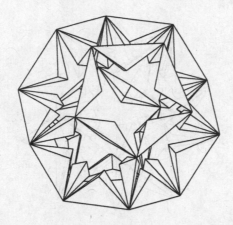

Figure 7.30. Icosahedron stellation g_1.

Figure 7.31. Icosahedron stellation $\mathbf{F}g_2$.

Figure 7.32. Icosahedron stellation f_1 (*dextro*).

Figure 7.33. Icosahedron stellation $\mathbf{f_2}$.

Bertrand's enumeration of star polyhedra

More than forty years after Cauchy's solution to Poinsot's problem, Joseph Bertrand considered it from a complementary point of view. Whereas Cauchy examined face-planes and reduced the problem to looking for regular stellations of the Platonic solids, Bertrand concentrated on the vertices of star polyhedra and ended up searching for regular polygons inside the Platonic solids. He claimed that his method is easier to visualise than the stellation process and I am inclined to agree.

The enumeration is based on a lemma which is dual to Cauchy's lemma: 'face' and 'kernel' are replaced by 'vertex' and 'convex hull', respectively. The *convex hull* of a set of points is the smallest convex polyhedron that contains them. The vertices of the hull always belong to the given set of points. The convex hull of a non-convex polyhedron is the just convex hull of its vertices. For example, the convex hull of the stella octangula is a cube.

Lemma. The vertices of a regular star polyhedron coincide with those of a Platonic solid.

PROOF: Let P be a regular star polyhedron and let Q be its convex hull. The vertices of P all lie on a sphere so all of them are also vertices of Q.

Now, P can be superposed on itself in many ways. We can rotate P so that a given vertex of P can be aligned with any designated vertex of Q. Furthermore, after the two vertices have been matched, there are at least three ways that P can be superposed on itself (because the vertex must be at least 3-valent). Therefore, all the solid angles of the convex hull are congruent, and each solid angle can be superposed on itself in at least three ways. A corollary of Euler's formula implies that the vertices of Q are 3-valent, 4-valent or 5-valent. These last two facts imply that each of the plane angles forming the solid angle can be superposed on any of the others, and similarly for the dihedral angles. Hence, each solid angle of the convex hull is formed from equal plane angles equally inclined to one another. Moreover, all the edges of Q have the same length. Since all the solid angles of Q are congruent, the plane angles in the corners of a face of Q are all equal. Therefore, the faces of Q are regular polygons and the convex hull is a Platonic solid. ∎

The next step in Bertrand's enumeration of the regular star polyhedra is to search for regular polygons inside the Platonic solids. This step, called *facetting*, is easier to visualise than the stellation process. The possibilities are illustrated in Figure 7.34. There are no facetted forms of the tetrahedron since the only regular polygons spanned by these vertices are the original faces. The vertices of a cube span a triangle (Figure 7.34(b)) and four of these triangles form a regular tetrahedron. The vertices of an octahedron bound three squares (Figure 7.34(c)) but these do not form a polyhedron since they do not share common sides.

Potentially the most fruitful case is the dodecahedron. Its vertices span five distinct regular polygons besides the original pentagons. Some of its vertices can be taken as the corners of a square (Figure 7.34(d)). This is clear from Euclid's construction which circumscribes a dodecahedron about a cube. There are also two kinds of triangular face (7.34(e)–(f)). The triangles of one kind do not join up, the others form tetrahedra. Another subset of the dodecahedral vertices are the corners of a pentagon and pentagram (7.34(g)–(h)). Only the star pentagon can be used to form a polyhedron—the great stellated dodecahedron. The convex pentagons do not have sides in common.

All the polygons in an icosahedron give rise to star polyhedra. As Poinsot noted, an inscribed pentagon is a face of the great dodecahedron (7.34(i)). A pentagram with the same vertices gives the small stellated dodecahedron (7.34(j)). There is also a triangle which is a face of the great icosahedron (7.34(k)).

This exhausts the possibilities. Hence there are only four regular star polyhedra.

Regular skeletons

Poinsot worked on his star polyhedra in the first decade of the nineteenth century. This was part of the period when people were trying to find a basis for Euler's formula, and, although the term 'polyhedron' was interpreted in a variety of ways, it was always thought of in terms of a surface—either hollow or as the boundary of a solid. This may go some way to explain why Poinsot is not consistent in his use of the term 'polygon'. When he is considering polygons for their own sake, he is happy to think of them as collections of line segments in the plane, but when it comes to building polyhedra, he uses the conventional filled-in kind of polygon for the faces, and uses his star polygons only to describe the vertex figures.

The two star-faced polyhedra that he discovered do not arise out of a detailed analysis like the other pair. He simply states that they exist, and they can be formed by extending the sides of pentagons in polyhedra already known. Since he mentions the number of times that they cover an inscribed sphere, we must assume that he thought of them as being constructed from some kind of filled-in polygons. Had he been consistent, Poinsot would have constructed his polyhedra from the polygons he discussed in the first part of his paper. In that case, a polyhedron would still be built up by attaching polygonal faces to each other, gluing their sides together to form edges, but the result would be a skeletal polyhedron—an edge framework. However, it is only recently that mathematicians have discarded the psychological crutch of filled-in polygons and started to investigate such skeletal polyhedra in any depth.

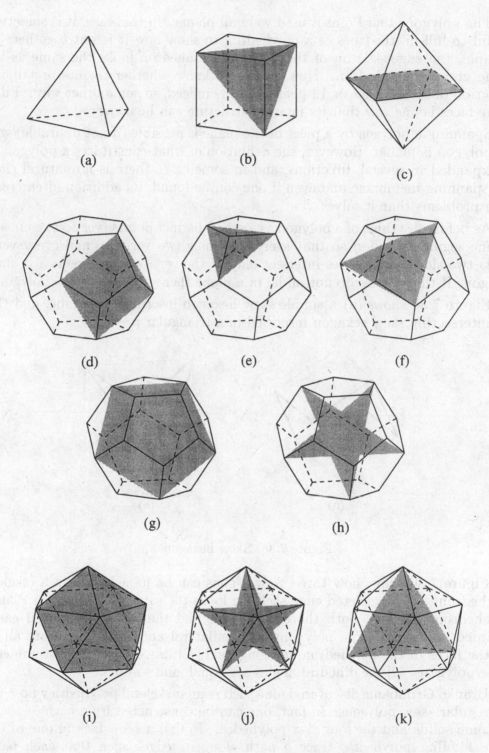

Figure 7.34. Facets of the Platonic solids.

The polygons that Poinsot used were all planar. In this case, it is sometimes helpful to 'fill-in' the faces of a polyhedron to show how it is put together. For example, the edge-skeleton of the Platonic icosahedron looks the same as that of the great dodecahedron. How are we to know whether to interpret it as a collection of 20 triangles or 12 pentagons or, indeed, in some other way? Filling in the faces is one way that its internal structure can be conveyed.

Spanning a polygon by a piece of the plane is possible (if not desirable) when the polygon is planar. However, the definition of what constitutes a polygon can be expanded in several directions, and in some cases there is no natural choice of a spanning membrane and even if one can be found, its addition often creates more problems than it solves.

As before, we think of a polygon as a set of distinct points (vertices) connected by line segments (sides) so that every line joins two vertices, and every vertex meets two sides. If the sides intersect only at the vertices then we have a *simple* polygon. If the vertices do not all lie in a plane then we have a *skew* polygon.

Figure 7.35 shows (a) a simple skew hexagon inscribed in a cube, and (b) a self-intersecting skew hexagon inscribed in a triangular prism.

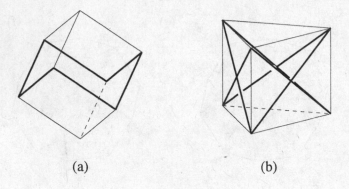

(a)   (b)

Figure 7.35. Skew hexagons.

Figure 7.36 shows how three skew 4-gons can be joined to form a (skeletal) polyhedron. The completed edge-skeleton looks the same as that of the Platonic tetrahedron. In fact, both the skew 4-gons and the polyhedron itself can be regarded as regular. The polygons are equilateral and their angles are all 60°. All the faces of the polyhedron are congruent, all its vertices are surrounded by three polygons, all the dihedral angles are equal, and so on.

Branko Grünbaum discovered nine such regular skeletal polyhedra whose faces are regular skew polygons. In fact, one can be constructed from each of the five Platonic solids and the four star polyhedra. To find a skew face in one of these more familiar polyhedra, trace a path along its edges such that each pair of adjacent edges are sides of a face of the polyhedron, but no three consecutive

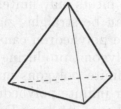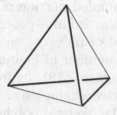

Figure 7.36.

edges are sides of a single face. A polygon constructed in this manner is called a *Petrie polygon* after its discoverer. The skew 4-gons in Figure 7.36 are Petrie polygons of the tetrahedron, and the skew hexagon in Figure 7.35(a) is a Petrie polygon of the cube.

Table 7.37 lists the properties of Grünbaum's nine regular skeletal polyhedra. The first three columns describe how many faces are used, how many sides each face has, and the angle at the corner of each face. The symbol 10/3 designates a skew polygon with ten sides which, when viewed from 'above', looks like the star decagon formed by connecting every third point on a circle. The fourth column states how many faces come together at each vertex of the polyhedron. The symbol 5/2 indicates that the vertex figure is a pentagram. The last column gives the Platonic or star polyhedron from which the skew-faced polyhedron is derived.

A further generalisation of a polygon would be to allow an infinite number of vertices. In this case, we can impose the restriction that the polygon is locally

No. faces	No. sides	Angle	Valence	Relative
3	4	60°	3	tetrahedron
4	6	90°	3	cube
4	6	60°	4	octahedron
6	10	108°	3	dodecahedron
6	10	60°	5	icosahedron
10	6	36°	5	small stellated dodecahedron
10	6	108°	5/2	great dodecahedron
6	10/3	36°	3	great stellated dodecahedron
6	10/3	108°	5/2	great icosahedron

Table 7.37. The composition of Grünbaum's regular skeleta.

finite in the sense that every bounded portion of space meets only finitely many sides of the polygon. The regular infinite polygons can be straight, zig-zag or helical. Using only a single kind of polygon, an infinite polyhedron can be constructed either from an infinite number of bounded polygons, producing objects like tilings and honeycombs (Figure 2.19), or from infinite polygons producing tubes or lattices. Grünbaum listed examples of regular infinite polyhedra along with his other examples of regular skeletal polyhedra. Later, Andreas Dress performed a systematic enumeration and showed that this empirically discovered list was complete.

We could also relax the condition that vertices are distinct points. This means that different vertices can occupy the same position in space, not that a polygon can visit the same vertex more than once. Although we defined a polygon to have distinct vertices, Poinsot did not make this explicit in his definition. He, Cauchy, and many others seem to have regarded it as an implicit condition. Grünbaum noticed this in 1990. More recently he remarked:

> The Original Sin in the theory of polyhedra goes back to Euclid, and through Kepler, Poinsot, Cauchy, and many others continues to afflict all work on this topic. ··· The writers failed to define what are the 'polyhedra' among which they are finding the regular ones. [c]

Frustrated by the use of implicit assumptions, the inconsistent use of terms like 'polygon' and 'polyhedron', and the imposition of unnecessary restrictions in the work of many authors, he developed extremely general definitions of polygon and polyhedron. His definitions are abstract and describe 'probably the most general type of object one may wish to call a polyhedron'. Starting from this baseline, the more familiar ways of interpreting what is meant by 'polyhedron' can be recovered by adding extra conditions. Grünbaum calls the polyhedral surface described in Chapter 5 an *acoptic* polyhedron—one which does not cut itself. A polyhedron all of whose faces are planar is called *epipedal*.

At last there is now some means of describing which particular family of polyhedra one is studying. This is important when trying to enumerate polyhedra with particular properties. Even trying to list the regular polyhedra is problematic unless you know which set of polyhedra you are investigating. Although this sounds obvious, authors rarely state in sufficient detail (if at all) the kinds of polyhedra that they are interested in. This means that results are unclear and proofs are incomplete. As Grünbaum remarks, the complete enumeration of regular polyhedra can still be considered an outstanding problem.

From *Perspectiva Corporum Regularium* by Wenzel Jamnitzer, 1568.

Symmetry, Shape and Structure

> *Symmetry is to the geometry of polyhedra what number theory is to arithmetic.*[a]
>
> A. Badoureau

A friend once remarked on seeing my collection of models of some of the stellations of the icosahedron (Figures 7.15–7.33) that they were like three-dimensional snowflakes. Apart from their white colour, what had triggered this analogy was their high degree of regularity and order. Snow crystals often have six similarly shaped spikes radiating out from a central point, equally spaced around it. This remarkable hexagonal symmetry is the same in many snowflakes even though they have such a diversity of forms, and it reflects the underlying order of their atomic structure. All the stellations of the icosahedron are also highly symmetrical and, although each stellation is unique and differs from each of the others, the nature of this symmetry is the same in every case. To appreciate what it means to say that two polyhedra have the same type of symmetry it is necessary to find a way to describe and quantify this aesthetic quality of polyhedra.

What do we mean by symmetry?

To the Greeks symmetry meant balanced and well proportioned. It was an ideal that signified perfection. In a geometric context symmetry was associated with commensurability. It also implied a regularity of form, and the harmonious relationship exhibited between the different parts of a whole. In the nineteenth century attempts to explain the physical properties and shapes of crystals led to the development of a more precise notion of symmetry. The scientific terminology captured the intuitive notions expressed in the Greek ideal and provided a basis for a mathematical theory of symmetry. This made it possible to compare the symmetry properties of two objects and clarify the meaning of statements like 'A dodecahedron is more symmetrical than a square-based pyramid'. We begin our discussion of symmetry with a fundamental question: What distinguishes a symmetrical object from an irregular one?

Suppose that you have a model of a symmetrical polyhedron (such as one of the Platonic solids) which is well constructed so that there are no blemishes on the faces to distinguish one from another. The model can be moved to a new position which cannot be distinguished from its original one. If your eyes were closed while the motion took place, when you opened them again, examining the polyhedron in its surroundings would not enable you to tell whether the model had been rotated.

A symmetrical polyhedron is characterised by the fact that it looks the same from different viewpoints. Alternatively, a polyhedron is symmetrical if it is possible to perform certain operations (such as rotation) which change the positions in space of individual faces but which leave the polyhedron in a position that is indistinguishable from its original position.

The number of different but indistinguishable positions in which a polyhedron can be placed is a measure of the amount, or degree, of symmetry which it possesses. For example, a regular tetrahedron can be placed in 12 different positions: any of the four faces can be placed on a table, and any of the three edges of this base face can be placed at the front. A hexagonal dipyramid can also be placed in 12 different positions since it can rest on any of its 12 faces in a unique way. Although these two polyhedra have the same *amount* of symmetry, they do not appear to have the same *kind* of symmetry. The dipyramid has two apices which differ from its other vertices imparting a definite direction to it, whereas the tetrahedron is more isotropic with all its vertices having an equal status. The symmetry of a polyhedron needs to be qualified as well as quantified.

To describe the different *kinds* of symmetry, it is helpful to investigate the operations which carry a polyhedron into its indistinguishable positions. Such an operation is called a *symmetry* of the polyhedron. A symmetry is determined by its effect on the polyhedron; it is not the action of moving the polyhedron or the route taken which are important, but rather the relationship of its initial and final positions. Using 'symmetry' as a noun in this way is justified by the fact that the more symmetrical a polyhedron is in the everyday sense, the more symmetries it has in this technical sense.

The symmetries of a polyhedron are rules which describe the relationships between its various parts. It has been said that the aesthetic appeal of symmetrical objects results from the psychological process of discovering these rules. They imply an orderly structure, the existence of an organising principle. In the following sections, the different kinds of symmetry operation are described, and the ways in which they can be combined to produce different systems of symmetries are investigated.

Rotation symmetry

The action of picking up a model of a polyhedron and repositioning it so that it appears not to have been moved is an example of a *direct* symmetry operation. The polyhedron is physically carried to a different but indistinguishable position. In later sections *indirect* symmetries will be encountered; the effect of these cannot be seen by manipulating a model but require the aid of a mirror.

The actual motion of a polyhedral model during a direct symmetry operation may be very complicated and difficult to describe, but since it is the difference between the initial and final positions which characterises a symmetry, the complexity of the motion is irrelevant—it is the effect which is important. To describe a symmetry we can choose a particular motion that produces the desired effect to represent the symmetry operation. The conventional motion is that of simple rotation.

The notion of rotation is a natural one. The Earth rotates around an axis which joins the north and south poles; a wheel rotates about an axle. In these physical examples of rotations there is always a part which remains fixed during the motion. For example, a point on the axle does not leave its position whereas a point in the wheel describes a circle as it rotates around. A direct symmetry *operation* for a polyhedron is a rotation about an axis through a certain angle. The symmetry *element* associated with a symmetry operation is the set of points not affected by the operation—the points which remain fixed. For a rotation, the symmetry element is the axis. (Later, we shall see that the symmetry element for an indirect symmetry is a plane or a single point.)

Mathematically, a rotation is described by giving a straight line (which forms the axis) and an angle. For example, the motion that the Earth makes in one hour about its axis is a rotation of 15°. (Of course, it has also revolved about the sun during that time as well.) Leonhard Euler showed that every direct symmetry of a polyhedron can be achieved by a rotation about some axis. A polyhedron can have more than one axis of rotational symmetry; Euler showed that in such circumstances all the axes must intersect at a point in the centre of the polyhedron.

A rotational symmetry of a polyhedron carries it from its original position to a second position. Because this new position is indistinguishable from the original one, the symmetry operation can be applied again, and the polyhedron rotated to a third position—also indistinguishable from the original one. Indeed, if the symmetry is rotation by 180° then it will be identical to the original position. It is often helpful to regard the act of moving a polyhedron and placing it back in an identical position as a symmetry operation. Such an operation is called the *identity* symmetry.

The fact that a symmetry can be repeatedly applied to a polyhedron provides an alternative method of describing the magnitude of a rotation. Rather than specifying the size of the angle, we can state the number of times that the rotation needs to be repeated to return the polyhedron to its starting position. For example, if a rotation of 90° is applied four times, the total angle through which the polyhedron is rotated by the combined rotation is 360°—a complete turn. Thus a rotation of 90° is also called a 4-fold rotation because a four-fold application of the symmetry is the same as the identity symmetry. In general, a rotation of $(\frac{360}{n})°$ is called an n-fold rotation.

An axis of n-fold rotational symmetry is sometimes called an n-fold axis. Care has to be taken when labelling axes in this way because the same axis can be associated with different rotations. A 4-fold axis must also be a 2-fold axis, for example. It is the convention that the largest possible value of n is chosen to label the axis.

Now that the terminology of rotational symmetry has been established let us see what it means and how it is applied in practice.

Systems of rotational symmetry

We already know that the number of distinct but indistinguishable positions in which a polyhedron can be placed is a measure of the *degree* of symmetry. Identifying the axes of rotational symmetry in a polyhedron gives us a way of distinguishing different *kinds* of symmetry. Thus a hexagonal prism and a regular tetrahedron have different types of symmetry even though they have the same degree of symmetry.

In order to gain a good understanding of the various kinds of rotational symmetry it is helpful to locate the positions of the axes on models of polyhedra. In simple cases, where there is a single axis, a picture may suffice but, as the number of axes increases, it becomes more difficult to imagine the way in which they are interrelated. In the case of the Platonic solids, in which there are many axes, the rotational symmetries form a complex, strongly interconnected system.

In the following examples, a variety of polyhedra are examined and the combinations of rotation axes that occur are recorded. You may find it helpful to rotate models of polyhedra between your fingertips as you read the text.

Cyclic symmetry.

The simplest system of rotational symmetry that a polyhedron can have is exhibited by the pyramids; they have a single axis of rotational symmetry. A hexagonal-based pyramid, for example, has an axis which passes through the apex and the centre of the base (Figure 8.1). The rotations about this axis which carry the pyramid onto itself are all the multiples of $(\frac{360}{6})°$, namely 60°, 120°, 180°, 240°, 300°, and 360°—the last symmetry being the identity. This kind of symmetry is called

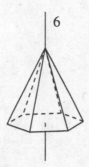

Figure 8.1. A rotation axis in a cyclic system.

cyclic. The pyramid is said to have 6-fold cyclic symmetry, or just C_6 symmetry for short. If the pyramid has a regular n-gon as its base then it has a single axis of n-fold rotational symmetry and has C_n symmetry.

Dihedral symmetry.

The triangular prism is another polyhedron which, like the hexagonal pyramid, has six rotational symmetries . However, unlike the pyramid, the prism has more than one axis of rotational symmetry. In fact, it has four such axes. One axis passes through the centres of the two triangular faces (as shown on the left of Figure 8.2). A rotation of 120° about this axis carries the prism to a different but indistinguishable position. Repeating this operation twice more returns the prism to its starting position as three rotations of 120° are the same as a rotation of 360° or the identity symmetry. Therefore, this axis is a 3-fold axis.

The other rotational symmetries of the prism turn it over so that the two triangular faces are interchanged. The rotations which invert the prism in this way have axes that pass through the centre of one of the rectangular faces and the midpoint of the opposite edge (Figure 8.2). There are three such axes and they are all 2-fold axes.

These then are the four axes of symmetry of a triangular prism. The 3-fold axis is called the *principal* axis of the system, and the three *secondary* 2-fold axes lie in a plane perpendicular to it. It is the principal axis which gives the directional nature to the symmetry of polyhedra such as prisms and dipyramids. This type of symmetry is called *dihedral* and is denoted D_n where n is the order of the principal axis. In the example given here, the prism has D_3 symmetry since the principal axis is 3-fold. In the general case of a prism with a regular n-gon base, the principal axis is an axis of n-fold rotational symmetry and there are n secondary axes of 2-fold symmetry lying in a plane perpendicular to it which are all equally spaced around it at intervals of $(\frac{180}{n})°$.

A closer study reveals that when n is odd (as for the triangular prism) all of the secondary axes are equivalent. However, when n is even, they separate out

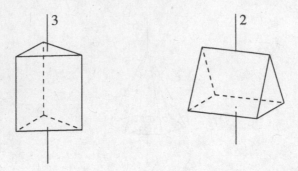

Figure 8.2. Principal and secondary axes in a dihedral system.

into two sorts. This can be seen in a hexagonal-based prism: half of the 2-fold axes join the centres of opposite faces, the others join the midpoints of opposite edges (Figure 8.3). When n is odd any secondary axis can be carried on to the position of any other by a rotation about the principal axis, but when n is even the two sorts of secondary axis cannot be interchanged in this way.

A special case of dihedral symmetry occurs when the principal axis is a 2-fold axis. In such circumstances it is not sensible to speak of a 'principal' axis since all the three axes are 2-fold, and they are mutually perpendicular—there is nothing to distinguish one axis from another. Examples of polyhedra with this type of symmetry are shown in Figure 8.4. Another example is a prism on a rectangular base whose width, length and height measurements are all different. This type of symmetry is usually classified along with the other dihedral types of symmetry and is labelled D_2. However, it should be remembered that in this case all of the axes have equal status.

Besides D_2, there are other types of symmetry where there is no preferred axis. These types of symmetry are collectively known as *spherical* (or *polyhedral*)

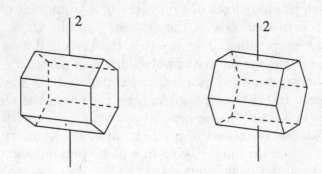

Figure 8.3. When n is even, the secondary axes in D_n can be separated into two kinds.

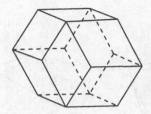 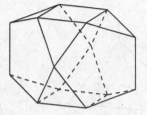

Figure 8.4. Polyhedra with D_2 symmetry.

types.[1] The Platonic solids are examples of polyhedra with spherical symmetry and we now turn our attention to some of these.

Tetrahedral symmetry.

The regular tetrahedron has seven axes of rotational symmetry—four 3-fold axes and three 2-fold axes (Figure 8.5). Each 3-fold axis passes through the centre of a face and the opposite vertex. Each 2-fold axis passes through the midpoints of opposite edges. Any polyhedron which has this system of rotational symmetries is said to have *tetrahedral* symmetry. This type of symmetry is denoted by the label T.

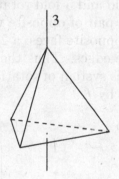 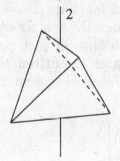

Figure 8.5. Rotation axes in the tetrahedral system.

Octahedral symmetry.

The regular octahedron has three sets of rotation axes (Figure 8.6). Firstly, there are three mutually perpendicular axes of 4-fold symmetry. Each of these passes through two opposite vertices. There are also four 3-fold axes each passing through the centres of a pair of opposite faces. Lastly there are six axes of 2-fold symmetry. These axes pass through the midpoints of pairs of opposite edges.

---

[1]Some mathematicians find it convenient to include the cyclic and dihedral cases as spherical systems. In the more restricted meaning used here, only the D_2 type is spherical; the term 'prismatic' is applied to the other systems .

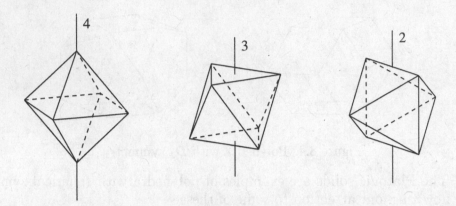

Figure 8.6. Rotation axes in the octahedral system.

Any polyhedron which has this system of rotational symmetries has *octahedral* symmetry. This system is denoted by O.

Icosahedral symmetry.

The regular icosahedron has axes of 2-fold, 3-fold and 5-fold rotational symmetry (Figure 8.7). A 5-fold axis passes through each pair of opposite vertices; a 3-fold axis passes through the centres of each pair of opposite faces; a 2-fold axis passes through the midpoints of each pair of opposite edges. Thus there are six 5-fold axes, ten 3-fold axes and fifteen 2-fold axes. This system of rotational symmetries is called the *icosahedral* system and is denoted by I.

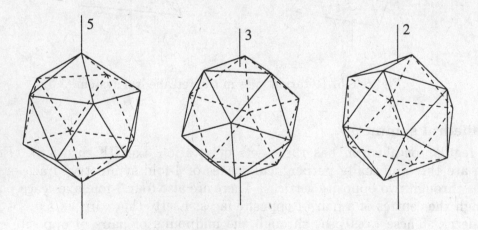

Figure 8.7. Rotation axes in the icosahedral system.

How many systems of rotational symmetry are there?

So far, every polyhedron that has been examined has given rise to a new type of symmetry. The pyramids, prisms, and some Platonic solids all have different systems of rotational symmetry. It is quite surprising to find that these are the only systems of rotational symmetries which a polyhedron can have. Any polyhedron, if it has rotational symmetry at all, must have one of the five types described above: cyclic, dihedral, tetrahedral, octahedral, or icosahedral. A cube, for example, has the same symmetry as an octahedron. So do some of the Archimedean solids such as the snub cube and the rhomb-cub-octahedron. Kepler's star-faced polyhedra have icosahedral symmetry. Miller's solid has dihedral symmetry.

Before proving that there are no other systems of rotational symmetry, it will be helpful to study the way in which the rotations in a particular system are related to each other. The tetrahedral system is used as an example.

The two points where the Earth's axis pierces its surface are called poles. By analogy, the set of points where the axes of rotational symmetry pierce a polyhedron are also called *poles*. Each axis punctures the polyhedron at two points so there are twice as many poles as axes. The poles can be split up into different kinds. Firstly, a pole which lies on an n-fold axis is called an n-pole. The tetrahedron, for example, has 2-poles and 3-poles. All the 2-poles are the same but the 3-poles can be separated into two sets: four 3-poles are situated at the face-centres, four more 3-poles lie at the vertices. Two poles are said to be *equivalent* if there is a rotational symmetry of the polyhedron which carries one pole to the other. So, for example, all the 2-poles of a tetrahedron are equivalent, but its 3-poles fall into two equivalence classes as no symmetry carries a vertex onto a face-centre.

The number of poles in an equivalence class is related to the total number of symmetries and the kind of pole. The number of 2-poles in the tetrahedral system is $\frac{12}{2} = 6$. The number of poles in each of the two equivalence classes of 3-poles is $\frac{12}{3} = 4$. In general if a polyhedron has a total of N rotational symmetries (including the identity symmetry) then an equivalence class of n-poles contains $\frac{N}{n}$ poles. To see that this is the case, place a point close to an n-pole and apply each symmetry to the point in turn marking the image point on the polyhedron. In Figure 8.8 this has been done with the 3-poles at the face-centres of a tetrahedron. This results in a total of N points being placed on the polyhedron arranged in groups of n around each pole in the equivalence class. Thus there are $\frac{N}{n}$ poles.

Knowledge of the number of poles in each equivalence class enables the number of rotational symmetries to be calculated. Ignoring the identity symmetry, there are $(n-1)$ rotations associated with an n-pole. These are the rotations of

$$\frac{1}{n} \times 360°, \quad \frac{2}{n} \times 360°, \cdots, \quad \frac{n-1}{n} \times 360°.$$

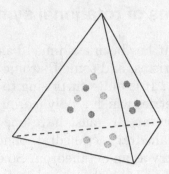

Figure 8.8.

For a 3-pole these are 120° and 240°. In the tetrahedral system the sum

(number of 2-poles) × (number of rotations associated with each 2-pole)
+ (number of 3-poles) × (number of rotations associated with each 3-pole)

equals twice the number of (non-identity) rotations because each rotation has been counted twice—once at each pole on its axis. Now, the number of n-poles in an equivalence class is $\frac{12}{n}$, and the number of rotations associated with an n-pole is $(n-1)$. So, by separating the term concerning 3-poles into its two classes of poles, the above sum becomes

$$\left(\frac{12}{2}\right) \times (2-1) \ + \ \left(\frac{12}{3}\right) \times (3-1) \ + \ \left(\frac{12}{3}\right) \times (3-1)$$

which evaluates to 22—twice the number of non-identity rotations.

Repeating this analysis in the general case leads to an equation relating the number of rotations in a system to the types of poles (and hence types of rotations) that the system can contain. By finding all the solutions of this equation, all the different systems of rotational symmetry can be deduced.

Theorem. The system of rotational symmetries of a polyhedron must be one of the following types: cyclic, dihedral, tetrahedral, octahedral or icosahedral.

PROOF: Consider a general system of rotational symmetries and assume that it contains a total of N rotations. Choose a polyhedron which exhibits this type of symmetry and examine the equivalence classes of poles. An equivalence class of n-poles contains $\frac{N}{n}$ poles, and associated to each n-pole there are $(n-1)$ non-identity rotations. Adding up all of these rotations associated to the poles gives twice the number of non-identity rotations since each is counted twice. Thus

$$2(N-1) \ = \ \sum_{\text{poles}} \frac{N}{n}(n-1)$$

which can be rewritten as

$$2 - \frac{2}{N} = \sum_{\text{poles}} \left(1 - \frac{1}{n} \right). \tag{A}$$

If $N = 1$ then the only symmetry in the system is the identity symmetry. So we can assume that N is at least 2. This means that

$$1 \leqslant 2 - \frac{2}{N} < 2.$$

The values of n in the summation must all be at least 2 (why?), hence

$$\tfrac{1}{2} \leqslant 1 - \frac{1}{n} < 1.$$

Suppose that the system of rotations contained only one equivalence class of poles. Then the right hand side of (A) would be less than 1, which contradicts the fact that the left-hand side of (A) is at least 1. Therefore, there must be at least two classes of poles.

Also, the system of rotations cannot contain four or more equivalence classes of poles for then the left-hand side of (A) is less than 2 while the right-hand side of (A) is

$$\sum_{\text{poles}} \left(1 - \frac{1}{n} \right) \geqslant \frac{1}{2} + \frac{1}{2} + \frac{1}{2} + \frac{1}{2} = 2.$$

Thus the possibilities are narrowed down to either two or three equivalence classes of poles.

Suppose that the system contains two equivalence classes of poles—one of p-poles and one of q-poles. Then substituting these values into (A) gives

$$2 - \frac{2}{N} = \left(1 - \frac{1}{p} \right) + \left(1 - \frac{1}{q} \right).$$

Rearranging gives

$$2 = \frac{N}{p} + \frac{N}{q}.$$

Now, the equivalence class of p-poles contains $\frac{N}{p}$ poles and hence $\frac{N}{p}$ is an integer not equal to zero. Similarly $\frac{N}{q}$ is a non-zero integer. Therefore $\frac{N}{p} = \frac{N}{q} = 1$ and hence $N = p = q$. Thus the system has two poles (one in each equivalence class) and, therefore, a single axis. Both poles are p-poles so the axis is p-fold, and the system of symmetry is cyclic, C_p.

Now consider the remaining case when there are three equivalence classes of poles. Suppose these are p-poles, q-poles and r-poles. Then substituting these into (A) gives

$$2 - \frac{2}{N} = \left(1 - \frac{1}{p}\right) + \left(1 - \frac{1}{q}\right) + \left(1 - \frac{1}{r}\right).$$

Rearranging gives

$$1 + \frac{2}{N} = \frac{1}{p} + \frac{1}{q} + \frac{1}{r}.$$

If all of p, q and r were at least 3 then

$$\frac{1}{p} + \frac{1}{q} + \frac{1}{r} \leqslant \frac{1}{3} + \frac{1}{3} + \frac{1}{3} = 1.$$

However, $1 + \frac{2}{N}$ is always greater than 1, so at least one of p, q, or r must be 2. Choose $r = 2$ and assume also that $p \geqslant q$. Thus

$$1 + \frac{2}{N} = \frac{1}{p} + \frac{1}{q} + \frac{1}{2}. \tag{B}$$

Rearranging this expression gives

$$(p - 2)(q - 2) = 4\left(1 - \frac{pq}{N}\right) < 4 \tag{C}$$

which implies that $(p - 2)(q - 2)$ is 0, 1, 2 or 3.

In the first case, when the product is zero, $q = 2$. Substituting this into (B) gives $\frac{1}{p} = \frac{2}{N}$, which implies $N = 2p$. Thus this system has three equivalence classes of poles, one of p-poles, and two others of 2-poles. Each class of 2-poles contains $\frac{N}{2} = p$ poles. This is the dihedral system D_p.

The other solutions to (C) are $(p, q) = (3, 3), (4, 3)$ and $(5, 3)$. These solutions correspond to the systems T, O and I respectively. There are no other possibilities. ■

Reflection symmetry

The symmetry operations discussed in the previous sections were all direct symmetries—they could all be physically applied to a model. The symmetries to be considered now are *indirect* and require a mirror to visualise the results. The polyhedron shown in Figure 8.9 does not have any rotational symmetry but it is not asymmetric. It has *bilateral* symmetry—the sort of symmetry seen approximately in a human face. This means it can be split into two halves which are mirror images of each other. The plane which separates the polyhedron in this way is called a *plane of symmetry* or a *mirror plane*. The polyhedron has mirror or *reflection* symmetry.

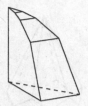

Figure 8.9. A polyhedron with bilateral symmetry.

Most everyday objects have reflection symmetry and objects without it, such as a corkscrew, can look twisted. Polyhedra which do not possess reflection symmetry are called *cheiral* (from the Greek word for 'hand'). They come in two forms which are related like a left and right hand—they are mirror images of each other. Such a pair of polyhedra are called *enantiomorphs*. The snub cube and the compound of five tetrahedra are two examples of cheiral polyhedra.

Bilateral symmetry is the simplest type of reflection symmetry as there is only one mirror plane. This type of symmetry is given the label C_s. (The subscript comes from 'Spiegel', the German word for mirror.)[2] When a polyhedron has more than one mirror plane, it must have rotational symmetry as well because the line in which the mirrors meet acts as a rotation axis. If a polyhedron is reflected first in one mirror plane and then in another, it is changed from being right handed to left handed and back to right handed again, so the composite effect must be a rotation. Thus all the types of symmetry (except C_s) which contain reflection symmetries must also contain one of the systems of rotational symmetries described above.

Prismatic symmetry types

The types of symmetry described in this section have a principal axis of rotation and their rotational symmetry type is either cyclic or dihedral. Examples of polyhedra exhibiting each of these types of symmetry can be made by decorating prisms with various patterns. Thus these symmetry types are collectively known as *prismatic*. The decorations can take the form of selective augmentation or truncation, making incisions or cutting out notches, or simply by painting on a design.

Patterns are occasionally seen on the faces of crystals. The facets are corrugated having tiny ridges and channels running across them. Such markings are called striae or striations. Examples can be seen on some facets of the pyrite crystals shown in Plate 2. These patterns reveal the symmetry of the underlying structure of the crystals.

---

[2]Several alternative labelling schemes are compared in Appendix 1.

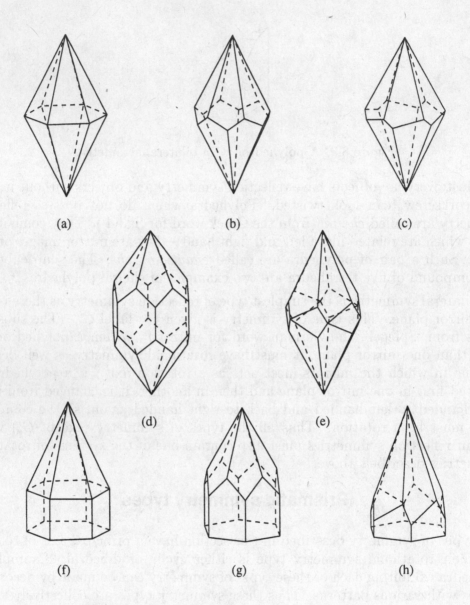

(a)   (b)   (c)

(d)   (e)

(f)   (g)   (h)

Figure 8.10. Polyhedra with prismatic symmetry.

The best way to gain a good understanding of the different symmetry types is to study a set of models. A set of small hexagonal prisms is easily made from thin card and patterns can be drawn on their surfaces so that they exhibit the various types of symmetry. Careful examination of such models allows the symmetries in each system to be identified, and the various systems to be compared with one another so that their differences become apparent. The reader is strongly encouraged to make such a set.

Symmetry type D_{nh}.

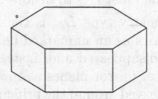

Figure 8.11.

First, the set of symmetries of an unmarked prism will be examined. A hexagonal-based prism will be used as a specific example (Figure 8.11). Its rotational symmetry type is dihedral since it has a principal axis of 6-fold rotational symmetry and six secondary axes of 2-fold rotational symmetry. Assume that the prism is resting on a hexagonal face so that the principal axis is vertical. Then there is a mirror plane which lies horizontally and cuts the prism in half. Other mirror planes are arranged vertically and they each contain the principal axis. The lines where the six vertical mirror planes meet the horizontal mirror plane are axes of rotational symmetry and form the 2-fold axes of the dihedral system. This is the complete system of mirror planes and rotation axes of an unmarked prism.

The label given to this system is D_{6h}. The D_6 refers to the type of rotational symmetry, and the subscript h refers to the existence of a horizontal mirror plane. The horizontal mirror cannot be the only mirror plane and the symmetry elements named in the label force the existence of the vertical mirrors. Another polyhedron which has this symmetry type is the hexagonal dipyramid (see Figure 8.10(a)).

Symmetry type D_{nv}.

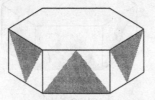

Figure 8.12.

If a hexagonal prism is decorated with chevrons that point up and down on alternate faces round the prism, as shown in Figure 8.12, then some of the symmetry elements of the unmarked prism are destroyed. The horizontal plane is no longer a plane of symmetry, neither are the vertical planes which pass through opposite edges. The 2-fold axes which join opposite face centres have been destroyed as well. The principal axis has been reduced to an axis of 3-fold rotational symmetry. This system of symmetry elements is labelled D_{3v}. The system of rotational symmetries is D_3 and the subscript v indicates the existence of vertical mirror planes. (Some authors use the label D_{3d} to denote this system.)

A polyhedron with symmetry type D_{6v} is the hexagonal antiprism. It has the same rotational symmetries as an unmarked hexagonal prism. However, the horizontal mirror plane has disappeared and, instead of the 2-fold rotation axes being contained in the vertical mirror planes as they are in the D_{nh} system, the axes and the planes are interleaved around the principal axis. Another polyhedron with D_{6v} symmetry is the hexagonal isosceles trapezohedron (see Figure 8.10(b)).

Symmetry type D_n.

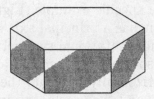

Figure 8.13.

If a hexagonal prism is decorated as shown in Figure 8.13 then all the reflection symmetries are destroyed but the complete system of rotational symmetries remains intact. This system of pure rotational symmetry is labelled D_6. The hexagonal scalene trapezohedron (Figure 8.10(c)) has this symmetry type.

Symmetry type C_{nv}.

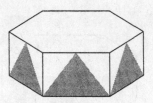

Figure 8.14.

The hexagonal prism can be decorated as shown in Figure 8.14 where all of the chevrons point upwards. In this case, the 2-fold rotations cease to be symmetries. This means that the rotational symmetry type has been reduced from dihedral to cyclic. The vertical mirror planes are still symmetries but the horizontal plane is not. (If both vertical and horizontal mirror planes were present then the 2-fold axes would also be present.) This type of symmetry is labelled C_{6v}. The C_6 shows the rotational symmetry type and the v refers to the vertical mirror planes. An elongated hexagonal pyramid has this type of symmetry (Figure 8.10(f)).

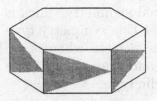

Figure 8.15.

Symmetry type C_{nh}.

If the pattern shown in Figure 8.15 is applied to the prism then, again, the rotational symmetry is reduced to the cyclic system. This time the vertical reflection symmetries are destroyed and the horizontal mirror plane remains. This system is labelled C_{6h}. Figure 8.10(e) shows a polyhedron with this symmetry type.

Symmetry type C_n.

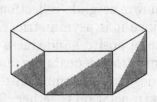

Figure 8.16.

When the prism is decorated as shown in Figure 8.16, all the symmetries except the axis of 6-fold rotational symmetry are destroyed. This system is just C_6. See Figure 8.10(h) for an example.

Problem. Identify the symmetry types of the polyhedra in Figures 8.10 (d) and (g).

Problem. Identify the symmetry types of the five octagonal-based prisms which are formed by applying each of the following patterns in turn to their rectangular faces:

Do any of the other letters of the alphabet give a symmetry type which is different from each of these?

Compound symmetry and the S_{2n} symmetry type

What is the symmetry type of the hexagonal-based prism decorated with a pattern like the one shown in Figure 8.17? It has no planes of reflection symmetry and only one axis of rotational symmetry which is 3-fold. So, in the above list, it has symmetry type C_3. Yet this symmetry type recognises only the fact that alternate faces have the same pattern whereas, in fact, the same motif appears on all the rectangular faces, and there seems to be a higher degree of symmetry than the straightforward cyclic type.

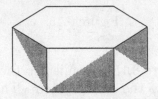

Figure 8.17.

There is an analogy between this kind of symmetry of a solid and the symmetry of a linear pattern known as glide-reflection. Suppose that we choose a scalene triangle as a motif since it is asymmetric. Then reflection symmetry repeats the motif but with the opposite handedness (see Figure 8.18). A second kind of symmetry (called translation) repeats the motif in an equally spaced linear pattern. These two symmetries can be combined to produce a single compound operation—a reflection and a translation together. To carry one copy of the motif onto the next one along, you translate it and then reflect it. This motion is called a *glide-reflection*. It is the symmetry of a line of footprints on a beach.

Now, let us take a section of the linear pattern with glide-reflection symmetry and join the ends together to form a loop (see Figure 8.19). If the section contains six copies of the triangle then the resulting loop will have the same symmetry as the decorated hexagonal prism above. The translational symmetry of the strip pattern has been converted into rotational symmetry of the loop, and the glide line of the strip pattern is now a plane. The symmetry which carries one face onto the next is a combination of a rotation about the principal axis by 60° followed by a reflection in a horizontal mirror plane. This operation is a compound symmetry and is called a *rotation-reflection*. The principal axis is called an axis of rotation-reflection, and in this example it is a 6-fold axis since the operation has to be carried out six times before the polyhedron returns to its original position. The label given to this type of symmetry is S_n. The example here has S_6 symmetry.

For this type of symmetry, n must be an even number otherwise it is the same as a C_{nh} type of symmetry. When n is 2, the strip has only two motifs on it, and there is no rotational symmetry. In this case, the symmetry is often described by a different operation, which is called *central inversion* or *reflection in a point*.

To see why the latter name is appropriate consider what happens when an object is reflected in a plane. The distance from the mirror plane to a point in

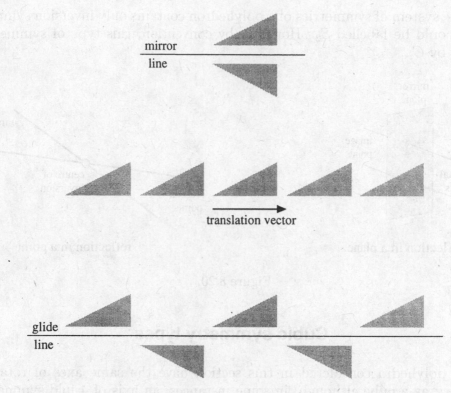

Figure 8.18. A glide-reflection is a compound symmetry made up of a translation and a reflection.

the object equals the distance from the mirror plane to the point's reflection in the mirror image (Figure 8.20). In an S_2 symmetry, there is a central point which functions in the same way as a mirror plane: the distance from the centre to a point in the object equals the distance from the centre to the point's image. This symmetry operation need not occur in isolation and is included in other, more complex, systems of symmetry. The cube, for example, has inversion symmetry. The regular tetrahedron, on the other hand, does not.

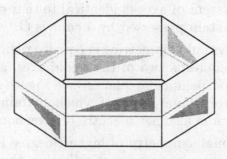

Figure 8.19.

If the system of symmetries of a polyhedron contains only inversion symmetry then it could be labelled S_2. However, by convention, this type of symmetry is denoted by C_i.

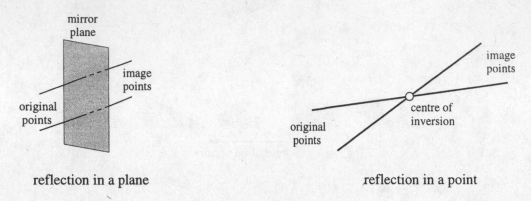

reflection in a plane   reflection in a point

Figure 8.20.

Cubic symmetry types

All the polyhedra considered in this section have the same axes of rotational symmetry as a cube although, in some instances, an axis of 4-fold symmetry is reduced to a 2-fold axis. As in the case of prisms, each of the types of cubic symmetry can be illustrated by decorating a cube with an appropriate pattern. It will be helpful to make some small cubes and apply patterns to them so that the differences in symmetry can be clearly seen.

Symmetry type O_h.

Firstly, the complete set of symmetry elements of an undecorated cube will be described. There are three axes of 4-fold rotational symmetry which join the centres of opposite faces; four axes of 3-fold rotational symmetry which join diagonally opposite vertices; and six 2-fold rotation axes joining the midpoints of opposite edges. This system of axes is identical to that of the octahedron, so the rotational symmetry system possessed by a cube is O.

The system of reflection symmetries is equally rich. There are three mirror planes, each of which contains two of the 4-fold axes, and which together form a system of mutually perpendicular planes (see Figure 8.21(a) and (b)). There are also six other mirror planes, each of which contains two of the 3-fold axes (Figure 8.21(c)). The cube also has a centre of inversion.

The axes of rotational symmetry of highest degree in this system are 4-fold. If one of these primary axes is placed vertically then there is a plane of reflection symmetry which lies horizontally. This system of symmetry is given the label O_h.

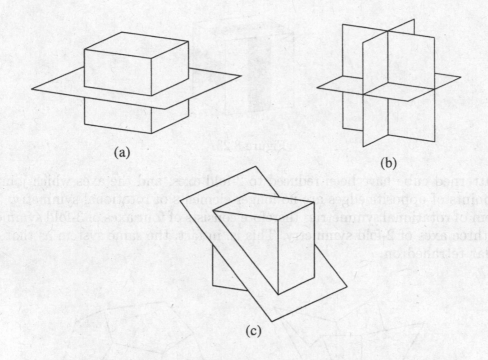

(a)

(b)

(c)

Figure 8.21. The reflection planes of a cube.

Symmetry type O.

If the pattern shown in Figure 8.22 is applied to the faces of the cube so that all the 3-fold axes of rotational symmetry are preserved, then the whole system of rotational symmetry is preserved as a consequence. However, all of the reflection symmetries are destroyed. So the label of this symmetry type is O. The snub cube has this type of symmetry, as does the octahedron covered with the tessellation of birds (Figure 2.5).

Figure 8.22.

Symmetry type T_h.

Suppose that the cube is decorated with the pattern shown in Figure 8.23. Again, the pattern respects the axes of 3-fold rotational symmetry. The other rotational symmetries have been altered or destroyed: the axes of 4-fold symmetry of the

Figure 8.23.

unpatterned cube have been reduced to 2-fold axes, and the axes which join the midpoints of opposite edges are no longer elements of rotational symmetry. The system of rotational symmetries therefore consists of four axes of 3-fold symmetry and three axes of 2-fold symmetry. This is, in fact, the same system as that of a regular tetrahedron.

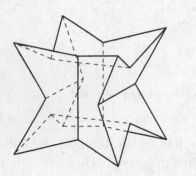
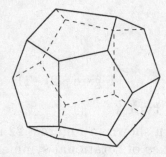

Figure 8.24. Polyhedra with T_h symmetry.

The decorated cube still retains some reflection symmetry. The three mutually perpendicular mirror planes are still planes of reflection symmetry; the other planes do not function as symmetry elements. This symmetry type is labelled T_h. The two dodecahedra shown in Figure 8.24 are examples of polyhedra with this symmetry type.

Polyhedra with this symmetry type also have inversion symmetry. This feature can be used to distinguish easily between T_h and the following symmetry system which does not contain a centre of inversion.

Symmetry type T_d.

A cube decorated as shown in Figure 8.25 also has the tetrahedral system of rotational symmetries. In this case, however, it is the orthogonal planes of reflection symmetry that are destroyed and the others which are retained. This symmetry type is labelled T_d. It is the complete symmetry type of a regular tetrahedron.

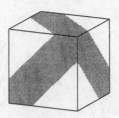

Figure 8.25.

Symmetry type T.

When the pattern shown in Figure 8.26 is used to decorate the cube, all the reflection symmetries are destroyed. The only remaining symmetries are the set of rotations of a tetrahedron. This symmetry type is labelled T. A polyhedron with this system of symmetry can be formed by suitably twisting and deforming an icosahedron to produce a 'snub tetrahedron' (Figure 8.27).

Icosahedral symmetry types

The analysis of the symmetry types of polyhedra having the icosahedral symmetry system is much easier than the preceding analyses of the prismatic and cubic systems. In fact, there are only two types. One of these contains planes of reflection symmetry and is labelled I_h; the other contains rotational symmetries only and is labelled I. The dodecahedron and the four Kepler–Poinsot star polyhedra have symmetry type I_h. The snub dodecahedron, the compound of five tetrahedra (Plate 12), and the icosahedron decorated with fish motifs (Figure 2.5) have symmetry type I.

Problem. The system of rotational symmetries of an icosahedron (I) was described above. Find the planes of reflection symmetry of a regular icosahedron and thus describe the system I_h.

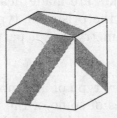

Figure 8.26.

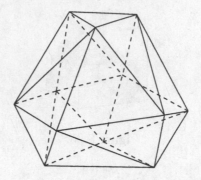

Figure 8.27. A 'snub tetrahedron'.

Determining the correct symmetry type

The above list of symmetry types of polyhedra is complete except for one further type—the system containing no symmetry operations at all. In such a situation the polyhedron is called *asymmetric*. The symmetry type is labelled C_1 since an axis of 1-fold symmetry implies that a polyhedron has to be rotated through a complete turn of 360° before its position is indistinguishable from its starting position. Every line meeting an asymmetric polyhedron is an axis of 1-fold symmetry and these are the only 'symmetries' it has.

So, a polyhedron can have one of 17 types of symmetry. (The prismatic types are actually families containing an unlimited number of closely related systems.) The 17 types of symmetry are

$$C_1, \quad C_i, \quad C_s,$$
$$C_n, \quad C_{nv}, \quad C_{nh}, \quad D_n, \quad D_{nv}, \quad D_{nh}, \quad S_n,$$
$$T, \quad T_d, \quad T_h, \quad O, \quad O_h, \quad I, \quad I_h$$

With so many classes, it is important to be able to correctly identify the symmetry type of a polyhedron in a straightforward manner. A simple scheme for doing this is shown in Figure 8.28. By answering a series of simple yes/no questions about the polyhedron in question, a path is traced through the decision tree which (provided no mistakes are made) leads to the correct symmetry type being obtained. When one of the prismatic types is obtained and the principal axis has n-fold rotational symmetry, the n part of the label should be replaced appropriately.

Problem. Find the symmetry type of any models which you have made. Also, try to find the symmetry types of some of the polyhedra depicted in this book. For example: the Archimedean solids, the deltahedra, and other regular-faced polyhedra (Chapter 2), and the rhombic polyhedra (Chapter 4).

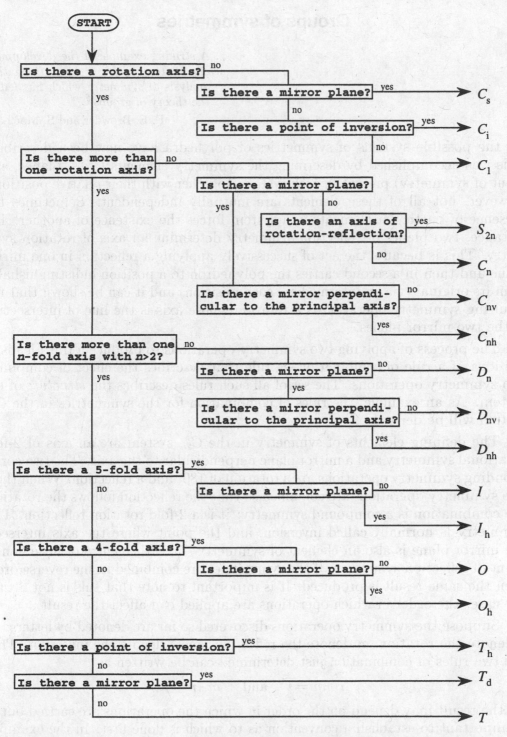

Figure 8.28. A decision tree to determine the symmetry type of a polyhedron.

Groups of symmetries

*A striking example of the development
of a Platonic vision is the mathematical
analysis of symmetry which has led to
the theory of groups.[b]*

F. E. Browder and S. Mac Lane

All the possible systems of symmetries of polyhedra have now been described. This was accomplished by describing the symmetry elements (axes, planes, and point of symmetry) present in each system, together with their relative positions. However, not all of these elements are mutually independent. Sometimes the presence of particular elements in a system forces the existence of another. For instance, two planes of reflection symmetry determine an axis of rotation symmetry. This is because the act of successively applying a reflection in one mirror plane and then in a second carries the polyhedron to a position indistinguishable from its original (via an intermediate such position) and it can be shown that the resulting symmetry operation is a rotation whose axis is the line of intersection of the two mirror planes.

The process of applying two symmetry operations to produce a third leads to the idea of a rule of combination—a rule that describes the effect of combining two symmetry operations. The set of all such rules describes the *structure* of the system. As an example, the rules of combination for the symmetries in the C_{2h} system will be derived.

The defining elements of symmetry in the C_{2h} system are an axis of 2-fold rotational symmetry and a mirror plane perpendicular to the axis. The two corresponding symmetry operations are a rotation of 180° and a reflection. When these two symmetry operations are combined so that the reflection follows the rotation, the combination is a compound symmetry: it is a 2-fold rotation-reflection. This symmetry is normally called inversion, and the point where the axis intersects the mirror plane is also an element of symmetry—it is the centre of inversion or point of reflection symmetry. If the symmetries are combined in the reverse order then the same result is produced. It is important to note that this is not always the case: the order in which operations are applied can affect the result.

Suppose the symmetry operations discovered so far are denoted by letters: let r denote the rotation, m denote the reflection, and i denote the inversion. Then the two rules of combination just determined can be written as

$$r \cdot m = i \quad \text{and} \quad m \cdot r = i.$$

As the result may depend on the order in which the operations are carried out, it is important to establish a convention as to which is done first. In the examples here, the leftmost operation is applied first. (The convention chosen varies from one author to another.)

Other rules of combination can be worked out. For example, what are the results of $r \cdot i$, $i \cdot m$ and $m \cdot m$? In fact, none of these combinations leads to new elements of symmetry; the resulting operation is always r, m, i or the identity symmetry. If the reflection is applied to a polyhedron twice in succession then the polyhedron returns to its original position. Denoting the identity symmetry by 1, this rule can be written

$$m \cdot m = 1.$$

The symbol 1 is chosen because of the analogy with combining numbers using multiplication. Multiplying any number by 1 has no effect. Likewise, combining any symmetry with the identity symmetry leaves it unchanged. Thus

$$1 \cdot m = m, \qquad r \cdot 1 = r, \qquad i \cdot 1 = i.$$

All of these rules of combination can be summarised in a table which gives the result of combining any two operations. Such a table is very like the multiplication tables which school children use. However, there is one important difference: the product of two numbers does not depend on their order whereas changing the order of two symmetry operations can produce different results. We shall use the convention that the symbols at the side of the table indexing the rows are the first operation and the symbols across the top of the table indexing the columns are the second. The table below shows the rules for C_{2h}.

C_{2h}	1	r	m	i
1	1	r	m	i
r	r	1	i	m
m	m	i	1	r
i	i	m	r	1

A table of rules can be worked out for any symmetry system. Two more examples deal with the cases of the S_4 system and the D_2 system.

In the S_4 system there is a single defining symmetry element: an axis of rotation-reflection symmetry. Let s denote the symmetry operation of rotating a polyhedron by 90° about this axis then reflecting in a mirror plane perpendicular to the axis. If this operation is performed twice in succession, it is equivalent to rotating the polyhedron by 180° about the axis. If r denotes a 2-fold rotation about the axis then $s \cdot s = r$. The combination of r and s is a rotation of 270° about the axis followed by a reflection and this operation is denoted by t. Thus $r \cdot s = t$. Applying s for a fourth time results in a rotation of 360° so $t \cdot s = 1$. The complete set of rules for these operations is shown in the table.

S_4	1	s	r	t
1	1	s	r	t
s	s	r	t	1
r	r	t	1	s
t	t	1	s	r

The symmetry elements in the D_2 system are three axes of 2-fold rotation which are mutually perpendicular. These axes can be labelled X, Y and Z as shown below. The operation of rotating a polyhedron by 180° about the X-axis can be denoted x. Likewise, y and z denote the 2-fold rotations about the Y and Z axes respectively. The combination table for this system is then

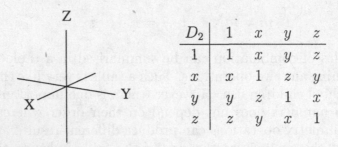

D_2	1	x	y	z
1	1	x	y	z
x	x	1	z	y
y	y	z	1	x
z	z	y	x	1

These tables, which provide a description of the structure of a symmetry system, are examples of the abstract mathematical concept called a *group*. A group is formed from a set of objects and a procedure which tells you how to combine any two objects to produce another. In our case, the objects of the group are symmetry operations and the combination procedure is 'apply the two symmetries one after the other'. For a set and its associated combination procedure to form a group, it must have several specific properties.

Firstly, the combination of any two objects must result in another object in the set. For example, the set of reflection symmetries of a cube do not form a group since the composition of two reflections is not a reflection but a rotation. The set of all rotational symmetries of a cube is a group since the composition of any two rotations always results in another rotation.

Secondly, there must be an object which acts like the identity symmetry. This object is usually denoted by 1. Then, if g is any other object in the group, both of the following rules must be true:

$$g \cdot 1 = g \quad \text{and} \quad 1 \cdot g = g.$$

The third property that a group must have is exhibited in the groups of symmetries by the fact that any symmetry operation can be undone. The effect of a symmetry can be neutralised by applying a second symmetry so that their combined result is the identity symmetry. For example, in the S_4 system, the operation denoted by s is undone by the operation t since $s \cdot t = 1$ (check the table). This property is expressed abstractly by requiring that, for every object g in the group, there exists an object h so that $g \cdot h = 1$.

The last property of a group allows lengthy expressions involving many compositions to be evaluated in any order. (This is not the same as allowing the order of the objects to be changed.) Using the S_4 system as an example again, the expression $r \cdot s \cdot t$ can be evaluated in two different ways. These are

(i) Evaluate the left-hand composition first: $(r \cdot s) \cdot t$

$$r \cdot s = t \qquad \text{and then} \qquad t \cdot t = r.$$

(ii) Evaluate the right-hand composition first: $r \cdot (s \cdot t)$

$$s \cdot t = 1 \qquad \text{and then} \qquad r \cdot 1 = r.$$

Both give the same result as required.

These four requirements on the objects of a set are the axioms of a group. When studying groups in this abstract sense, it is only the structure which is taken into account. The objects are not interpreted as any physical operation which can actually be carried out. All that is used is the knowledge of the way in which they combine. The tables which express this structure are called *group tables*. In this abstract sense, the structures of the C_{2h} and the D_2 systems of symmetries are the same. Their group tables have the same pattern and thus the same abstract structure. The groups are said to be *isomorphic*, meaning 'same form'. However, this structure is interpreted in two different ways. In the D_2 case, all of the objects are realised as rotations, but in the realisation as C_{2h}, indirect symmetries also appear.

The other group table found so far is that of the S_4 system of symmetries. This group is abstractly different from those of the D_2 and the C_{2h} systems. The patterns in the tables are different. This can be easily checked by noticing that in the D_2 group table, the composition of every symmetry with itself is the identity, whereas in the S_4 system there are only two such symmetries. This property of the patterns in the group tables shows that the groups are abstractly different and that the symmetry systems have different structures.

In groups with a large number of objects, the structure can be very complex. For example, the group table of the symmetry system I_h is an array of 120×120 elements. The abstract structure of the dihedral symmetry system D_3 is shown below. The objects of this group can be interpreted as the rotations of a triangular prism: the 2-fold rotations correspond to p, q and r while a and b are rotations round the principal axis. Notice that this group contains examples of composition where the order of the objects is important. For instance, $a \cdot p = r$ but $p \cdot a = q$.

D_3	1	a	b	p	q	r
1	1	a	b	p	q	r
a	a	b	1	r	p	q
b	b	1	a	q	r	p
p	p	q	r	1	a	b
q	q	r	p	b	1	a
r	r	p	q	a	b	1

People were using groups long before they had been formally defined. In the problems being studied, such as the symmetries of polyhedra, the axioms of a group are automatically satisfied. It was only after much experience of these and other similar structures that the four properties listed above were distilled out as their characteristic features and abstract groups were defined axiomatically. The term *group* was first used to describe such structures in 1869 by Camille Jordan (1838–1922), although he used only the first property, that of closure of the system, to define a group. He was studying groups of symmetries and the other three axioms follow automatically from the way that symmetries combine. He did notice that every symmetry had an 'equal and opposite' symmetry associated with it, that is, an inverse. In 1854, Arthur Cayley recognised the need for the fourth property (associativity) and the existence of an identity. He defined a group using abstract symbols and expressed rules of combination as a group table. In 1856, William Rowan Hamilton gave one of the first *presentations* of a group—a method of describing a group without writing out its complete group table. He gave a presentation of the icosahedral group which can be written on a single line (a considerable saving of space and effort when compared to the 60×60 group table) and called his system 'icosian calculus'. The four properties listed above, which are the modern axioms of a group, were first published in 1882 independently by both Walter von Dyck and Heinrich Weber.

The analysis of abstract groups is called group theory and the area now forms a very important branch of mathematics. It has applications in many disciplines including subatomic particle physics, molecular bonding schemes in chemistry, classification of patterns and ornamental designs, the description of different kinds of geometry, and crystallography.

Crystallography and the development of symmetry

The development of the mathematical theory of symmetry groups spanned the nineteenth century and much of the motivation came from investigations into the nature of crystals.

Crystals have always attracted man's attention. The flatness of their surfaces, their precise geometrical shapes, their translucency and light refracting properties distinguish them from the other amorphous forms found in nature. The Greeks confused the clear crystals of quartz with glaciers, thinking them to be water which had been congealed by the intense cold in the mountains. The Greek word 'krystallos' means ice.

Until the seventeenth century the only crystals known were those visible to the naked eye, and crystals were thought to be rare and something of a curiosity. The invention of the microscope enabled naturalists to see that many fine grained minerals contain crystals. The early chemists watched solids crystallising out of solutions, the plane faces and symmetrical shapes appearing spontaneously. Observations like these hinted at an orderliness to the structure of crystals.

One of the first to suggest that internal structure accounts for the external form was Robert Hooke (1635–1703). In his *Micrographia, or some Physiological Descriptions of Minute Bodies* (1665) he suggested that crystals were stacks of small spherical particles packed closely together. The fact that crystals fracture easily in certain directions leaving flat faces led Christiaan Huygens (1629–1695) to propose that these cleavage planes are natural divisions between sheets of particles. Domenico Guglielmini (1655–1710), a professor of mathematics at Bologna and Padua, observed that the directions of the cleavage planes were always the same in any given substance and he believed that the fundamental units from which a crystal is built must themselves be miniature crystals with plane faces.

In 1772, Jean-Baptiste Louis Romé de l'Isle (1736–1790) published his *Essai de Cristallographie*. In his view, the chief characteristic for classifying minerals was their external geometrical shape, and his essay contains detailed descriptions of over 100 crystal forms. In a later work (1783) he expanded this list to more than 450. Precise measurements were taken of each crystal described and this led to the discovery of the constancy of interfacial angles.

René Just Haüy (1743–1822) also wrote on the structure of crystals. In an essay that appeared in 1784 he put forward ideas which also arose from the contemplation of cleavage planes. There is a story that he accidentally dropped a large calcite crystal which belonged to a friend who thereupon presented him with the resulting fragments. Haüy noticed that the cleavage planes were not the same as the face planes and that if the crystal were split apart sufficiently, a small core would be left, all of whose faces were parallel to cleavage planes. In the case of calcite, this core is rhombohedral in shape. He went on to show how units of this shape could be stacked up to produce faces with different slopes by supposing that rows were omitted in a regular fashion in successive layers. The idea that a crystal was built from tiny identical building blocks arranged in an array that can be finished off in different ways accounted for the variety of external forms in crystals of a single mineral. At one time the existence of different forms of the same crystal had been used as an argument against internal structure. The diagram reproduced in Figure 8.29 is from Haüy's *Traité de Mineralogie* published in 1801. It shows how small cubes can be stacked to produce shapes with the form of a rhombic dodecahedron and a pyritohedron. The faces are stepped because the ratio of the size of the solid to the size of the building block is large. In real crystals the blocks are submicroscopic and the faces are apparently planar. The importance of this building block theory of crystal structure has led to Haüy being called the 'Father of crystallography'.

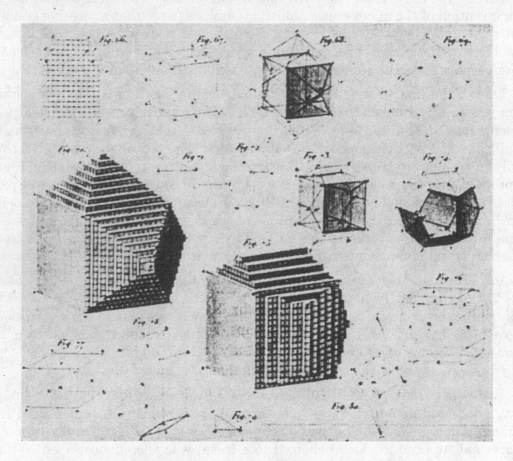

Figure 8.29. Diagram from Haüy's *Traité de Mineralogie* showing how various crystal shapes can be formed by stacking up many small congruent blocks.

At first, Haüy had just six building blocks but later the set was enlarged to eighteen. He also allowed his blocks to be modified so that new faces appeared but he imposed a law of symmetry to restrict the possibilities: parts of the initial form that were *indistinguishable to the eye* had to be modified in the same fashion. This gave Haüy an intuitive understanding of some of the different symmetry types of solid objects. Mathematicians of the time had not ventured beyond simple mirror symmetry.

A problem with Haüy's theory is that the fundamental units formed by cleaving a crystal cannot always be stacked up to fill space in the way that cubes can. This led to debates about the reality of Haüy's building blocks. Some suggested replacing each block by a point at its centre of gravity which resulted in a lattice of points distributed through space in a pattern that repeated itself over and over in all directions. Others replaced the blocks by spheres, reviving Hooke's idea. Gabriel Delafosse (1796–1878) proposed that Haüy's blocks be replaced by 'polyhedral molecules' which had spherical atoms at their vertices. These molecules were to be arranged in a lattice (like Haüy's blocks) and their shape would determine the external form and physical properties of the crystal.

Understanding this new structure required a detailed analysis of the symmetries of the molecules themselves, of the lattice structure, and of their interrelationships. These problems were attacked by Auguste Bravais (1811–1863). In a paper that appeared in 1849 he defined axes, planes and centres of symmetry and distinguished between principal and secondary axes. By considering the various combinations of symmetries, he enumerated the list of symmetry types of polyhedra presented earlier in the chapter apart from one exception: he missed the type S_{2n} when n is even. (His list of symmetry operations did not include rotation-reflections. When n is odd, S_{2n} can be generated from rotations and inversion.)

The mathematician and astronomer August Ferdinand Möbius (1790–1868) also investigated the possible systems of symmetries which a polyhedrally shaped object can have. His definition of symmetry is the same as that used today: a figure is symmetrical if it is congruent to itself in more than one way. He entered his work for a prize offered by the Paris Academy in 1861 for 'work perfecting some aspect of the theory of polyhedra'. Among the other entrants was Eugène Charles Catalan, who enumerated some semiregular polyhedra including the Archimedean solids. None of the entries was judged to be of sufficient quality to be awarded the prize.

The first enumeration of the symmetry types that the external form of a crystal can have was achieved earlier than this. The mineralogist Johann Friedrich Christian Hessel (1796–1872) classified the 32 kinds of symmetry that a crystal can have in 1830, but his work went unnoticed until Leonhard Sohncke (1842–1897) drew attention to its importance. It was republished in the 1890's as part of a collection of papers on crystallography.

In 1850 Bravais continued his own investigations and studied the symmetries of points arranged regularly in space. He found seven different lattice systems, now called the cubic, hexagonal, tetragonal, rhombohedral (or trigonal), orthorhombic, monoclinic, and triclinic systems. He followed this with a finer classification which produced fourteen different types of lattice.

The lattice structure of a crystal restricts the kinds of rotational symmetry that can appear to 2-fold, 3-fold, 4-fold and 6-fold. This *crystallographic restriction* is implicit in Haüy's work. A consequence is that crystals cannot have icosahedral symmetry since this system contains forbidden 5-fold symmetry axes. The rotational symmetries of a cube are not excluded and cubic crystals occur in nature. Of the infinite families of symmetry types in the prismatic systems of symmetry, only those whose principal axis is 2-fold, 3-fold, 4-fold or 6-fold can occur as the symmetry type of a crystal. Hence there are only finitely many possibilities for the symmetry types of the external forms of crystals. These 32 symmetry types are called the *crystal classes*.

Crystallographers separate the 32 crystal classes into seven kinds corresponding to Bravais' original classification of lattices. Each lattice is built up by repeating a fundamental unit, and the shape of the building block determines the type of lattice. Haüy's diagram shows how cubic blocks can produce crystals having cubic symmetry types O_h (the cube and rhombic dodecahedron) and T_h (the pyritohedron). The other kinds of building blocks and the related symmetry groups are shown in Table 8.30. One is a square-based prism with rectangular sides. This is a building block for the tetragonal systems. If the fundamental units are shaped like bricks in which the height, width and length measurements are all different then the orthorhombic symmetry types are created. If the brick is sheared in one direction it becomes the building block for the monoclinic systems, and if it is sheared in two directions to produce a skew parallelopiped then it is the fundamental unit for the triclinic systems. The two kinds of symmetry which contain 3-fold rotational symmetry are the rhombohedral (or trigonal) and the hexagonal systems. The building blocks for these systems are a rhombohedron and a rhombus-based prism respectively.

The variety of external forms which crystals of the same species can exhibit means that classification systems based on geometrical shape are not very useful. However, all crystals of a given species have the same internal structure and this makes a better basis for classification. For this reason investigations into the possible structures continued in the second half of the nineteenth century.

In his classification of patterns in space, Bravais had assumed that every molecule in the lattice had the same orientation so that they were all aligned in the same direction. This restriction was relaxed and replaced by the condition that each molecule had the same relation to the pattern as a whole, irrespective of its orientation. This meant that direct symmetries in general needed to be considered, not just translations.

Louis Poinsot studied screw motions, another type of direct symmetry operation. A screw is a compound operation involving a rotation followed by a translation along its axis. Building on Bravais' and Poinsot's work, Camille Jordan investigated systems of direct symmetries in a group theoretical manner (although groups were not yet formally defined). He examined the ways that ro-

Building block	System	Group	Related groups
	cubic	O_h	O, T_h, T_d, T
	tetragonal	D_{4h}	$D_4, D_{2v}, C_{4v}, C_{4h}, S_4, C_4$
	orthorhombic	D_{2h}	D_2, C_{2v}
	monoclinic	C_{2h}	C_2, C_s
	triclinic	C_i	C_1
	rhombohedral	D_{3v}	D_3, C_{3v}, S_6, C_3
	hexagonal	D_{6h}	$D_6, C_{6v}, D_{3h}, C_{6h}, C_6, C_{3h}$

Table 8.30.

tations, translations and screw symmetries could be combined. Sohncke reworked Jordan's ideas in the less abstract form of repeating patterns in space. He completed the list of groups extending it from 59 to 66 (two of these were later found to be the same).

Indirect symmetries such as rotation-reflection and rotation-inversion had been neglected in these classifications and the resulting crystallographic theory was unable to account for certain directional properties of crystals. Some crystals become polarised when they are subjected to changes in temperature or pressure—two opposite faces acquire positive and negative charges. The pyro-electric property was known since at least 1757 when Franz Aepinus found that tourmaline became polarised after being placed in hot water. These pyro-electric and piezo-electric properties led Pierre Curie (1859–1906) to study crystal symmetry, and he drew attention to the omission of indirect symmetries.

The inclusion of indirect symmetries raises the number of groups of symmetries that a three-dimensional repeating pattern can have to a total of 230.

The enumeration of these 'space groups' was carried out in the early 1890's by the Russian Evgraf Stephanovich Fedorov (1853–1919), who extended Sohncke's work, and simultaneously by the German mathematician Arthur Moritz Schoenflies (1853–1928), who, under the influence of Felix Klein, followed the group theoretical approach of Jordan. Both men missed a few cases but when they became aware of each other's work they corresponded to check that their lists agreed. They finally produced a list of 230 groups. The same problem was being worked on independently by the Englishman William Barlow (1845–1934), who published his classification in 1894. Like Sohncke and Fedorov, he thought in terms of lattices and he made models by arranging gloves on a rack to help him visualise the regular patterns in space. He also studied the ways that spheres can be packed together in an economical manner and found five dense or close packings which he believed represented the structures of some crystals.

This achievement was the climax of the golden age of theoretical crystallography—a complete mathematical classification of crystal structures based on the assumption that crystals are built of units stacked up in a repetitive manner. The theory was written up by Harold Hilton and published as *Mathematical Crystallography and the Theory of Groups of Movements* in 1903. Near the close of his book Hilton remarks

> *The geometrical theory of crystal-structure now seems to be fairly complete; it is probable that further advance is to be expected on the physical or mechanical side.*[c]

At the turn of the century, the physical composition of crystals was undecided although it was frequently stated that all crystals possessed a periodic structure. Another unsolved problem at that time was to determine the nature of the radiation discovered by Wilhelm Conrad Röntgen in 1895. It had been suggested that these x-rays were very short wavelength electromagnetic radiation. Max von Laue (1879–1960), who had been working on the interference patterns formed when light passes through a diffraction grating, realised that if both assumptions were correct then a similar phenomenon should occur if x-rays were passed through a crystal. Walter Friedrich and Paul Knipping tested his idea experimentally by irradiating a copper sulphate crystal. The photographic plate placed behind the crystal revealed a set of regularly ordered dark patches. This was the first x-ray diffraction pattern and provided the key experimental evidence for a repeating structure on the atomic scale. The announcement of success was made in 1912 and the significance of this discovery was recognised with the award of a Nobel prize in 1914.

©1996 P. R. Cromwell

Counting, Colouring and Computing

A mathematician, like a painter or a poet, is a maker of patterns. ⋯ The mathematician's patterns, like the painter's or the poet's, must be beautiful; the ideas, like the colours or the words, must fit together in a harmonious way. Beauty is the first test: there is no place in the world for ugly mathematics.[a]

G. H. Hardy

One way of enhancing the aesthetic qualities of polyhedral models is to paint them. Besides adding to their visual appeal, colour can be used to highlight various relationships between the component parts of a polyhedron since the eye is drawn to elements of like colour. This technique has been used to colour the models shown in some of the colour plates. In many of the models of the Archimedean solids (Plates 3 and 4) faces of like shape have the same colour showing how each type of face is distributed over the polyhedron. In each of the compound polyhedra (Plates 11–16) all the visible parts of a component are coloured the same and each individual component has a separate colour. Likewise for the star polyhedra (Plates 8–9) where each face is monochromatic: even though the accessible parts of a face are physically disconnected, they are linked by colour and are seen as parts of a planar whole.

The faces of a polyhedron can also be embellished with markings or patterns. Some of the oldest examples of this are a pair of Roman icosahedral dice found in Egypt, now in the British Museum, which date from the Ptolemaic dynasty. The tessellations of the octahedron with bird motifs and of the icosahedron with fish shown in Figure 2.5 are further examples of decorated polyhedra. However, in the present investigation these kinds of pattern will be ignored; we shall concentrate on colourings in which each face is coloured in a single colour. Furthermore, we shall restrict attention to 'spherical' polyhedra, that is, to polyhedral surfaces[1] which can be continuously deformed into a sphere. It is sometimes necessary to appeal to this property to show that particular kinds of colouring are possible, as for instance when Euler's formula is applied.

Colouring the Platonic solids

The first mathematical question relating to coloured polyhedra appeared in 1824 in Joseph Diaz Gergonne's journal *Annales de Mathematique, Pures et Appliquées*. It concerned the number of different ways each of the Platonic solids can be coloured if all of their faces receive different colours. Implicit in this question is the problem of deciding whether two colourings are different.

As an illustration we shall examine the possibilities of colouring a cube with six colours, each face receiving a different colour. The six faces of the cube can be labelled 'top', 'bottom', 'left', 'right', 'front', and 'back'. When colouring the cube, the faces are to be coloured in this order. We can choose from six colours for the top face. There is a choice of five unused colours remaining for the bottom face, four for the left face, and so on until only one colour remains unused and this is applied to the back face. Therefore the number of possibilities for assigning the colours to the faces is

$$6 \times 5 \times 4 \times 3 \times 2 \times 1 = 720.$$

This result is not yet the final answer to our question. For the purposes of calculation the faces of the cube were labelled so that they could be distinguished but, when studying a model of coloured cube, all the faces are alike. Apart from colour, there is nothing to distinguish between them. If you were handed two coloured cubes and asked whether they were coloured in the same way, you would turn them and try to match the patterns. Because the cube is symmetrical, the same pattern of colours can occur in several orientations, and rotating one of the cubes would allow you to see all the possibilities.

The above calculation did not take into account the symmetry properties of the cube. The answer of 720 is too large because colourings which one intuitively thinks of as being the same have been counted more than once. The cube has 24 rotational symmetries so each particular pattern of colours can be placed in any of 24 orientations. Each pattern will appear in the list of 720 colourings 24 times over. Therefore, the number of different colourings of the cube using six colours is actually $\frac{720}{24} = 30$.

This method can be applied to the other Platonic solids. Suppose that the polyhedron has F faces and that each face is given a unique label. Colour each

---

[1]See the definition of 'polyhedron' in Chapter 5.

face choosing from a palette of F colours so that no two faces have the same colour. Then the number of distinct colourings of the labelled polyhedron is

$$F \times (F - 1) \times (F - 2) \times (F - 3) \times \cdots \times 3 \times 2 \times 1.$$

This expression is usually written in the shorthand notation $F!$ (read 'F factorial').

At this point it is necessary to decide which colourings count as being the same. This became hidden in the discussion of the cube and the definition will be isolated here since it is used throughout the chapter.

Suppose that two models of a polyhedron have been coloured. Then the two colourings are the same if one of the models can be rotated and repositioned so that it looks identical to the other with corresponding faces having like colour. This can be restated as follows:

Definition. Two colourings of a polyhedron are equivalent if there is a rotational symmetry of the polyhedron which carries one colouring onto the other.

The same pattern of colours can appear in various orientations and the number of these different positions is equal to the number of rotational symmetries of the polyhedron. Arthur Cayley observed that, in the case of a Platonic solid, the number of rotational symmetries equals twice its number of edges.[2] In the cube example, there are 12 edges and 24 symmetries. If E denotes the number of edges of a polyhedron, each colouring can occur in $2E$ different orientations. Each of these is counted as a distinct colouring when the faces of the polyhedron are labelled, so every pattern is repeated in the list $2E$ times over. To find the number of different colourings of the unlabelled polyhedron we divide the number of colourings of the labelled polyhedron by the repeat frequency. Thus the number of different colourings of a Platonic solid using a different colour for each face is

$$\frac{F!}{2E}.$$

Problem. The 30 differently coloured cubes form the pieces of a puzzle.[3] The aim is to select any cube from the set then find eight of the remaining 29 cubes which can be put together to form a cube twice as high as the original pieces which is coloured the same as the chosen cube on the outside, and so that the touching faces of adjacent cubes have the same colour.

---

[2] The Platonic solids are special in this respect—they are edge-transitive as we shall see in Chapter 10.

[3] This puzzle, together with its solution, is discussed in A. Ehrenfeucht, *The Cube Made Interesting*, translated from the Polish by W. Zawadowski, Pergamon Press 1964, pp53–58.

How many colourings are there?

In the previous section a particular case of the question 'How many colourings are there?' was considered, namely, 'How many different ways are there to colour a polyhedron with F faces if no two faces have the same colour?'. The answer was shown to be $\frac{F!}{n}$, where n is the number of rotational symmetries of the polyhedron.

The more general problem of finding the number of different ways to colour a polyhedron where the colour of each face can be freely chosen from a given set of colours is more difficult. For example, how many indistinguishable colourings of a square-based pyramid are there using only black and white? Ignoring the symmetries of the pyramid, we see that there can be at most $2^5 = 32$ such colourings since the pyramid has five faces and each of them can be either black or white independently. In fact, there are only 12 indistinguishable colourings and the situation is simple enough that they can all be found by hand. The four triangular faces can be coloured either all white, three white and one black, two white and two black, one white and three black, or all black. The case where there are two faces of each colour produces two types of pattern according to whether opposite faces have the same colour or not. Each of the other cases gives rise to a single type of pattern. All six pattern types are shown in Figure 9.1. Each can occur whether the square face is black or white, so there is a total of 12 different colourings.

Figure 9.1. Overhead views of six coloured pyramids.

The problem of determining the number of different colourings becomes more difficult as the number of faces of the polyhedron increases. A large number of colours, or a high degree of symmetry of the polyhedron also complicates the problem. For instance, try to imagine how many different colourings of a cube there are which use at most five colours. Fortunately, we do not have to make an exhaustive search and list all the possibilities to answer this kind of question. There is a technique which allows us to *calculate* the number of different colourings of any polyhedron with any number of colours. By applying this method, we shall see that the cube has 800 such colourings.

A counting theorem

Trying to determine the number of ways a polyhedron can be coloured is an example of a more general class of problems that concern the number of ways in which something can happen. Counting the number of chemical isomers of a molecule is another example. This kind of problem can often be solved by using group theory.

The method described below is an application of an abstract result in group theory generally known as Burnside's Lemma or Burnside's Counting Theorem after William Burnside (1852–1927), although the theorem originates in work by G. Frobenius. The theorem is not stated in its general context here, neither is a proof given for the special case of colourings. Rather, the black and white colourings of the square-based pyramid are analysed in detail to show how the resulting formula is derived and illustrate why the method works.

For the purposes of the following discussion it is necessary to make a distinction between colourings of labelled and unlabelled polyhedra. The term *colouring* will be used when referring to labelled polyhedra (those in which all the faces can be distinguished regardless of symmetry), and *colouring type* when referring to unlabelled polyhedra. Thus a colouring is a particular assignment of colours to labelled faces, and colouring type refers to the pattern so produced; different colourings may have the same colouring type.

When asking how many different ways a polyhedron can be coloured we want to know the number of colouring types. Burnside's theorem gives a formula for the number of colouring types in terms of the numbers of colourings with particular properties. The latter quantities are straightforward to calculate and this makes the formula effective.

Example. Re-analysing the pyramid.

To simplify the situation slightly, assume that the base of the square-based pyramid has been coloured white. Each of the other four faces can be either black or white independently. This results in a set of $2^4 = 16$ possible colourings of the pyramid. We now make a list of the symmetries which each of these coloured polyhedra possesses. The (direct) symmetries of the pyramid comprise the identity and three rotations about the axis joining the centre of the base to the apex. Two of the rotations are 4-fold, the other is 2-fold. A symmetry operation in this set is a symmetry of a coloured pyramid if it carries the coloured pyramid to a position indistinguishable from its original one. This requires that the faces are carried onto faces of like colour.

The list of the 16 colourings and their symmetries is given in Table 9.3. The first column assigns a number to each colouring. The next column describes the colours of each of the four triangular faces. These faces of the pyramid can be

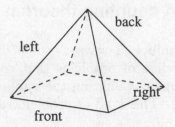

Figure 9.2.

labelled 'front', 'back', 'left', and 'right' as shown in Figure 9.2. The symbol in the second column of the table gives the (initial letters of the) colours of the faces in the cyclic order: 'front', 'left', 'back', 'right'. Thus the colouring in the sixth row is a pyramid whose 'front' and 'back' faces are coloured black, and whose 'left' and 'right' faces are white. The third column indicates the symmetries possessed by each particular colouring. The symmetries are labelled 1 for the identity, r_2 for the 2-fold rotation, r_4 for the 4-fold rotation clockwise, and r_4^{-1} for the 4-fold rotation counter-clockwise. A tick ($\checkmark$) under a particular symmetry indicates that a colouring possesses that symmetry. The horizontal lines divide up the colourings into the six colouring types found previously, and the final column describes each of these types.

The aim now is to calculate the number of ticks in the table. This will be done in two ways: first by adding the totals of each column, and second by adding the totals of each horizontal division. Since these two calculations must give the same result, they can be equated and a formula deduced for the number of types of colouring.

The number of ticks in each column is the number of colourings which have a particular symmetry. Every colouring has the identity symmetry; four colourings have 2-fold rotational symmetry; only two colourings have 4-fold rotational symmetry. The total number of ticks in the table is

$$\begin{aligned} &\quad \text{(the number of colourings with symmetry 1)} \\ +\ &\quad \text{(the number of colourings with symmetry } r_2) \\ +\ &\quad \text{(the number of colourings with symmetry } r_4) \\ +\ &\quad \text{(the number of colourings with symmetry } r_4^{-1}). \end{aligned}$$

The sum can be written in a concise form using the summation notation:

$$\sum_{s \text{ in } C_4} \text{(the number of colourings with symmetry } s)$$

where the summation variable s ranges over all the symmetries of the polyhedron. This sum is quite easy to calculate for a given polyhedron and any set of colours: it is this that makes the resulting formula useful.

	Colouring	1	r_2	r_4	r_4^{-1}	Colouring type
1	(W,W,W,W)	✓	✓	✓	✓	all white
2	(W,W,W,B)	✓				
3	(W,W,B,W)	✓				three white and
4	(W,B,W,W)	✓				one black
5	(B,W,W,W)	✓				
6	(B,W,B,W)	✓	✓			two white, two black,
7	(W,B,W,B)	✓	✓			opposite faces the same
8	(B,B,W,W)	✓				
9	(B,W,W,B)	✓				two white, two black,
10	(W,W,B,B)	✓				opposite faces different
11	(W,B,B,W)	✓				
12	(B,B,B,W)	✓				
13	(B,B,W,B)	✓				one white and
14	(B,W,B,B)	✓				three black
15	(W,B,B,B)	✓				
16	(B,B,B,B)	✓	✓	✓	✓	all black

Table 9.3. The symmetries of the different colourings of a pyramid.

The second part of the analysis consists of calculating the number of ticks in each horizontal division, then adding the results. For this we need an important result from group theory which is known as the Orbit–Stabiliser Theorem.[4] Its technicalities need not concern us since its consequences are clearly visible in the table: in each of the horizontal divisions which correspond to a colouring type there are precisely four ticks. The reason that four occurs here is that the pyramid has four symmetries. More generally, the theorem implies that

$$\begin{pmatrix} \text{number of colourings with} \\ \text{a particular colouring type} \end{pmatrix} \times \begin{pmatrix} \text{number of symmetries} \\ \text{of that colouring type} \end{pmatrix}$$

$$= \begin{pmatrix} \text{number of symmetries of} \\ \text{the uncoloured polyhedron} \end{pmatrix}.$$

In the pyramid example, there are two colourings which have the colouring type 'two black, two white, opposites the same', namely colourings 6 and 7. Each of

---

[4]This theorem is discussed in Appendix 2.

these colourings has two symmetries: 1 and r_2. The product $2 \times 2 = 4$ is the number of symmetries of the pyramid.

As the number of ticks in each horizontal division of the table is constant, and the number of divisions equals the number of colouring types, the total number of ticks in the table equals

$$\begin{pmatrix} \text{number of symmetries of} \\ \text{the uncoloured polyhedron} \end{pmatrix} \times \begin{pmatrix} \text{number of types} \\ \text{of colouring} \end{pmatrix}.$$

This expression contains the quantity that we want to know (the number of types of colouring) together with a quantity which is easily calculated (the number of symmetries of the polyhedron). Equating this result to the previous one, and writing G for the symmetry group of the polyhedron gives

$$\begin{pmatrix} \text{number of types} \\ \text{of colouring} \end{pmatrix} = \frac{\displaystyle\sum_{s \text{ in } G} \begin{pmatrix} \text{number of colourings} \\ \text{with symmetry } s \end{pmatrix}}{\begin{pmatrix} \text{number of symmetries of} \\ \text{the uncoloured polyhedron} \end{pmatrix}}.$$

This is the required formula. It is not difficult to calculate both of the quantities on the right-hand side and the following examples show how this is done.

Applications of the counting theorem

Example.   The number of colourings of a square-based pyramid.

In this first example, the square-based pyramid is analysed again, this time considering the general case of how many colourings with n colours are distinguishable. We need to determine the number of colourings which possess each type of symmetry. Every colouring has the identity symmetry, and the total number of colourings is n^5 since there is a choice of n colours for each of the five faces of the pyramid. How many colourings have 2-fold rotational symmetry? This symmetry places a restriction on the arrangement of the colours on the triangular faces: the front and back faces must be of the same colour, and so must the left and right faces. There are three choices to be made: a colour for the base, a colour for the front–back pair, and a colour for the left–right pair. Each of the three choices is made from n colours so there are n^3 colourings which have 2-fold symmetry. If a colouring has 4-fold rotational symmetry then all of the triangular faces must have the same colour. There are two choices to be made: a colour for the base of the pyramid, and a colour for the other four faces. Thus there are n^2 colourings with 4-fold symmetry. Both of the 4-fold symmetries contribute the same number of colourings to the summation. Substituting these values into the above formula, and recalling that the pyramid has four symmetries gives that the number of different colourings of a square-based pyramid using n colours is

$$\frac{n^5 + n^3 + n^2 + n^2}{4}$$
$$= \; {}^{1}\!/_{4}\, n^2(n^3 + n + 2).$$

When n is 2 this evaluates to 12, which is the same result as before.

Example. The number of colourings of a regular tetrahedron.

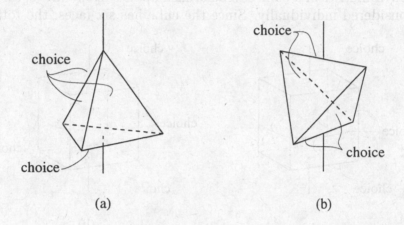

Figure 9.4.

Firstly, we need to calculate the number of colourings which have a given symmetry of the tetrahedron. Again, the number of colourings using at most n colours will be found. The regular tetrahedron has 12 direct symmetries: the identity, eight 3-fold rotations, and three 2-fold rotations.

The number of colourings with the identity symmetry is n^4 since there are four faces and there is a choice of n colours for each face. A 3-fold rotation axis passes through the centre of one face (which will be called the base) and the opposite vertex. For the colouring to have 3-fold symmetry, the three faces adjacent to the base must all be the same colour. So there are only two choices of colour: one choice of colour for the base, and one for the other three faces (Figure 9.4(a)). As each of the choices can be made independently from the set of n colours, there are n^2 colourings with 3-fold symmetry.

The 2-fold axes pass through the midpoints of opposite edges. Therefore, both faces adjacent to an edge which meets the axis must have the same colour. The faces separate into two pairs, and again there are two choices of colour, one for each pair (Figure 9.4(b)). So there are n^2 colourings with 2-fold symmetry. The formula now gives the number of indistinguishable ways to colour a regular tetrahedron using at most n colours:

$$\frac{1}{12}\left(n^4 + 8n^2 + 3n^2\right)$$
$$= \frac{1}{12}\,n^2\left(n^2 + 11\right).$$

Problem.   Find the five different colourings of the tetrahedron using black and white.

Example.   The number of colourings of a cube.

The 24 symmetries of a cube can be broken up into five families, each of which must be considered individually. Since the cube has six faces, the total number

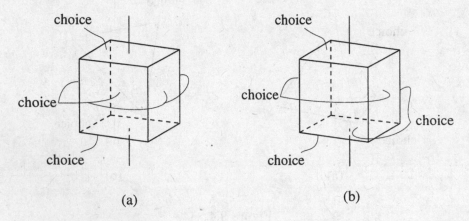

(a)   (b)

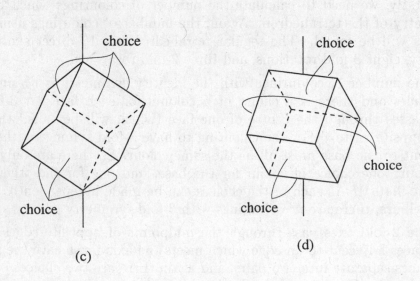

(c)   (d)

Figure 9.5.

of colourings (all of which have the identity symmetry) is n^6. There are six 4-fold rotations. An axis of one of these symmetries joins the centres of opposite faces. Assume that the axis is vertical. There are three choices of colour: one each for the top and bottom faces, and one for the other four faces which must all be the same colour (Figure 9.5(a)). So n^3 colourings have this 4-fold symmetry. Each of the three 4-fold axes is also an axis of 2-fold rotational symmetry. There are now four choices of colour: one each for the top and the bottom; the four other faces divide into two pairs of opposite faces, and a colour can be chosen for each pair (Figure 9.5(b)). There are eight 3-fold rotations whose axes are the diagonals of the cube joining opposite vertices. With this axis in a vertical position, the two choices of colour are: one for the upper three faces, and one for the lower three (Figure 9.5(c)). Thus there are n^2 colourings preserved by this 3-fold symmetry. Lastly, there are six more 2-fold rotations to consider whose axes join the midpoints of opposite edges. With one of these axes vertical, the three independent choices of colour are: one for the two upper faces which meet the axis, one for the two lower faces meeting the axis, and one for the remaining pair of opposite faces (Figure 9.5(d)). Thus there are n^3 colourings which have this 2-fold symmetry. Substituting all these data into the formula gives that the number of different ways to colour a cube in at most n colours is

$$\tfrac{1}{24}\,(n^6 + 6n^3 + 3n^4 + 8n^2 + 6n^3)$$
$$= \tfrac{1}{24}\,n^2\,(n^4 + 3n^2 + 12n + 8).$$

As mentioned above, substituting $n = 5$ into this formula shows there are 800 colourings using at most five colours.

Problem. Find the ten different colourings of the cube using only black and white.

Problem. The adventurous reader who wants to try out this technique may like to verify that the number of ways to colour a regular octahedron with at most n colours is

$$\tfrac{1}{24}\,n^2\,(n^6 + 17n^2 + 6).$$

Proper colourings

The only restriction placed on the colourings in the previous sections was that each face be coloured in a single colour. This allowed the whole polyhedron to be painted in a single colour which does not show up any structural features of the polyhedron. In this section, an additional requirement is forced upon the colouring: faces that share a common edge must have different colours. One result of this is that the outlines of the faces stand out and their individual shapes show up. A colouring which has this property is called a *proper* colouring.

As before, the question can be asked 'How many different proper colourings of a polyhedron are there which use a given number of colours?'. As might be expected, this extra restriction on the behaviour of a colouring substantially reduces the number of possibilities. Among the 57 colourings of a cube which use at most three colours, there is a unique proper colouring. Out of almost 60 million colourings of the icosahedron using at most three colours, only 144 are proper colourings. Some of them are illustrated schematically in Figure 9.6 and the others can be obtained from these by permuting the colours and taking their mirror images. The first seven are unchanged by swapping pairs of colours but do change when the colours are cycled. Reflections also give new colourings so these patterns contribute 42 colourings. For the next two patterns, swapping two colours or taking a reflection achieves the same result. This gives another 12 colourings. The next pattern is the only one which has mirror symmetry. All six permutations of the colours give different colourings. For the remaining seven patterns, all permutations of the colours and reflections give different colourings— 84 in all.[5]

It may seem surprising that, in some situations, it is impossible to find a proper colouring of a polyhedron with the given number of colours. For example, the tetrahedron cannot be properly coloured with three or fewer colours since each of its four faces touches all of the others, forcing every face to have a different colour. With four colours, however, there are two enantiomorphic proper colourings. This shows that four colours are necessary to colour the tetrahedron. Rather than investigate how many proper colourings are possible using a given number of colours, it is perhaps more pertinent to ask whether a proper colouring is possible at all.

We have already seen that the tetrahedron requires four colours. The cube cannot be properly coloured with two colours since three faces meet at a vertex and each of them must be coloured differently. However, three colours are sufficient. For the regular octahedron, only two colours are required, and it can be properly coloured in the manner of a chessboard.

It would be nice to be able to determine how many colours are necessary to properly colour a given polyhedron *by inspection*, that is to say, just by checking whether the polyhedron has particular properties. An example of this was used above to show that the cube could not be properly coloured with two colours—its faces meet wrongly at the vertices. This observation can be extended to provide a complete characterisation of properly 2-colourable polyhedra. This means that there is a property which is shared by all polyhedra that can be properly coloured in two colours and which only they possess.

---

[5]For further information on how to enumerate these colourings see W. W. R. Ball and H. S. M. Coxeter, *Mathematical Recreations and Essays* (thirteenth edition), Dover, New York 1987, pp239–242.

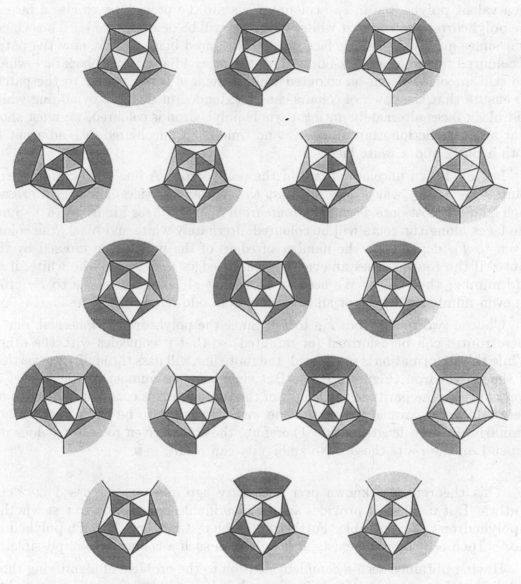

Figure 9.6. Schematic diagrams showing the proper colourings of the icosahedron.

Theorem. A polyhedron can be properly coloured with two colours if, and only if, every vertex is surrounded by an even number of faces.

PROOF: Suppose that a polyhedron is properly coloured with two colours. Then the colours must alternate on the faces around any vertex. This implies that there is an even number of faces around each vertex.

To show the converse, it is sufficient to describe a method for colouring an even-valent polyhedron in two colours. To start the procedure, choose a face of the polyhedron and colour it white. This face will be denoted by F_W. Those faces that share an edge with this face must be coloured black so that now the patch of coloured faces has a black border. All the faces adjacent to a black face which are still uncoloured can be coloured white leaving a white border to the patch. To ensure that the patch of colours can be extended in this way by adding white and black faces alternately until the whole polyhedron is coloured, we must show that a contradiction cannot arise: at no time is an uncoloured face adjacent to both a black and a white face.

Let F_U be an uncoloured face of the polyhedron. A line on the polyhedron joining F_W to F_U which does not pass through any vertices, and which crosses each edge at most once, is called a *route* from F_W to F_U (see Figure 9.7(a)). Since the faces along the route will be coloured alternately white and black, the colour given to F_U depends on the number of edges of the polyhedron crossed by the route: if the route crosses an even number of edges then F_U will be white, if an odd number then black. We need to show that all routes from F_W to F_U cross an even number of edges, or all routes cross an odd number of edges.

Choose two routes from F_W to F_U. Since the polyhedron is spherical, one of these routes can be deformed (or rerouted) so that it coincides with the other. While this deformation is performed, the route line will pass through some vertices of the polyhedron (Figure 9.7(b)). But since an even number of edges meet at every vertex, the parity of the length of the route remains constant: if the length were odd going around the vertex one way, it would also be odd via the other; similarly for even length paths. Therefore, the colour given to face F_U does not depend on the route chosen. No ambiguity can result.   ∎

This theorem was known over a century ago and was discussed by Peter Guthrie Tait in 1880. It provides an easily verifiable prescription to test whether a polyhedron is 2-colourable. Furthermore, the test is conclusive: if a polyhedron passes then a 2-colouring exists; if it fails then such a colouring is impossible.

Having obtained such a complete solution to the problem of identifying those polyhedra that can be properly 2-coloured, it is natural to progress to a slightly more complex situation and ask whether it is possible to characterise the 3-colourable polyhedra.

Clearly, a polyhedron which is properly coloured in two colours can also be properly coloured in three colours just by recolouring any one of the faces with a third colour. Hence, any even-valent polyhedron can be properly 3-coloured. It is also clear that the even-valent polyhedra do not exhaust all the possibilities. The cube, for example, can be properly coloured in three colours. Perhaps it is too optimistic to expect that a simple criterion can be found which will determine whether or not any polyhedron can be properly coloured with three colours. However, in two families of polyhedra a characterisation has been found. These are the 3-valent polyhedra, and the polyhedra whose faces are all triangles. These families are sometimes called the *simple*, and the *simplicial* polyhedra respectively.

Theorem. A 3-valent polyhedron can be properly coloured with three colours if, and only if, each face has an even number of sides.

PROOF: Suppose that a 3-valent polyhedron has been properly coloured with three colours red, green and blue. Choose a red face and look at the faces which are adjacent to it. They must be either blue or green. As three faces meet at every vertex, the colours must be alternately blue and green around the boundary of the red face. Hence, the red face has an even number of sides. A similar argument can be applied to every face.

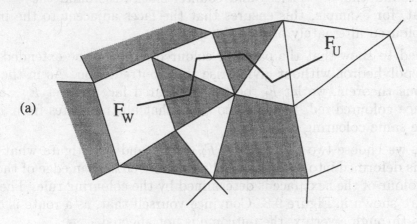

(a)

F_W F_U

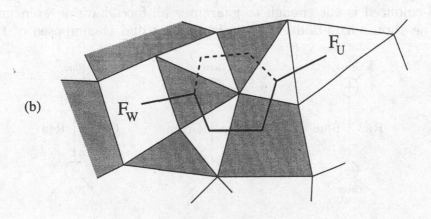

(b)

F_W F_U

Figure 9.7.

The converse is proved by describing a colouring procedure and then showing that it cannot lead to inconsistencies. The first stage in the colouring procedure is to label the vertices of the polyhedron. The current situation is the reverse of that considered in the preceding theorem: here every face is bounded by an even number of sides; previously, an even number of edges met at every vertex. Just as it was possible to colour the faces in two colours so that every edge separated the two colours, so a similar argument shows (in the current context) that it is possible to label the vertices with (say) '+' and '−' signs so that every edge connects a vertex labelled '+' to one labelled '−'.

With the vertices labelled in this fashion, the faces can be coloured as follows. Choose a face and colour it red. The colours of the remaining faces are determined from the faces already coloured and the '+' and '−' labels of the vertices by applying the following rule: the colours appear in the order red-green-blue clockwise round the '+' vertices and counter-clockwise round the '−' vertices. Notice that, for example, this ensures that the faces adjacent to the initial red face are coloured alternately blue and green.

We need to show that the patch of coloured faces can be extended to cover the whole polyhedron without giving rise to a contradiction. As in the proof of the previous theorem, we let F_U be an uncoloured face and let F_R denote the starting face coloured red. We need to show that all the routes from F_R to F_U lead to the same colouring of F_U.

Again, we choose two routes from F_R to F_U and investigate what happens when one is deformed into the other. When a route crosses an edge of the polyhedron the colour of the next face is determined by the colouring rule. The possible patterns are shown in Figure 9.8. Convince yourself that, as a route is deformed and passes through a vertex, the outcome is not affected. ∎

If the polyhedron is not 3-valent then the hypothesis that the polyhedron can be 3-coloured is not enough to guarantee all faces have an even number of sides. The rhomb-icosi-dodecahedron is 4-valent and is composed of triangles,

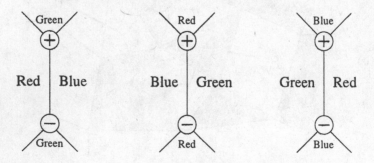

Figure 9.8.

squares, and pentagons. If each type of face is coloured differently then a proper 3-colouring results. However, not all of the faces have an even number of sides.

The condition of 3-valency is also important for the converse. There are examples of polyhedra, all of whose faces have an even number of sides, which cannot be properly coloured with three colours. An example is shown in Figure 9.9. It can be viewed as a hexagonal prism from which two wedges have been shaved off. Its faces have either four or six sides but it is not 3-valent. Suppose the base is coloured red. The six vertical faces must be coloured alternately (in blue and green, say) if only three colours are to be used. One of the top faces can be coloured red, but there is no colour remaining for the last face since it meets faces of all three colours.

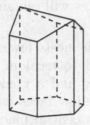

Figure 9.9. This polyhedron cannot be properly 3-coloured.

The other family of polyhedra to be considered are those with triangular faces. To prove the following theorem, which characterises which of these polyhedra are 3-colourable, we need to know how many different routes there are on a polyhedron from one face to another. Recall that a route is a line on the polyhedron joining two faces which does not pass through any vertex, and which crosses each edge at most once. It follows from work done in the early 1930's by Hassler Whitney that on any convex polyhedron[6] there are at least three distinct routes between any pair of faces. (Two routes are *distinct* if the only faces they have in common are their end faces.)

Theorem. Let P be a polyhedron all of whose faces are triangular. Then P can be properly coloured with three colours if, and only if, P is not a tetrahedron.

PROOF: We have already seen that a tetrahedron requires four colours to colour it properly. So we need to show that every other triangular-faced polyhedron can be properly coloured with three colours. In the proofs of the two preceding theorems, a method of colouring was described which was then shown to produce a proper colouring. A different approach will be used in this proof. Here, we will

---

[6] Convexity is not the critical point here. What we are using is the fact that the edge-skeleton of the combinatorial dual is 3-connected.

start with a proper colouring which contains too many colours (in this case four) and then show how to rearrange them so that their number can be reduced.

The faces of P can be coloured with the four colours red, green, blue, and yellow to produce a proper colouring of P as follows: start at any face and apply any colour to it; this forms the initial patch of colour. This patch can be extended by colouring a face which adjoins the patch. Since the face is triangular, it meets at most three different colours and so there is always a fourth colour which can be used to colour it. In this way the patch of coloured faces can be extended to cover the whole polyhedron.

The aim now is to rearrange the colours on some of the faces of P while maintaining the properness of the colouring, so that the number of yellow faces is reduced. Eventually no yellow faces will remain and P will be properly coloured with three colours.

If there is a yellow face which meets fewer than three colours then there is a third colour which can be used to recolour the yellow face. For example, if a yellow face meets two red faces and a green one, it can be recoloured blue. By using this observation, the yellow faces can be made to wander around on P so that they migrate towards a central place. Eventually a single yellow face remains. This then wanders to a place where it can be recoloured. The details of this process will now be described.

Suppose that at least two faces of P are coloured yellow, and choose a route joining two of them. A route can be described by listing the faces which it passes through in order. So suppose our chosen route is $F_1 \to F_2 \to F_3 \to \cdots \to F_n$, where F_1 and F_n are yellow. The first two faces can be recoloured so that F_1 is not yellow: simply uncolour the faces F_1 and F_2, then F_1 can be recoloured in a non-yellow colour since it meets only two other colours, and F_2 can then be recoloured, possibly with yellow. This process is illustrated in Figure 9.10. The effect of this is to shorten the length of the route between the two yellow faces, or to remove one of them. This trick can be repeated so that the face F_2 which has just been coloured yellow can be recoloured non-yellow and the (possibly) yellow face migrates to F_3 again reducing the distance between the two yellow faces. At the last stage, the face F_{n-1} can be coloured non-yellow since it is adjacent to a yellow face and at most two other colours. Thus the number of yellow faces of P can be reduced by one. This process can be repeated with any pair of yellow faces until eventually a solitary yellow face remains.

Now there is a single face of P coloured yellow. Let F be a face of P which is not adjacent to the yellow face, and let A, B, C be its three neighbouring faces. At least two of these neighbours must have the same colour because the three non-yellow colours are used up by the face F itself and two of its neighbours. Suppose that faces B and C have the same colour. Let F_Y be the yellow face of P. From the remark that preceded this theorem we know that there are

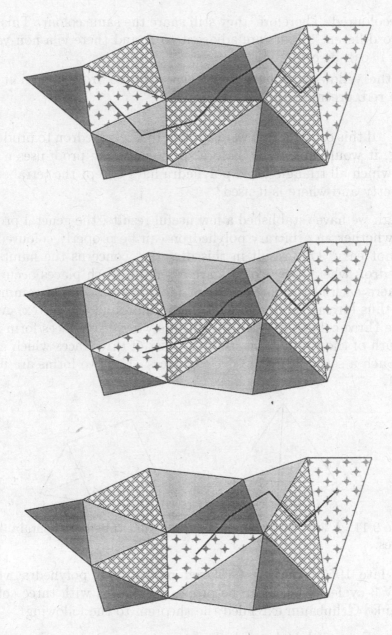

Figure 9.10.

three distinct routes on P which connect F_Y to F, one passing through each of A, B, and C. The route which we need is the one passing through A, that is, $F_Y \rightarrow \cdots \rightarrow A \rightarrow F$. The same procedure that was used above can be repeated on this route so that the yellow face migrates along the route towards F. This time, when we reach the end, there is certain to be a non-yellow colour with which to label F. Its two neighbours, B and C, did not appear on the route and so have not been recoloured. Therefore, they still share the same colour. This means that at most two different colours are adjacent to F and there is a non-yellow colour with which it can be coloured.

All of the yellow faces have now been removed from P and it is properly coloured in red, green and blue.   ■

Problem.   If this construction were applied to a tetrahedron to produce a proper 3-colouring, it would fail. Why? More specifically, the proof uses a property of polyhedra which all triangle-faced polyhedra have except the tetrahedron. What is the property and where is it used?

Although we have established a few useful results, the general problem of determining whether an arbitrary polyhedron can be properly coloured with three colours is not easy. One result in this direction concerns the number of places in a polyhedron where three colours are essential. Such places occur at 3-valent vertices where each of the three incident faces has an edge in common with the other two, thus forcing three colours to be used around the vertex. Another situation where three colours are essential occurs when three faces form a triangular tube. In both of these situations, there is a set of three faces which are mutually adjacent. Such a set is called a 3-cycle of faces. The two forms are illustrated in Figure 9.11.

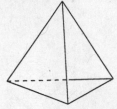
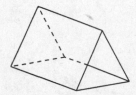

Figure 9.11. A 3-valent vertex and a triangular tube both contain 3-cycles of faces.

In the late 1950's Herbert Grötzsch proved that polyhedra which do not contain any 3-cycles of faces can be properly coloured with three colours. Later (1963) Branko Grünbaum extended the theorem to the following.

Theorem.   Every polyhedron which contains at most three 3-cycles of faces can be properly coloured with three colours.   ■

In some respects this is the best result possible. A further increase in the number of 3-cycles allowed is not possible since the tetrahedron has four 3-cycles (one at each vertex) and it cannot be properly coloured in three colours. On the other hand, this is not the complete set of properly 3-colourable polyhedra since the cube has six 3-cycles and is still properly 3-colourable.

How many colours are necessary?

As the number of available colours is increased further, one might think that the problem of determining which families of polyhedra can be properly coloured would get still more difficult. In fact, this is not the case. If your paint-box is sufficiently large then you can properly colour every polyhedron placed before you. It is a simple consequence of Euler's formula that a palette containing six colours is large enough to properly colour any polyhedron.

Theorem. Every polyhedron can be properly coloured with at most six colours.

PROOF: A colouring of the faces of P is a topological property. Therefore, no relevant information is lost if P is continuously deformed, and we can assume that P is a coloured network on a sphere. We shall prove the theorem for such spherical networks.

Let F be the number of faces of P, and suppose that we have established that all spherical networks with fewer than F faces can be properly coloured with at most six colours.

One of the corollaries of Euler's formula is that every polyhedron contains at least one face with fewer than six sides. For the purposes of illustration, assume that P contains a pentagonal face. There is a polyhedral network, Q, which has $F - 1$ faces, and which is the same as P except near this face where the two networks differ as shown in Figure 9.12. Since Q has fewer than F faces, it can be properly coloured with at most six colours (by hypothesis). The corresponding faces of P can, therefore, also be coloured using at most six colours leaving just the pentagonal face uncoloured. Since this face meets at most five different colours, there is always a spare colour for it. Hence P can be properly coloured with six or fewer colours.

This analysis has reduced the problem of colouring a network with F faces in six colours to colouring a network with $F - 1$ faces in six colours. If we can solve the latter (less complicated) problem then we can solve the former. The

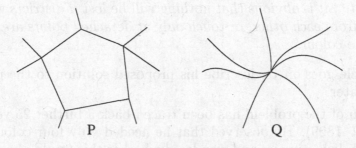

Figure 9.12.

argument can be repeated to show that Q can be properly 6-coloured *if* we can colour all networks with $F - 2$ faces in six colours. Proceeding in this fashion we can reduce the complexity of the problem (number of faces) as far as we wish. Once we reach the situation where 'We can colour P if we can colour a network with six faces' the problem is solved because every polyhedral network with six or fewer faces can be properly coloured using at most six colours: each face can be painted a different colour.

(This strategy is called the inductive method of proof. In this case we used induction on the number of faces.) ∎

This theorem raises a new question: if six colours are sufficient to colour every polyhedron, how many are necessary? Determining how many colours are required is a problem with a history dating back to the middle of the nineteenth century. It has been known since the 1890's that any polyhedron can be properly coloured with five colours. But trying to decide whether every polyhedron can be properly coloured with four colours turned out to be very complex. It became one of the best known unsolved problems in mathematics, and in the end was resolved only with the aid of a computer.

The four-colour problem

The problem of finding proper colourings of polyhedra is a particular case of a more general topological problem, namely, colouring the faces of a polyhedral network on a sphere. Another case of this problem is colouring maps on a globe, and indeed, this is the original setting of the four-colour problem. Alfred Bray Kempe (1849–1922) writing in *Nature* in 1880 described it like this:

> *The problem is to show how the districts of a map may be coloured with four colours, so that no two districts which have a common boundary or boundaries shall be of the same colour. The object of this colouring being to make the division of the map into districts clear without reference to boundary lines which may be confused with rivers, etc., it is obvious that nothing will be lost if districts which are remote from each other, or touch only at detached points are coloured the same colour.[b]*

Kempe's article goes on to describe his proposed solution to the problem which we shall see later.

The origin of the problem has been traced back a further 25 years to Francis Guthrie (1831–1899). He observed that he needed only four colours to colour a map of the English counties and wondered whether this would always be the case, whatever the map. He wrote to his younger brother, then at university in London, who raised the problem with Augustus De Morgan (1806–1871). De Morgan could not solve it and wrote of it in a letter to William Rowan Hamilton in the autumn of 1852. This is the earliest documented reference to the four-colour problem.

A major source of confusion and misunderstanding of this problem stems from the fact that no arrangement of five faces exists so that each is adjacent to the other four, thus forcing five colours to be used. De Morgan proved that it is impossible to construct five mutually adjacent faces. Prior to this, in the 1840's, Auguste Ferdinand Möbius teased his students by asking them to divide a kingdom into five parts, each having a border with the other four, so that the land could be shared between the king's five sons.

To see why this fact is not sufficient to solve the problem, consider an analogous situation involving only three colours. The presence of four mutually adjacent faces in a polyhedral network prevents or obstructs attempts to properly colour the network in three or fewer colours. The obstruction is localised and is due to the arrangement of only four faces of the whole network. A pentagonal-based prism does not have any set of four mutually adjacent faces and thus has no local obstructions to proper colouring in three colours. However, four colours are still required to achieve a proper colouring. In this case there is a global obstruction to proper colouring in three colours—the nature of the polyhedron as a whole prevents a proper colouring in fewer than four colours. De Morgan and Möbius knew that five mutually adjacent faces cannot exist. So there are no polyhedral networks that contain a *local* obstruction to proper colouring with four colours. This does not prove the four-colour problem is true in every case because there may be an example of a *global* obstruction.

Problem. Convince yourself that a pentagonal prism cannot be properly coloured with only three colours.

The four-colour problem received little attention until it was publicised by Arthur Cayley. At a meeting of the London Mathematical Society in 1878, he asked whether anyone could solve it. This prompted Kempe to work on the problem and the following year he proposed a solution. His argument is similar to the proof of the 6-colour theorem above, though he describes a procedure by which the colours can be rearranged so that fewer than six colours are required.

Kempe's argument.

The structure of Kempe's proof is the same as that in the preceding theorem: he uses induction on the number of faces. The tetrahedron, with four faces, provides a foundation for the induction since it can clearly be properly coloured with four colours. Given a polyhedron with F faces, we assume that every polyhedron having fewer than F faces can be properly coloured with four colours. In the same way as before, we can colour all but one of the faces using at most four colours. The problem is to show that the colouring can be extended to cover the last remaining face.

If the uncoloured face has fewer than four sides then it is clear that this is possible.

Suppose that the uncoloured face has four sides. If the four adjacent faces do not use all four colours then there is a colour left over which can be used to colour the final face. If all four colours do appear on the adjacent faces then Kempe showed how to rearrange some of the colours to overcome this problem.

Suppose that the colours surround the uncoloured face in the order red, blue, yellow, green. Kempe's idea was to look at areas coloured in two colours. In this example, areas coloured red-or-yellow and areas of blue-or-green. This can be thought of as red–yellow continents and islands, and blue–green oceans and lakes. These two-coloured regions are now called *Kempe chains*. It is possible that the blue face and the green face adjacent to the uncoloured face both belong to the same sea, or that the red face and the yellow face adjacent to the uncoloured face both belong to the same landmass but, since the network is drawn in a sphere, it is impossible for *both* of these to happen simultaneously. So assume, for example, that the red face and the yellow face are in different islands. Then the colours red and yellow can be interchanged on one of the islands without destroying the properness property of the colouring. When this is done the uncoloured face is adjacent only to blue, green and yellow faces and so red can be used to colour the final face. This process is illustrated in Figure 9.13. The square face in the upper diagram meets all four colours. The faces on the left and right of this square belong to the same Kempe chain so interchanging the colours in this chain will not help. However, if the colours in the Kempe chain above the square are interchanged, a colour becomes available to colour the square (as shown in the lower diagram).

This trick can be applied when the uncoloured face has five sides. However, this only reduces the number of different colours adjacent to the uncoloured face to four. Kempe assumed that the trick could be repeated on different pairs of colours so that the uncoloured pentagon was adjacent to at most three different colours. However, there is a flaw in this assumption. It went unnoticed for more than a decade, but in 1890 Percy John Heawood (1861–1955) found an example for which Kempe's method fails. Kempe's argument does show, however, that five colours are always sufficient to colour any polyhedral network on a sphere.

∎

Although Kempe's proof turned out to be incomplete, it contains the basis of the technique which was used to solve the problem. The idea is to contract a subset of faces, colour the remaining faces (by induction), replace the contracted faces, then show how to extend the colouring to the whole network. This process

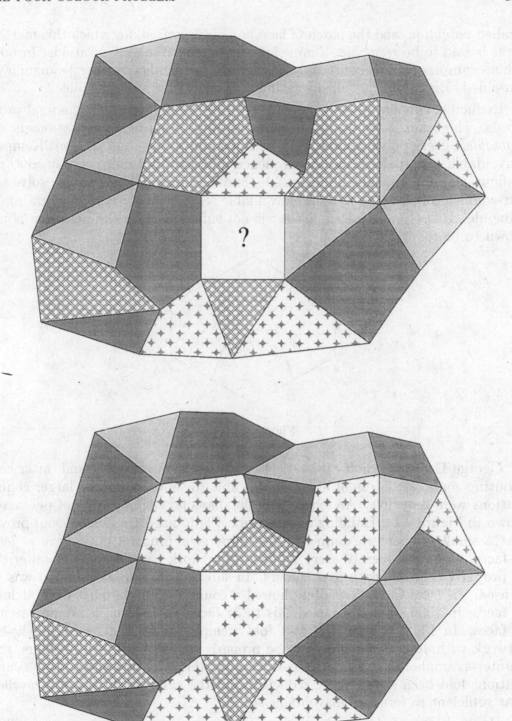

Figure 9.13.

is called *reduction*, and the patch of faces, or *configuration*, for which this method works is said to be *reducible*. Kempe found two reducible configurations. In both of his examples the configuration consists of a single face: either triangular or four-sided. His argument failed to show that pentagons are reducible.

Reducible configurations are one of the twin pillars on which the final proof stands. The other is the notion of unavoidability. A set of configurations is *unavoidable* if every polyhedral network must contain at least one of them. Kempe's unavoidable set, derived as a consequence of Euler's formula, contained three configurations: a triangle, a quadrangle, and a pentagon. One way to solve the four-colour problem is to find an unavoidable set in which every configuration is reducible. Unfortunately Kempe's set is not sufficient since the pentagon is not known to be reducible.

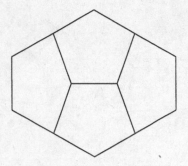

Figure 9.14.

George David Birkhoff (1884–1944) analysed Kempe's work and, after contributing some new ideas of his own, was able to show that some larger configurations were reducible. An example is the diamond configuration of pentagons shown in Figure 9.14. Building on this work, Philip Franklin (1898–1965) proved that a smallest counter-example to the four-colour problem must have at least 26 faces, and consequently every polyhedral network with 25 faces or fewer can be properly coloured with four colours. In subsequent years this bound was increased. In 1926 C. N. Reynolds showed a counter-example must have at least 28 faces; in 1936, Franklin raised this to 32 faces; in 1938 C. E. Winn reached 36 faces. In 1968, Oystein Ore and Joel Stemple proved that every polyhedral network with 40 or fewer faces can be properly coloured with four colours, so a counter-example has at least 41 faces. Through this collective effort many configurations had been shown to be reducible, but the number of these was nowhere near sufficient to form an unavoidable set.

Heinrich Heesch began working on the four-colour problem in the 1930's. He became convinced that it could be solved by finding an unavoidable set of reducible configurations. In the 1950's he estimated that such a set would be quite large, possibly containing 10 000 configurations, but that each configuration would be limited in size. At the time, this strategy seemed to offer little prospect of producing a solution because of the huge amount of calculation involved. However, the arrival of computers and their rapid development meant that this kind of attack became feasible.

In surveying the known techniques for proving that a configuration was reducible, Heesch noticed at least one procedure was mechanical enough to be performed by machine. His student Karl Dürre wrote a program to test the reducibility of configurations. As with many techniques, when the program was successful, the configuration was reducible, but a failure signified only that the method was not powerful enough, not that no way of reducing the configuration existed. During the testing of many configurations, Heesch noticed that configurations containing particular patterns were never reduced by the program. He found three such obstacles to reduction and all of them are easily described. No configuration which contains one of these obstacles has yet been reduced.

Heesch also developed a technique for producing unavoidable sets. The method is based on the idea of moving charge around an electrical network. A *discharging procedure* is an algorithm for redistributing initial charges in a network. Charge cannot be created or destroyed in the process and the algorithm must stop, and not circulate charge forever. By analysing the possible outcomes, the places where charge accumulates can be used to construct an unavoidable set of configurations. Changing the algorithm produces different sets.

In 1970, Wolfgang Haken noticed ways to improve Heesch's discharging procedures. He hoped that such improvements might be sufficient to solve the four-colour problem. His planned attack on the problem is outlined schematically in Figure 9.15. Before embarking on such a programme it is helpful to have some kind of indication that such a strategy is likely to succeed. What if the unavoidable set is very large? What if checking for reducibility takes so long that the project would take decades or centuries to complete? What if an unavoidable set of reducible configurations does not exist and the process never stops?

In 1972 Haken was joined by Kenneth Appel and they began computer experiments to try to answer some of these questions and to search for an effective discharging procedure. After some months of experimenting, gathering information about the likely sizes of configurations in an unavoidable set and the size of the set itself, they decided to prove that their method of attack had a reasonable chance of success. This involved arguing from certain assumptions that a discharging procedure which produced an unavoidable set of reducible configurations was 'overwhelmingly likely' to exist. Their first observation was that for configurations with a given perimeter, the likelihood of reducibility increases with the complexity of the interior. This means that when configurations are sufficiently large they are almost certain to be reducible. This makes it very unlikely that the process can run forever. The next step was to show that 'suf-

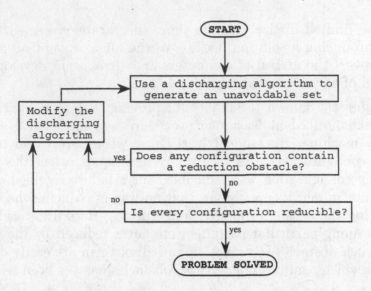

Figure 9.15.

ficiently large' configurations are actually small enough to be manageable and that, therefore, the process would finish in a reasonable amount of time.

In 1975 their attention turned to developing computer programs that would test configurations for reducibility. By 1976 they were ready to start constructing a suitable set of configurations. The discharging procedure used to generate the unavoidable set was implemented by hand. This meant that the procedure was more flexible and could be modified as required. Over the first six months of 1976 the discharging procedure was refined until at last an unavoidable set of reducible configurations was produced. The final set contained almost 2000 configurations.

What is proof?

When Appel and Haken announced their proof it caused much controversy in the mathematical community. The fame of the problem and its apparent simplicity have led many mathematicians to try to settle it. The announcement of a proof led people to expect that some brilliant insight had been discovered, and many were dismayed and disappointed to find that it involved analysing hundreds of cases by computer. Also people became suspicious when they realised that computer calculations were essential to the proof. Although the discharging procedure can be performed by hand and the generation of the unavoidable set can be checked, most of the configurations are so complicated that proving reducibility can be achieved only with the use of a computer. Appel and Haken themselves remarked 'It does not seem possible to check the reducibility computations themselves by hand.'

This raises philosophical questions concerning the nature of proof. A mathematical proof is supposed to leave no room for doubt, to deduce guaranteed conclusions from an agreed starting point. Prior to the introduction of computers, the arguments used in proofs could always be verified by other mathematicians; each step in the deduction from hypothesis to conclusion could be checked for its logical validity. When many of these steps are performed by a computer, and the calculations can be verified only by other computers, a mathematician has to ask himself whether his faith that computers behave as they ought is equal to his faith in his own reasoning ability.

Three other examples of computer-assisted proof have been mentioned in earlier chapters. Both V. A. Zalgaller's enumeration of convex polyhedra with regular faces (Chapter 2) and J. Skilling's enumeration of uniform polyhedra (Chapter 4) used a computer to search for new possibilities. In both cases, a previously known list was shown to be complete. Unlike the four-colour problem, neither of these problems was particularly well-known and so these confirmations did not attract much attention. Maksimov's proof that all polyhedra with fewer than nine vertices are rigid (Chapter 6) also made use of a computer.

Another use of computers in mathematics is to test large numbers for primality. In a computational sense there is no difference between a computer calculation which shows that a large number is prime and another which shows that a configuration is reducible. The difference lies in the interpretation of the results. Both types of calculation can be seen as establishing certain facts whose truth depends on the correct working of the computer. In the first case, these facts are treated as data from which one can draw opinions or conjectures—about the distribution of prime numbers, for instance. The data do not constitute a proof, merely an indication of what one might try to prove, so a small risk of error is tolerable. In the second case, an accumulation of computed facts *is* the proof.

It can be argued that the nature of the calculations involved in the proof of the four-colour theorem is so mechanical and repetitive that, if they could be performed by hand, a human attempting the task would, in all probability, make more errors than a computer. However, this does not eliminate the possibility of errors occurring, of either human or computer origin. At this point Appel and Haken's methodology provides a way out. If one accepts the assumptions underlying their probability argument that the method must succeed then there are many different proofs of the theorem depending on the choice of discharging algorithm and the resulting set of configurations which is used. So if there is an error in the published proof, there is almost certain to be another similar set which will suffice. The theorem is 'overwhelmingly likely' to be true.

Another objection to computer proof is that it violates the aesthetic quality of mathematics. Mathematicians are not concerned merely with facts and answers but with understanding; not only with truth but also with beauty. A good proof is supposed to expose the underlying reasons of *why* a result is true. When most of a proof is hidden away in a machine this psychological requirement is denied. A computer provides no insight, just a mechanical verification.

History has shown that the accepted standard of proof varies with time. It is probable that this new style of proof will become accepted as it is used to solve other problems in the future. In fact, such proofs may be necessary to solve some problems. At the turn of the century, it was generally believed that any mathematical problem could be solved by using sufficiently powerful techniques. But, in the 1930's, it was shown that there are true statements for which no proofs exist, and others whose proofs are so long that it is impracticable to write them down. It is possible that a conventional short topological proof of the four-colour theorem will be found. However, Appel and Haken estimate that over ten million man-hours of effort have been spent trying to find such a solution. It is conceivable that their method is the only way and that we are witnessing the birth of a new style of proof.

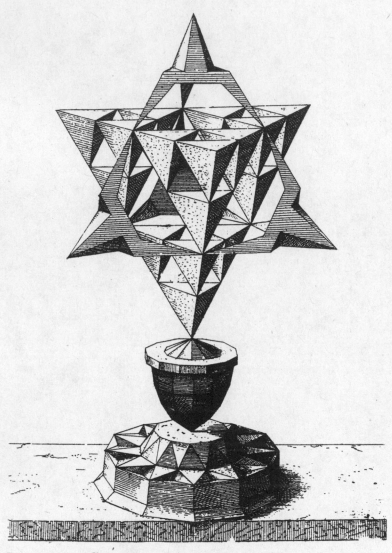

From *Perspectiva Corporum Regularium* by Wenzel Jamnitzer, 1568.

10 Combination, Transformation and Decoration

> *The whole is more than the sum of its parts.* [a]
>
> Aristotle

Some of the most fascinating polyhedral models to play with are compounds. As you turn them in your hands, the individual components catch your eye in turn, and your attention jumps from one to another as you study the intriguing way they are interlocked. The easiest examples to appreciate are those in which the individual components stand out clearly and are quickly recognised. For this reason compounds of regular polyhedra are particularly striking, especially when each component is painted in a different colour.

Examples of compound polyhedra have appeared in earlier chapters. Compounds of two tetrahedra, a cube and octahedron, and a dodecahedron with an icosahedron were known to Kepler (Chapter 4). Compounds of five and ten tetrahedra, and of five octahedra occur among the stellations of the icosahedron (Chapter 7). We will now construct more examples.

Making symmetrical compounds

A *compound* polyhedron is a set of distinct polyhedra, called the *components* of the compound, which are placed together so that their centres coincide. If the component polyhedra are of similar sizes then they will probably intersect each other, the faces of one component passing through the faces of the others. Although it is possible to make compounds of any polyhedra, we shall restrict attention to compounds of Platonic solids and in which all the components are the same. These can have a high degree of symmetry which makes them very attractive.

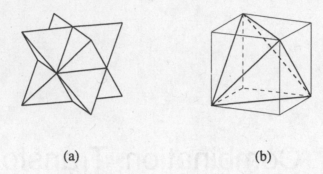

(a)   (b)

Figure 10.1. A tetrahedron can be inscribed in a cube in two ways.

Consider the compound of two tetrahedra shown in Figure 10.1. It can be inscribed in a cube so that the $4 + 4$ vertices of the two tetrahedra coincide with the eight vertices of the cube. This shows that a single tetrahedron can be inscribed in a cube in two different ways. The compound results from doing this in both ways simultaneously. This method of inscribing one polyhedron in another in several ways is a quick way to generate compound polyhedra.

Recall Euclid's construction of a dodecahedron. He started with a cube and erected a 'roof' on each face in such a way that the sloping parts on adjacent faces were coplanar and formed regular pentagons (Figure 10.2(a)). There are two ways to erect a roof on the face of a cube. Using both simultaneously produces the compound of two dodecahedra shown in Figure 10.2(b) and Plate 16.

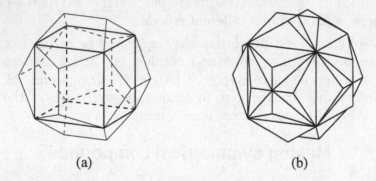

(a)   (b)

Figure 10.2. A dodecahedron can be circumscribed about a cube in two ways.

This idea of placing one polyhedron in another in different ways can also be used the other way round. Euclid's construction also shows that a cube can be inscribed in a dodecahedron. To form a compound, we do this in as many different ways as possible, simultaneously. How many cubes can be inscribed in a dodecahedron? Each of the 12 edges of the cube is a diagonal of one of the 12 faces of the dodecahedron. There are five choices for the diagonal of a pentagonal face, and therefore five different ways to inscribe a cube in a dodecahedron. All of these cubes taken together form a compound of five cubes. The compound is shown in Plate 13.

We know that a tetrahedron can be inscribed in a cube in two ways. Unfortunately, trying to reverse the situation does not produce a new compound: the cube can be circumscribed about a tetrahedron in only one way.

Symmetry breaking and symmetry completion

The question of how many times one polyhedron can be inscribed in (or circumscribed about) another polyhedron in a particular way is closely related to symmetry. More precisely, it is connected with the symmetry of three things: the inscribed polyhedron, the circumscribed polyhedron, and the compound of the two taken together. We need to introduce some terminology: the inscribed polyhedron will be called the *kernel*, the circumscribed polyhedron will be called the *shell*, and the compound of the kernel and the shell will be called the *amalgam*.

In the cube-in-dodecahedron example, the cube is the kernel and the dodecahedron forms the shell. The (rotational) symmetry groups of the kernel, shell and amalgam are O, I and T respectively. Notice that the amalgam has a lower degree of symmetry than either the kernel or the shell alone. Some of the symmetry has been destroyed. However, in the resulting compounds of five cubes and of two dodecahedra, the lost symmetry has been reinstated: the five cubes have icosahedral symmetry; the two dodecahedra have octahedral symmetry.

This is not a coincidence and the idea of reinstating destroyed symmetry can be used to create a variety of compounds. The cube-in-dodecahedron example will be examined in more detail so that the relationships between the symmetry groups of the kernel, shell, amalgam and their compounds become apparent. Then this process of symmetry completion will be used to generate more compound polyhedra.

Example. A cube in a dodecahedron.

symmetry groups: kernel   (cube) O
                    shell   (dodecahedron) I
                    amalgam T

The symmetries of the kernel can be divided into two sets: those that are also symmetries of the amalgam, and those that are not. The first set will form a group, in this case the tetrahedral group, T. The other set contains the symmetries of the kernel which are destroyed in the amalgam. This set is not a group since, for one thing, it does not contain the identity symmetry. If these destroyed symmetry operations are performed on the amalgam then the kernel will remain unchanged because these operations are symmetries of the kernel. However, they are not symmetries of the shell so the shell will be carried to new positions different from the original one. In our example, the rotations of 90°, which are symmetries of the cube, are not symmetries of the amalgam. They carry the dodecahedron to a second position. This compound of dodecahedra will then have the same symmetry as that of the kernel. Furthermore, the number of components can be worked out from the sizes of the groups involved:

$$\begin{pmatrix} \text{number of components in the} \\ \text{compound of shell polyhedra} \end{pmatrix} = \frac{(\text{number of symmetries of the kernel})}{(\text{number of symmetries of the amalgam})}.$$

This process can be applied using the symmetries of the shell in place of those of the kernel. This results in a compound of kernel polyhedra which has the same symmetry as the shell. The number of components is then:

$$\begin{pmatrix} \text{number of components in the} \\ \text{compound of kernel polyhedra} \end{pmatrix} = \frac{(\text{number of symmetries of the shell})}{(\text{number of symmetries of the amalgam})}.$$

The examples here have produced compounds of $\frac{24}{12}$ shells (that is two dodecahedra) and $\frac{60}{12}$ kernels (five cubes).

Example.   A tetrahedron in a cube.

symmetry groups:	kernel	(tetrahedron)	T
	shell	(cube)	O
	amalgam		T

The point to notice here is that the symmetry groups of the kernel and the amalgam are identical. Therefore, although applying the symmetries of the shell to the amalgam produces a compound of two tetrahedra, applying the symmetries of the kernel to the amalgam does not produce anything new.

Example.   A cube in an octahedron.

symmetry groups:	kernel	(cube)	O
	shell	(octahedron)	O
	amalgam		D_4

A cube can be inscribed in an octahedron as shown in Figure 10.3(a). Both the kernel and the shell have octahedral symmetry. However, they are arranged so that only one pair of 4-fold axes match up in the amalgam, the other 4-fold axes are reduced to 2-fold axes. This means that the amalgam has dihedral symmetry of type D_4. Some symmetries of both the shell and the kernel are destroyed in the amalgam and both sets can be used to generate compounds. Reinstating the symmetry of the shell gives a compound of three cubes (Figure 10.3(b)) and

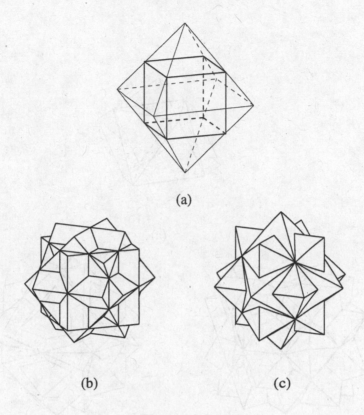

(a)

(b)   (c)

Figure 10.3.

completing the symmetry of the kernel produces a compound of three octahedra (Figure 10.3(c)). Both compounds have octahedral symmetry and are shown in Plate 14.

Example. An octahedron in a cube.

symmetry groups:	kernel	(octahedron)	O
	shell	(cube)	O
	amalgam		D_3

An octahedron can be inscribed in a cube as shown in Figure 10.4(a). The symmetry of the amalgam is again dihedral but is now of type D_3. Completing the symmetry of the shell gives a compound of four octahedra, and completing the symmetry of the kernel produces a compound of four cubes. These are shown in Figures 10.4(b) and (c), and in Plate 15.

In the above examples, the amalgam of the two polyhedra is formed by inscribing one polyhedron in another. The fact that the two polyhedra share such a close relationship is not essential to the process of generating new compounds. All that we are doing is using a set of rules (the symmetries) to repeat a given object. One of the polyhedra is a template for the components of the compound.

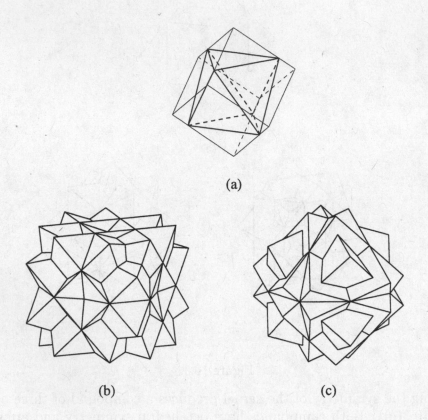

(a)

(b)   (c)

Figure 10.4.

and the other acts merely as a reference marker to indicate how the template is
situated with respect to the symmetry elements which are used to generate the
compound. We can choose any polyhedron to be the template and repeat it using
the symmetry operations in any symmetry group. However, the most interesting compounds arise when some of the symmetries of the template coincide with
those being used to generate the compound.

The following example illustrates some of these points. Figure 10.5 shows
two tetrahedra with a common vertex. The larger one is the template and the
symmetries of the other determine how it is to be repeated to generate a compound. The result is the compound of four tetrahedra shown in Plate 11. Models
of other compound polyhedra are shown in the plates. These include compounds
of five octahedra and five dodecahedra.

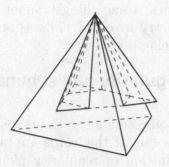

Figure 10.5.

Are there any regular compounds?

Do any of the compounds described above (or any others) deserve the prestigious label of 'regular'? Clearly, since all the components of these compounds are Platonic solids, all their faces are equal regular polygons and the same number meet at every vertex. Even so, some are still 'more regular' than others. This can be seen most clearly by examining the visible parts of the faces. In some compounds the same regions are visible whichever face is looked at. Others have more than one kind of face pattern. In the three octahedra, for example, all the faces are equivalent whereas the compound of three cubes has two kinds of face. These are shown in Figure 10.6—the shaded parts are visible. In this latter case, the faces do not all play the same role because we can distinguish the two kinds.

We can also consider the vertices. They are all similarly surrounded but are they all indistinguishable? Do they all play the same role? In the compound of four cubes there are two kinds of vertex: some are surrounded by three kite-shaped portions of the square faces, others by an 'L'-shaped piece and two triangles.

Figure 10.6.

The fact that, sometimes, some things cannot be distinguished hints that there is an aspect of symmetry involved. This is not surprising since regularity and symmetry are closely related concepts.

Regularity and symmetry

When considering the Platonic solids the terms 'regular' and 'symmetrical' are often regarded as synonyms but, if the words are taken in their technical sense, this is incorrect. The definition of regularity requires that the same number of congruent regular polygons surround every vertex. This is a local condition which specifies how an individual face is situated with respect to its neighbours. It restricts the way that faces can meet. On the other hand, symmetry operations consider the polyhedron as a whole; they take a global viewpoint.

The difference between local order properties and global symmetry becomes noticeable when the conditions are applied in a wider context. The rhomb-cub-octahedron and Miller's solid have the same local order properties: their solid angles are all congruent. However, their symmetry properties are different. In the rhomb-cub-octahedron all the vertices are equivalent, but in Miller's solid this is not the case: we can distinguish two kinds of vertex when we look at their position in the polyhedron as a whole.

The contrast between local and global viewpoints has recently become the focus of attention in various subjects. In crystallography, for example, symmetry groups have proved very useful in describing the internal structure of crystals, and they convey information about global properties. However, as a crystal grows, the molecules or ions joining the crystal align themselves according to the configuration of those molecules already belonging to it. The atomic forces which govern the building of a crystal are of a local nature. When confronted with the differences between the rhomb-cub-octahedron and Miller's solid it seems rather fortuitous that the resulting crystal structure has a strong global symmetry. Why should local ordering rules that are effective over a short range produce a structure with long range symmetries?

Returning to the case of polyhedra, it seems clear that the local and global viewpoints are different responses to the same problem. They are both ways of trying to express the same visual property of the Platonic solids—namely, that they look the same from different directions. When you turn a model of a Platonic solid in your hands then whenever you see it face first, the same image is presented to the eye no matter which face is at the front. Similarly, when viewed edge on, or vertex first, the picture does not depend on the edge or vertex that you choose to look at. This aesthetic quality can be expressed by saying that all the faces are congruent, that the dihedral angles are equal, and that all the vertices are similarly surrounded. It can also be expressed by noting that the model has a particularly strong kind of symmetry. The idea involved in this second case can be made precise using the notion of transitivity which unites the concepts of regularity and symmetry.

Transitivity

The notion of transitivity makes explicit the intuitive idea that certain objects (such as faces or vertices) are equivalent or indistinguishable, that they all look the same no matter which is focused on. The transitivity properties of a polyhedron describe the different directions from which the polyhedron looks the same. The word 'transitive' has the same root as the words 'transit' and 'transition', commonplace words that convey ideas of motion, of change of place, of passage from one state to another. This is appropriate because transitivity concerns the possibility of moving objects around.

A polyhedron is said to be *face-transitive* if, for any pair of faces, there is a symmetry of the polyhedron which carries the first face onto the second. Physically, this means that the polyhedron looks the same when viewed face on, no matter which face is presented to the eye. Every time the model is rotated so that it is seen face first, the rotation involved is a symmetry of the polyhedron so the initial and final positions are indistinguishable. (Face-transitive polyhedra are sometimes called *isohedral*.)

Examples of face-transitive polyhedra include the Platonic solids, and the dipyramids and trapezohedra (Figure 8.10(a)–(c)). Another thirteen examples, which include Kepler's two rhombic polyhedra, are shown in Figure 10.7. They are related to the Archimedean solids and were first described by Eugène Charles Catalan (1814–1894).

Polyhedra which are not face-transitive are easy to find since the faces of a face-transitive polyhedron must be all the same shape. Therefore the prisms and Archimedean solids cannot be face-transitive. The faces of the Siamese dodecahedron are all congruent equilateral triangles, but it is still not face-transitive. Its symmetry group (D_{2v}) contains only six symmetries: the identity, two reflections in perpendicular mirror planes, and three 2-fold rotations (about the principal axis, and each of the two secondary axes). So any one of the 12 faces can be carried to at most five other faces by the symmetry operations in this group.

Many non-convex polyhedra are face-transitive. The four regular star polyhedra are examples. The stellations of the dodecahedron are still face-transitive when we interpret them as spherical polyhedral surfaces each composed of sixty isosceles triangles. If the octahedron is augmented by regular tetrahedra then we see that the stella octangula can be interpreted as a spherical surface composed of 24 equilateral triangles. Other compounds give examples of non-convex face-transitive polyhedra when seen in this naive way. In this light the 'five tetrahedra'

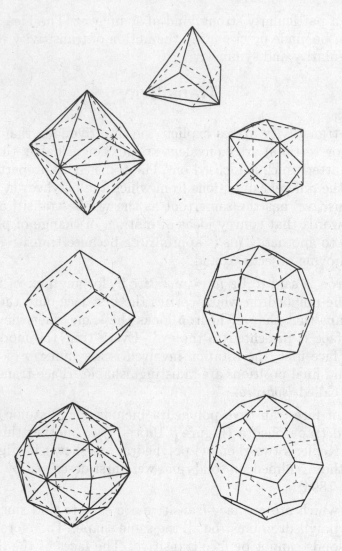

Figure 10.7. The Catalan solids.

is composed of sixty non-convex pentagons. Some of the other stellations of the icosahedron provide further examples of face-transitive polyhedra.

General properties of face-transitive polyhedral surfaces have been investigated by Branko Grünbaum and Geoffrey Shephard. They found that all such polyhedra must be *star-shaped* (that is, there must be a point in the interior from which it is possible to get an unobscured view of the whole surface). The faces of these polyhedra can be triangles, convex quadrilaterals, or star-shaped[1] pentagons. Some of Grünbaum and Shephard's examples are shown in Figure 10.8.

---

[1]Do not confuse star-shaped polygons and polyhedra with star polygons and star polyhedra in the Kepler–Poinsot sense.

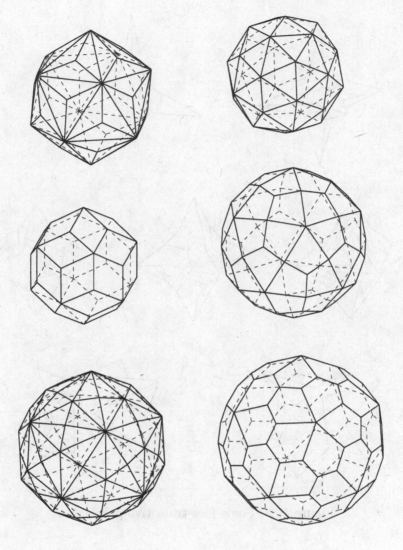

Figure 10.7 (*continued*).

A polyhedron is *vertex-transitive* (or *isogonal*) if any vertex can be carried to any other by a symmetry operation. This corresponds to the fact that a polyhedron looks the same when viewed with any of its vertices directed forwards. The Archimedean solids are all vertex-transitive. However, Miller's solid has symmetry group D_{4v} and these 16 symmetries are insufficient to carry a vertex onto each of the 23 others. The Greeks did not express the aesthetic quality they saw sufficiently well to ensure that all the semiregular polyhedra allowed under their definition looked the same from every vertex. The fact that Archimedes did not record Miller's solid among his list of semiregular polyhedra may indicate

Figure 10.8. Some face-transitive polyhedra.

that he was intuitively searching for vertex-transitive solids. However, his list did not include the prisms either and these are vertex-transitive. Perhaps he, like Kepler, did not find polyhedra with prismatic symmetry so pleasing on the eye.

Rather surprisingly, there are examples of vertex-transitive polyhedra which are not spherical. Unlike the face-transitive polyhedra, they cannot be deformed into a sphere but have tunnels running through them. These vertex-transitive polyhedra were discovered by Grünbaum and Shephard. Some can be obtained from prisms. Figure 10.9(a) shows a polyhedron which has the same vertices as the octagonal prism shown in Figure 10.9(b) but which has been 'twisted' so that each of the eight square faces has become a pair of triangles. Combining the squares of the prism with the sixteen triangles of the 'twisted prism' produces a vertex-transitive polyhedral surface which is topologically a torus.

Other examples can be constructed which have higher genus (more tunnels).
Figure 10.9(c) shows a polyhedron which is combinatorially equivalent to a snub
cube. Its convex hull, shown in Figure 10.9(d), has the same set of vertices and is
also equivalent to a snub cube but geometrically the two polyhedra are different.
If the square faces are removed from each polyhedron and the two remaining
pieces glued together, a vertex-transitive polyhedron of genus five is produced.
Other examples can be constructed which have genera 3, 7, 11 and 19.

A polyhedron is *edge-transitive* (or *isotoxal*) if any edge can be carried to any
other by a symmetry operation. This corresponds to the fact that the polyhedron looks the same when viewed edge on, from any edge. Edge-transitivity differs
from the two previous kinds of transitivity in that it cannot occur alone. Every
edge-transitive polyhedron has to be either face-transitive or vertex-transitive
(possibly both). For example, the rhombic dodecahedron is edge-transitive and
face-transitive. Examples of polyhedra which are both face-transitive and vertex-
transitive but not edge-transitive are the irregular tetrahedra known as sphenoids
(see Figure 10.10). These have isosceles or scalene triangles for faces. The com-

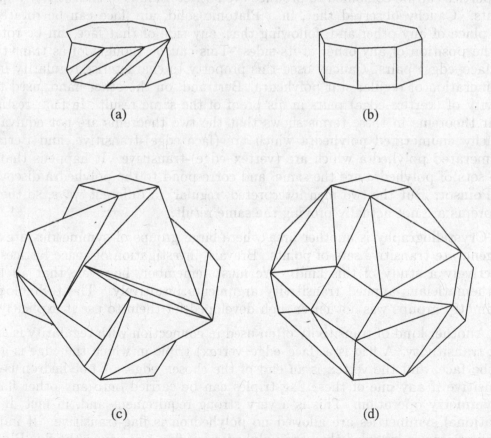

(a)   (b)

(c)   (d)

Figure 10.9. Constructions of two aspherical, vertex-transitive polyhedra.

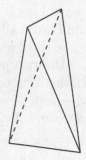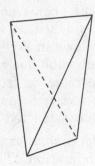

Figure 10.10. Isosceles and scalene sphenoids.

plete icosahedron (Plate 10) provides another example if its faces are interpreted as irregular star 9-gons which pass through each other.

The three basic elements of a polyhedral surface used to define transitivity properties can be combined to produce even stricter kinds of transitivity requirements. Cauchy observed that, in a Platonic solid, any face can be rotated to the place of any other and, following that, any side of that face can be rotated to the position of any other of its sides. This can be thought of as transitivity of (face–edge) pairs. Cauchy used this property to characterise regularity in his enumeration of regular star polyhedra. Bertrand, on the other hand, used transitivity of (vertex–edge) pairs in his proof of the same result. In fact, restating their theorems in these terms shows that the two theorems are not equivalent. Cauchy enumerated polyhedra which are (face–edge)-transitive, and Bertrand enumerated polyhedra which are (vertex–edge)-transitive. It happens that the two sets of polyhedra are the same, and correspond to the polyhedra discovered by Poinsot. But the two men interpreted 'regular' in different ways, so the two theorems are not actually proving the same result.

Crystallography is another area where large groups of symmetries are used to generate transitive sets of points. Bravais' investigation of space lattices was effectively a study of this kind. We must remember, however, that all these mathematicians applied transitivity arguments intuitively. The technology of symmetry groups was not far enough developed for them to use it explicitly.

Another kind of transitivity often used in connection with regularity is called flag-transitivity. A flag is a (face–edge–vertex) triple in which the edge is a side of the face, and the vertex is an end of the chosen edge. A polyhedron is flag-transitive if any one of these flag triples can be carried onto any other flag by a symmetry operation. This is a very strong requirement and, in fact, if only rotational symmetries are allowed no polyhedron is flag-transitive. If indirect symmetries are included then ten polyhedra are flag-transitive: the five Platonic solids, the four Kepler–Poinsot star polyhedra, and the stella octangula.

Polyhedral metamorphosis

Transitivity properties are particularly nice features of polyhedra. They make descriptions very easy to state. Instead of describing the positions of all the parts, we can give the position of just one of them and the rules for repeating it. A vertex-transitive convex polyhedron can be completely specified with just two pieces of information: the coordinates of one vertex and the symmetry group of the polyhedron. A cub-octahedron, for example, might be given by the pair $((1,1,0), O_h)$. Of course, the amount of information required has not been reduced: we still need to know three real numbers to position each vertex. The difference is in how the information is presented. A lot of information is encoded into the symbol for the symmetry group and this is what makes the description so concise. The repetition rules of the symmetry group describe how the set of vertices is related to the given point and, because we are assuming that the polyhedron is convex, the faces can then be filled in in a unique way.

This notation for describing particular polyhedra can also be used in reverse. The given information consists of a point in space and a symmetry group and these data can be used to generate a polyhedron: apply the symmetries to the given point to produce a set of vertices and form their convex hull to make a polyhedron. Which polyhedra can be created in this way? The seed point and the group can both be changed—the point can move continuously in space and there are several symmetry groups to choose from. How do these parameters affect the resulting polyhedron?

For the moment we shall restrict our attention to one symmetry group and investigate what effects the choice of seed point has. We shall concentrate on the group O_h. If the seed point is moved along a line that passes through the centre of symmetry then all that changes is the size of the polyhedron. Since changes of scale are not very interesting we can ignore how far away the seed point is from the centre of symmetry and just record its direction. This can be achieved by restricting the seed point to lie on a sphere centred at the centre of symmetry. Choosing different points on this sphere should give rise to different polyhedra.

The mirror planes of the O_h symmetry system intersect the sphere in great circles. They divide up the sphere into a network of 48 spherical triangles (shown in Figure 10.11) and these act as a reference grid to indicate where the symmetry elements are situated. Where two great circles intersect, a 2-fold rotation axis pierces the sphere. Similarly, the 3-fold and 4-fold axes occur where three or four great circles meet. If the seed point is chosen to lie on an axis then the polyhedron which results is either an octahedron (4-fold axis), a cube (3-fold axis) or a cub-octahedron (2-fold axis).

When the seed point lies inside a triangle we get a polyhedron which has 48 vertices—there will be one vertex in each triangle. For one special point inside

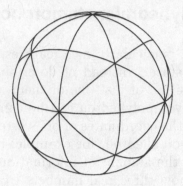

Figure 10.11.

each triangle, the generated polyhedron will have regular faces: it will be the great rhomb-cub-octahedron, an Archimedean solid. All the other polyhedra generated from points inside the triangle will be like the great rhomb-cub-octahedron in many respects (some examples are shown in the centre of Figure 10.12). They are all made up of six octagons, eight hexagons, and twelve rectangles, and every vertex is surrounded by one of each of the three kinds of face. The shapes of the faces change as the seed point is varied but the way that they are put together is preserved. All the polyhedra have the same combinatorial structure. Such polyhedra are said to be *isomorphic*.[2] This word is derived from the Greek for 'same form' and is used in many areas of mathematics when objects have a common structure.

There are other points on the parameter sphere which have not been investigated yet. These are the points that lie on exactly one great circle. The points on any one of the sides of a spherical triangle generate a family of polyhedra which are all isomorphic. The points on a line connecting a 2-fold axis to a 3-fold axis are isomorphic to a truncated cube, the points on a line connecting a 2-fold axis to a 4-fold axis are isomorphic to a truncated octahedron, and the points on a line connecting a 3-fold axis to a 4-fold axis are isomorphic to a rhomb-cub-octahedron. One point on each line generates a regular-faced Archimedean solid. Examples from these isomorphism classes are illustrated around the edge of Figure 10.12. The diagram is a schematic map of one of the spherical triangles. The polyhedra illustrated show what kinds of polyhedron are generated by choosing a seed point in various regions. Studying the differences in these figures and their relative positions gives a good idea of how the polyhedron varies with the choice of seed point.

A better way to get a feel for the variety of polyhedra and their interrelationships is to use a computer (if your machine and programming skills cope with the

---

[2]Note that some authors describe polyhedra which have the same net as isomorphic. In this book, such polyhedra are called stereo-isomers.

challenge). Interactive graphics provide an excellent tool for understanding these changes in a very direct way. Using a mouse to drag the seed point about inside a triangular region, you can watch as the polyhedron changes shape in response. You actually experience the metamorphosis of polyhedral forms.

What happens if we change the symmetry group? If we replace O_h by the group containing only the rotations of a cube, O, then many of the polyhedra that appear are the same as those generated by the O_h group. The point on the 3-fold axis still produces a cube, for example. In fact, all the points which lie on one or more of the great circles produce a polyhedron with mirror symmetry. The only differences between the O and O_h groups occur in the interiors of the spherical triangles. Instead of generating isomorphs of the great rhomb-cub-octahedron, isomorphs of the snub cube appear. Some examples are illustrated in Figure 10.13.

There are two enantiomorphic forms of the snub cube. Which one is produced depends on where the seed point is located. Half of the triangles give one form, and the rest give its mirror image. It is apparent from these examples that the spherical triangles in the parameter space are more than just a convenient reference system. Knowing how the polyhedra are related to the seed points in any one of these triangles is sufficient to understand the whole behaviour. Each triangle is called a *fundamental region* for the group O_h: each triangle can be carried to any other by a symmetry in the group, and together they cover the whole sphere. All the triangles are equivalent because they are all in the same transitivity class. For the group O there are two transitivity classes of triangles which are mirror images of each other. Taking one triangle from each class produces a fundamental region for O.

The analysis of the symmetry groups I_h and I is analogous that for O_h and O. Schematic diagrams showing the relative positions of the various polyhedra would exhibit the same general features. The three tetrahedral groups are more interesting. Figures 10.14–10.16 are schematic diagrams showing the kinds of polyhedron which are generated by seed points in a spherical triangle using the symmetry groups T, T_d and T_h. Notice the appearance of the Platonic icosahedron. It lies on the great circle connecting the cub-octahedron to the octahedron in Figures 10.14 and 10.16. The squares of the cub-octahedron are folded along a diagonal to produce a polyhedron isomorphic to the icosahedron but which has obtuse-angled isosceles triangles for some of its faces. As the seed point progresses towards the 4-fold axis these triangles become acute-angled and finally vanish to become an edge of the octahedron. Somewhere along this route, all the triangles are equiangular and the icosahedron is regular.

The spherical triangles in these figures are still the fundamental triangles shown in Figure 10.11. It is sufficient to study just one of these even though all the tetrahedral groups have fundamental regions which are larger than this. The fundamental regions for the groups T_d and T_h are formed from two of the triangles.

Seed points in the two triangles produce mirror-image polyhedra but because these polyhedra are not cheiral we cannot tell them apart. The rotation group T does produce cheiral polyhedra—what we might call 'snub tetrahedra'. Its fundamental region is made up of four triangles but again, if we ignore cheirality, it is sufficient to consider only one of them.

The space of vertex-transitive convex polyhedra

A mathematician who asks "what space you are working in?" is not referring to the size of your office.[b]

Ian Stewart

Traditionally, space is where the objects of geometry live and have extension. The structural properties of this Euclidean space are set out at the beginning of the *Elements*. For the modern mathematician, the word 'space' is not limited to the conventional notions of a three-dimensional space of points. It is used to refer to the collection of objects being considered, and the structure of the space reflects the relationships between these objects.

In the previous section, polyhedra were generated by choosing a seed vertex from a spherical triangle. The point chosen is a variable parameter and the triangle is the collection of all the possible values it can take. The triangle is an example of a *parameter space*. There is a one-to-one correspondence between the points in a triangle and the convex vertex-transitive polyhedra with a particular symmetry group. We can also interpret the triangle as a space of polyhedra: each 'point' (or element) in the space *is* a polyhedron. Furthermore, the structure of the space has a useful interpretation: polyhedra which are almost the same are very close together in this space. The distance between two points of the space (that is, two polyhedra) gives a measure of how different they are.

The idea of a space of polyhedra can be made very precise. Using a parameter-space as an example of how such a space can be constructed is somewhat misleading but it does give the flavour of what is involved. We want a space of polyhedra to have a structure which reflects the relationships between polyhedra. We can give a qualitative description of such a space which exhibits its primary features. This is rather like a map of an underground rail network. In this case, the important points which need to be recorded are things like the order of the stations along each track and the crossover stations where you can change lines. The actual physical layout of the tracks through the city is irrelevant to the passengers. All that they need to know is that nearby points on the map correspond to nearby points in the city. How can we make a schematic map of the space of polyhedra?

To simplify the problem greatly, we shall restrict our attention to convex

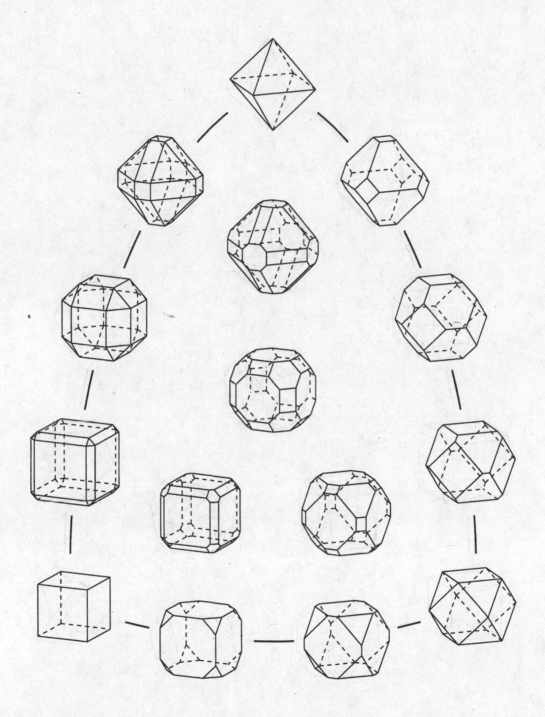

Figure 10.12. Polyhedra in the parameter space of the group O_h.

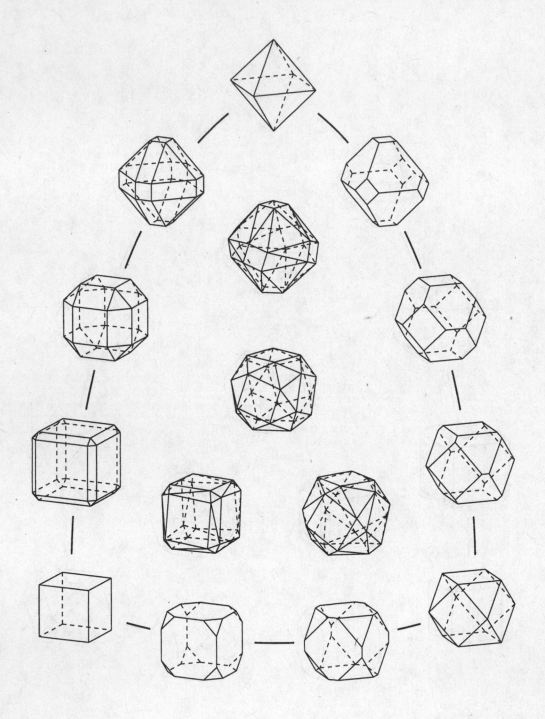

Figure 10.13. Polyhedra in the parameter space of the group O.

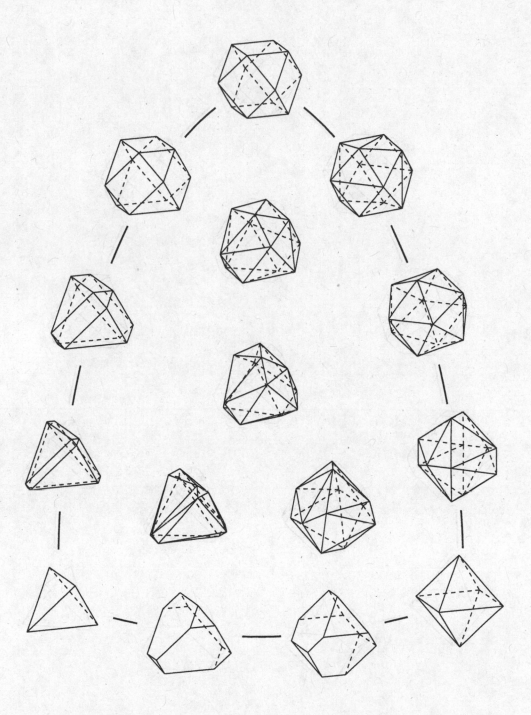

Figure 10.14. Polyhedra in the parameter space of the group T.

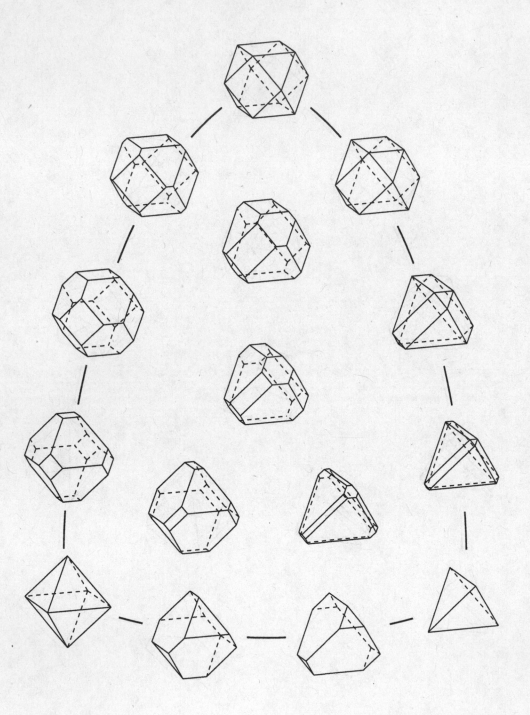

Figure 10.15. Polyhedra in the parameter space of the group T_d.

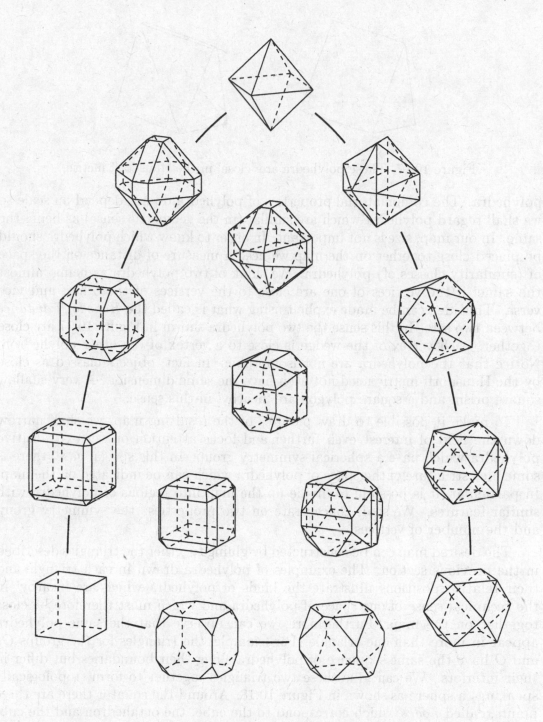

Figure 10.16. Polyhedra in the parameter space of the group T_h.

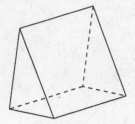 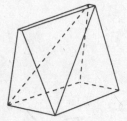

Figure 10.17. These polyhedra are 'close' in the Hausdorff metric.

polyhedra. The combinatorial properties of polyhedra do not depend on scale so we shall regard polyhedra which are similar (in the geometric sense) as being the same. In our map, size is not important. In order to know which polyhedra should be placed close together on the map we need a measure of distance on the space of (similarity classes of) polyhedra. We think of two polyhedra as being 'almost the same' if the vertices of one are close to the vertices of the other, and vice versa. This idea can be made explicit using what is called the *Hausdorff distance* between two sets. In this sense the two polyhedra shown in Figure 10.17 are close together: each vertex of the wedge is close to a vertex of the other polyhedron. Notice that the polyhedra are not isomorphic. In fact, objects classed as close by the Hausdorff metric need not even have the same dimension. A very shallow square prism and a square polygon are 'nearby' in this sense.

To make it possible to draw pictures of the resulting map, we shall narrow down the field of interest even further and focus attention on vertex-transitive polyhedra which have a spherical symmetry group. In this special case, there is some nice structure of the space of polyhedra which can be indicated on the map. In particular, it is possible to divide up the map into regions of polyhedra with similar features. We shall concentrate on two properties: the symmetry group and the number of vertices.

The desired map can be constructed by gluing together the triangles described in the previous section. The examples of polyhedra drawn in each triangle and their relative positions illustrate the kinds of polyhedra which are 'nearby' in the technical sense of our space of polyhedra and which must therefore be close together on the map. Furthermore, we can observe that the same polyhedra appear in more than one triangle. For example, the triangles for the groups O_h and O have the same sequence of polyhedra along their boundaries but differ in their interiors. We can glue these two triangles together to form (topologically speaking) a sphere as shown in Figure 10.18. Around the equator there are three points, called *nodes*, which correspond to the cube, the octahedron and the cub-octahedron. These have 8, 6 and 12 vertices respectively. All the other polyhedra on the equator have 24 vertices but they are not all the same. Some are isomorphic to the truncated cube or the truncated octahedron, others to the rhomb-cub-

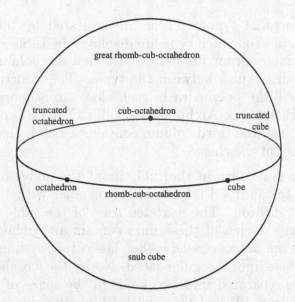

Figure 10.18. A map of the space of convex, vertex-transitive polyhedra
with octahedral symmetry groups (O and O_h).

octahedron. The three nodes divide the equator into three line segments—one line
corresponds to each isomorphism class. In the upper hemisphere are polyhedra
with 48 vertices and O_h symmetry; in the lower hemisphere are polyhedra with
24 vertices and symmetry group O.

A similar analysis of the polyhedra with icosahedral symmetry can be used
to construct another sphere for the groups I_h and I. Such a sphere is shown in
the upper-right part of Figure 10.20. The numbers on the nodes, lines and faces
match the polyhedra listed in the Table 10.21.

The triangles containing the polyhedra with one of the tetrahedral symmetry
groups contain examples from both the octahedral and icosahedral spheres. In
fact, the three triangles in Figures 10.14–10.16 can be joined into a large region
whose boundary is the same as the equator of the sphere of octahedral symmetries
(see Figure 10.19). This region also contains a point corresponding to the regular
icosahedron. So we must wrap this region around the equator of the octahedral
sphere and attach it to the icosahedral sphere at a single point. This completes
our map of the space of vertex-transitive convex polyhedra whose symmetry group
is one of the seven spherical groups.

The map is shown in Figure 10.20. The upper hemisphere of the octahedral
sphere which contains polyhedra with O_h symmetry (coloured grey in the figure)
has been flattened so that the tetrahedral part of the map can be attached to its
equator. The points, line segments, and regions of the map delineate its structure.
They mark places where there is a discontinuity in the space—either a jump in
the number of vertices or a switch of symmetry group. This division of the space

of polyhedra produces 24 types. These are indicated by the numbers on the diagram and the data about each type are displayed in Table 10.21. The number of vertices and symmetry group are listed in the first two columns but these alone are not enough to distinguish between the types. The structure of the space is richer than that. All the polyhedra in each class are isomorphic to each other and to one of the Platonic or Archimedean solids. The last column contains the name of that solid, and the third column contains a tick if the regular-faced one is actually a member of the class.

The first seven polyhedra in the table lie at the nodes of the map. These points correspond to the five Platonic solids together with the cub-octahedron and the icosi-dodecahedron. The next ten rows of the table deal with the line segments in the map. Seven of these lines contain an Archimedean solid. The other three classes are not covered by the classical discussions because none of the polyhedra on these lines is regular-faced. Notice the icosahedra with isosceles triangular faces are separated into two kinds in the space of polyhedra. Those with 'fat' triangles (row 10) and those with 'thin' triangles (row 11). The apex angle in a fat isosceles triangle is larger than 60°, in a thin triangle it is less than 60°. In this context the special angle is not the right angle which gives rise to the acute/obtuse classification of triangles in Euclid's *Elements* but the angle in an equiangular triangle. The last seven rows of the table contain the polyhedra in the regions of the map. Four of the regions contain a familiar Archimedean solid, the others contain more examples of polyhedra unknown in ancient times.

Notice how the icosahedron plays a crucial role in making the space connected. Because it is possible to connect any two points in the map by a line, it is possible

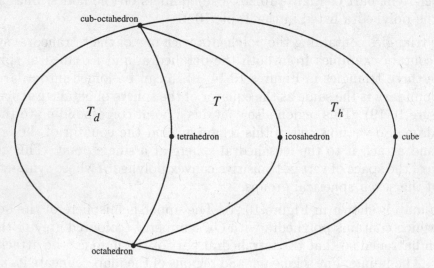

Figure 10.19. A map of the space of convex, vertex-transitive polyhedra with tetrahedral symmetry groups (T, and T_d and T_h).

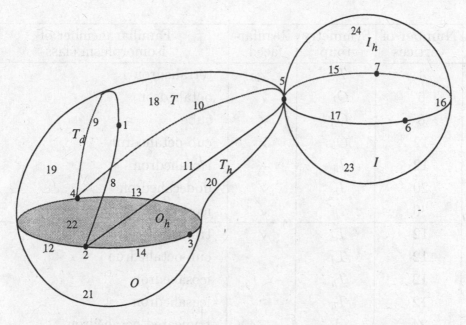

Figure 10.20. A map of the space of convex, vertex-transitive polyhedra with a spherical symmetry group.

to continuously deform any one of these polyhedra to any of the others in such a way that all the polyhedra it passes through are also convex and vertex-transitive.

A more detailed and rigorous study of this space was carried out by Stewart Robertson, Sheila Carter and Hugh Morton in 1970. They also analysed the prismatic groups. This is more awkward to describe since there are infinitely many prismatic groups. However, the map of the spherical groups is connected to the prismatic part of the map through the tetrahedron, the octahedron, and the cube.

Totally transitive polyhedra

We now return to the problem of finding compound polyhedra which are regular. The definition of a regular polyhedron can be recast using transitivity. The conditions of equal faces, equal dihedral angles, and congruent solid angles correspond to face-, edge-, and vertex-transitivity. A polyhedron that is face-transitive, edge-transitive and vertex-transitive is said to be *totally transitive*.

The totally transitive polyhedra include the Platonic solids and the four Kepler-Poinsot star polyhedra—the polyhedra we call regular. The advantage of the transitivity interpretation of regularity is that it applies to compounds. The transitivity properties of the examples of compounds described above are listed in Table 10.22. Five of them are totally transitive. The compound of two

	Number of vertices	Symmetry group	Regular-faced	Familiar member of isomorphism class
1	4	T_d	✓	tetrahedron
2	6	O_h	✓	octahedron
3	8	O_h	✓	cube
4	12	O_h	✓	cub-octahedron
5	12	I_h	✓	icosahedron
6	20	I_h	✓	dodecahedron
7	30	I_h	✓	icosi-dodecahedron
8	12	T_d	✓	truncated tetrahedron
9	12	T_d		cub-octahedron
10	12	T_h		icosahedron
11	12	T_h		icosahedron
12	24	O_h	✓	truncated octahedron
13	24	O_h	✓	truncated cube
14	24	O_h	✓	rhomb-cub-octahedron
15	60	I_h	✓	truncated icosahedron
16	60	I_h	✓	truncated dodecahedron
17	60	I_h	✓	rhomb-icosi-dodecahedron
18	12	T		icosahedron
19	24	T_d		truncated octahedron
20	24	T_h		rhomb-cub-octahedron
21	24	O	✓	snub cube
22	48	O_h	✓	great rhomb-cub-octahedron
23	60	I	✓	snub dodecahedron
24	120	I_h	✓	great rhomb-icosi-dodecahedron

Table 10.21. Classes of vertex-transitive polyhedra.

tetrahedra was first depicted in Pacioli's *Divina Proportione*. The compounds of
five and ten tetrahedra, of five cubes, and of five octahedra were first described
by Edmund Hess in 1876. The last example in the table, the compound of two
sphenoids, is obtained by stretching the stella octangula along one of its axes
(Figure 10.23). It is interesting in that it is face- and vertex-transitive but not

Components	Quantity	Vertex-trans	Edge-trans	Face-trans
tetrahedra	2	✓	✓	✓
	4			
	5	✓	✓	✓
	10	✓	✓	✓
cubes	3	✓		
	4			✓
	5	✓	✓	✓
octahedra	3			✓
	4			
	5	✓	✓	✓
dodecahedra	2			✓
	5			✓
sphenoids	2	✓		✓

Table 10.22. Transitivity properties of some compounds.

edge-transitive. The compound of four tetrahedra has no transitivity properties at all.

The remainder of this section will be spent proving that all the totally transitive polyhedra have been found. To do this correctly we need to state the kind of polyhedra we are interested in. Only when this is done can we begin to look among them for those that are totally transitive. Since the primary motivation

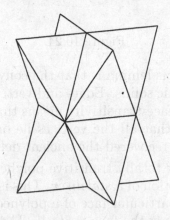

Figure 10.23. A compound of two sphenoids.

for introducing transitivity was to study compounds, we shall allow the polyhedra to have self-intersections where faces pass through each other. The faces
are to be planar polygons whose sides may pass through each other. Thus our
objects of study include the star polyhedra described in Chapter 7 together with
compounds of such polyhedra. With this understanding, the only possibilities
are the five Platonic solids, the four Kepler-Poinsot star polyhedra, and the five
compounds listed above.

Lemma.   The faces of a polyhedron that is edge- and vertex-transitive are regular polygons.

PROOF: As the polyhedron is vertex-transitive, all of its vertices lie on a sphere.
A face of the polyhedron can be thought of as the base of a pyramid whose apex
is the centre of this sphere (see Figure 10.24). All the sides of the base of the
pyramid are the same length because the polyhedron is edge-transitive. So, the
other faces of the pyramid are all congruent isosceles triangles. This forces the
pyramid to be a right pyramid on a regular polygon base.   ■

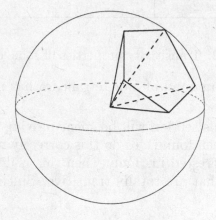

Figure 10.24.

A simple corollary of this lemma is that the convex totally transitive polyhedra are precisely the Platonic solids. Edge- and vertex-transitivity imply that the
faces are regular polygons, face-transitivity means that the faces are all equal, and
vertex-transitivity implies that all the vertices lie on a sphere and are similarly
surrounded. Thus we have recovered the ancient definition.

To enumerate the other totally transitive polyhedra we shall use some group
theory, in particular, the notion of a *stabiliser*. This is a sort of localised symmetry
group. The stabiliser of a particular face of a polyhedron is the set of symmetries
of the polyhedron which carry the face to itself. These symmetries leave the face
unchanged or 'stabilise' it. The stabiliser of a face is always a subgroup of the

symmetry group of the polyhedron and it always contains at least one symmetry: the identity.

Stabilisers of edges and vertices can be defined in the same way.

Theorem. There are fourteen totally transitive polyhedra: the five Platonic solids, the four Kepler–Poinsot star polyhedra, and five compound polyhedra.

PROOF: The preceding lemma showed that the faces of a totally transitive polyhedron must be regular polygons. Face-transitivity forces all the faces to be congruent.

The proof of the theorem proceeds in several steps and combines ideas from both Cauchy's and Bertrand's enumerations of the star polyhedra. The first step is to show that (at least) one of the following holds:

(i) there is a rotation which carries a face onto itself, or

(ii) there is a rotation which carries a vertex onto itself.

In the first case, we can show that the face-planes of the polyhedron bound one of seven possible convex kernels. Searching their stellation patterns for regular polygons produces candidates for totally transitive polyhedra. This is Cauchy's method. Case (ii) corresponds to Bertrand's method: totally transitive polyhedra can be produced by facetting any of seven convex solids.

STEP 1. The stabiliser of a face contains a rotation.
Since the polyhedron is face-transitive, the stabilisers of the faces are all the same size. (In fact, the subgroups are all conjugate.) Similarly, all the stabilisers of the edges are the same size. Let ϕ be the number of symmetries in the stabiliser of some face, and let η be the number of symmetries in the stabiliser of an edge. If we can show that ϕ is at least three then the stabiliser of a face must contain rotations which carry the face onto itself and therefore an axis of rotation must pass through the face.

We now apply the Orbit–Stabiliser theorem. This implies that the number of faces multiplied by the number of symmetries which stabilise a face equals the number of symmetries of the polyhedron. Similarly for the edges. If we write F and E for the numbers of faces and edges of the polyhedron then

$$\text{the number of symmetries of the polyhedron} \quad = \quad F \cdot \phi$$

and also

$$\text{the number of symmetries of the polyhedron} \quad = \quad E \cdot \eta.$$

Equating these two things and rearranging gives

$$\frac{\eta}{\phi} = \frac{F}{E}.$$

The ratio of faces to edges is easy to calculate. All the faces are regular n-sided polygons, for some $n \geqslant 3$, and two sides are brought together to form each edge. So $nF = 2E$. This implies that

$$\frac{\eta}{\phi} = \frac{F}{E} = \frac{2}{n}$$

$$= \frac{2}{3}, \frac{2}{4}, \frac{2}{5}, \frac{2}{6} \cdots.$$

When these fractions are in their lowest terms, all except one have a denominator bigger than two. Hence ϕ must be at least three in every case except when the faces are squares.

We shall see later that if the polyhedron had square faces and the stabiliser of a face did not contain a rotation then case (ii) would hold: its vertices would lie on rotation axes.

STEP 2. Every vertex lies on a rotation axis.
Since the polyhedron is vertex-transitive, the stabilisers of the vertices all contain the same number of symmetries. Let ψ be the number of symmetries in the stabiliser of a vertex. As in the case of faces, if we can show that ψ is at least three then the stabiliser must contain rotations which carry the vertex onto itself and therefore the vertex must lie on the axis of one of these rotations.

Writing V for the number of vertices of the polyhedron and applying the Orbit–Stabiliser theorem again we see that

the number of symmetries of the polyhedron $= V \cdot \psi$.

As before, equating this statement with the analogous one about edges, we deduce that

$$\frac{\eta}{\psi} = \frac{V}{E}.$$

All the vertices have the same valency (because the polyhedron is vertex-transitive). Assume that m edges meet at each vertex. Then $mV = 2E$. This implies

$$\frac{\eta}{\psi} = \frac{V}{E} = \frac{2}{m}$$

$$= \frac{2}{3}, \frac{2}{4}, \frac{2}{5}, \frac{2}{6} \cdots.$$

When these fractions are in their lowest terms all except one have a denominator bigger than two. Hence ψ must be at least three in every case except when the vertices are 4-valent.

There are two situations that are problematic. Step 1 did not deal with the case of square faces and step 2 has not dealt with 4-valent vertices. However, it is impossible to construct a polyhedron which has square faces *and* 4-valent vertices so both situations cannot occur together. Therefore, at least one of statements (*i*) and (*ii*) above is true.

STEP 3. There are seven possible kernels for stellating.
Suppose a rotation axis passes through every face. A symmetry which carries one face onto another must also carry one axis onto another. So, as the polyhedron is face-transitive, the axes piercing the faces must be of the same kind. That is, they must all be 2-fold, all be 3-fold, all be 4-fold, or all be 5-fold. The convex polyhedra which satisfy these properties are the upper seven shown in Figure 10.25. They are the five Platonic solids together with the rhombic dodecahedron and the rhombic triacontahedron. The systems of rotational symmetry in which all the axes of a given kind lie in a plane cannot occur. The remaining possibilities are D_2, T, O and I.

STEP 4. There are seven possible shells for facetting.
A symmetry which carries one vertex onto another must also carry one axis onto another. If the polyhedron is vertex-transitive, its vertices must all lie on axes of the same kind. The convex polyhedra which satisfy these properties are the five Platonic solids, the cub-octahedron, and the icosi-dodecahedron. They are the lower seven figures in Figure 10.25.

STEP 5. The enumeration.
Of the nine convex polyhedra shown in Figure 10.25, only the Platonic solids are totally transitive. The other four are sometimes called *quasiregular*. The rhombic solids are not vertex-transitive, and the Archimedean solids are not face-transitive. Notice that they are all edge-transitive.

We have shown that the faces of a totally transitive polyhedron must lie in the face-planes of one of the first seven solids, or can be found by facetting one of the last seven solids.

So to find the possibilities we can scan stellation patterns looking for regular polygons. This was done for the Platonic solids in Chapter 7. The stellation patterns for the two rhombic solids are shown in Figure 10.26. The rhombic dodecahedron does not have any totally transitive stellations. The square in the stellation pattern of the rhombic triacontahedron gives rise to the compound of five cubes.

The other method of generating totally transitive polyhedra is to facet one of seven solids. The Platonic solids were covered in Chapter 7. The cub-octahedron cannot be facetted: the only regular polygons it contains are equatorial hexagons and triangles. Besides equatorial 10-gons and pentagons, the icosi-dodecahedron contains triangles. Eight of these close up to form an octahedron (Figure 10.27), and inscribing all the triangles produces the compound of five octahedra. All the

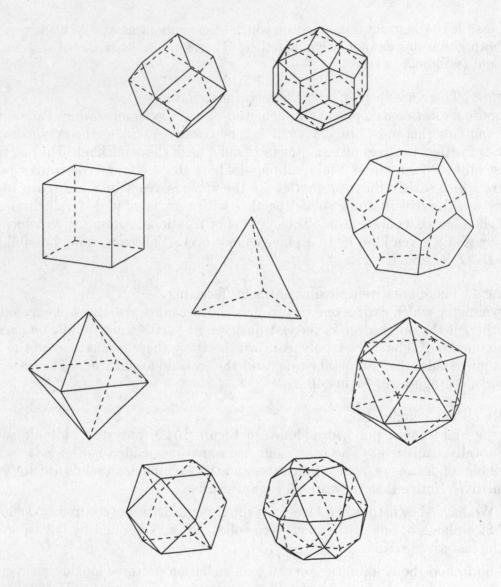

Figure 10.25.

totally transitive polyhedra must be generated by at least one of these methods. Curiously, they all appear on both lists.

The totally transitive polyhedra are listed in Table 10.28. The kernels of stellated forms are listed in the second column; the shells of the facetted forms are given in column three. ■

One last comment on this proof. Recall from Chapter 7 Poinsot's confusion over when a set of points is regularly distributed on a sphere. He thought that the vertices of the Platonic solids and also the midpoints of their edges should be

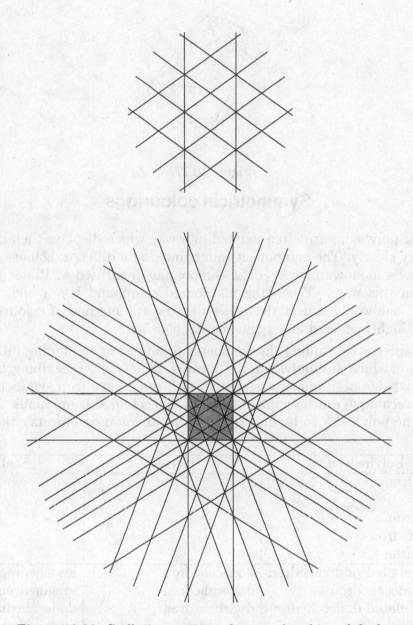

Figure 10.26. Stellation patterns for two rhombic polyhedra.

examples of such sets. We can now see that these are the places where the rotation axes of the polyhedra puncture the sphere. Only the 3-fold, 4-fold and 5-fold axes can be derived from the vertices. The 2-fold axes come from the midpoints of the edges or alternatively from the vertices of two of the quasiregular solids.

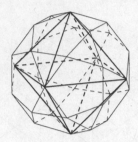

Figure 10.27.

Symmetrical colourings

Compound polyhedra are often painted in a way which displays their composite nature very clearly. The components are painted in a different colours, and each component is monochromatic. (The compounds illustrated in Plates 11–16 are coloured in this way.) When the uncoloured compound has a high degree of symmetry, one would expect that such a systematic method of colouring would produce a highly ordered and symmetrical colouring.

In Chapter 8 we studied the symmetry of an object by finding different positions from which the polyhedron appeared the same. Describing how these indistinguishable positions are related to each other gave us a symmetry group. A similar technique can be applied to study the coloured compounds. This time we want the polyhedra to be indistinguishable as coloured objects—the symme-

Polyhedron	Kernel	Shell
tetrahedron		
cube		
octahedron		
dodecahedron		
icosahedron		
small stellated dodecahedron	dodecahedron	icosahedron
great dodecahedron	dodecahedron	icosahedron
great stellated dodecahedron	dodecahedron	dodecahedron
great icosahedron	icosahedron	icosahedron
compound of two tetrahedra	octahedron	cube
compound of five tetrahedra	icosahedron	dodecahedron
compound of ten tetrahedra	icosahedron	dodecahedron
compound of five cubes	rhombic triacontahedron	dodecahedron
compound of five octahedra	icosahedron	icosi-dodecahedron

Table 10.28. The totally transitive polyhedra.

tries must preserve the pattern of colours. Such symmetries are called *colour-preserving* symmetries, and together they form a structure called the *colour-preserving group*.

It is important to differentiate between the symmetry group of a coloured object and the symmetry group of the underlying uncoloured polyhedron. For example, the compound of three cubes shown in Plate 14 has colour-preserving group D_2 but has octahedral symmetry as an uncoloured object. The number of rotational symmetries has been reduced from 24 to 4. The colour-preserving groups of the coloured compounds of five tetrahedra (Plate 12), of five cubes (Plate 13), and of four cubes (Plate 15) are all trivial: the only colour-preserving symmetry for these coloured polyhedra is the identity! The following examples show other ways in which the colour-preserving group can differ from the ordinary geometric symmetry group.

Example 1.

The simplest polyhedron, the regular tetrahedron, can be coloured with four colours, one per face. Only one of the twelve rotational symmetries of the tetrahedron, the identity, preserves the arrangement of the colours. So the colour-preserving group is the trivial group, C_1. The identity is always a colour-preserving symmetry.

Example 2.

At the other extreme, every symmetry of the uncoloured polyhedron can also be a colour-preserving symmetry. For example, if a low, rhombus-based pyramid is added to every face of a rhombic triacontahedron then a polyhedron with 120 faces is obtained (see Figure 10.29). It can be properly coloured with two colours. In this case, the group of colour-preserving symmetries is the whole icosahedral group, I, the same as the group of the underlying polyhedron.

Figure 10.29. An augmented rhombic triacontahedron.

Example 3.

The octahedron is another polyhedron which can be properly coloured with two colours. Here, the group of colour-preserving symmetries is the tetrahedral group. This can be seen by extending all the faces of either colour until they meet forming a tetrahedron. This shows that the tetrahedral group is a subgroup of the octahedral group. This relationship is often written as $T < O$.

Example 4.

A cube can be coloured with three colours so that opposite faces have the same colour. The colour-preserving symmetries are the 2-fold rotations about axes joining opposite face centres, so the colour-preserving group is D_2.

A pyritohedron is a dodecahedron with irregular pentagons for faces (Figure 10.30). Its symmetry group is the tetrahedral group. It can be coloured so that from a distance, it looks similar to the colouring of the cube described above: three colours are used and the faces of each colour fall into two pairs of adjacent faces on opposite sides of the polyhedron. The colour-preserving group is again D_2. These two examples show that $D_2 < O$ and also $D_2 < T$.

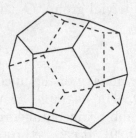

Figure 10.30.

Example 5.

A pyramid with a regular hexagon base can be coloured with two colours so that the colours alternate round the sloping sides, and either colour occurs on the base. The colour-preserving group is C_3 and the symmetry group of the uncoloured pyramid is C_6. Similar colourings of pyramids can be devised to show that $C_n < C_{np}$, where p is any integer.

Problem. A hexagonal dipyramid has symmetry group D_6. Find three colourings of this polyhedron which have the following colour-preserving groups: C_3, C_6, and D_3.

These examples illustrate that groups of colour-preserving symmetries are often much smaller than the geometric symmetry groups, even when the colours are distributed in a regular and systematic pattern. Clearly, the colour-preserving group does not capture our intuitive idea of what a symmetrical colouring is like.

The problem stems from the definition of what we mean by a symmetrical colouring. The definition uses geometrical symmetries such as rotations to try to describe a non-geometric situation. Geometrical symmetries are good at describing the symmetrical nature of *shape*, but they are not sufficient to describe the symmetry of a *pattern* formed by an arrangement of colours. The problem, then, lies not in the desire to use groups of symmetries to describe the degree and structure of the symmetry of an arrangement of colours, but in the use of *geometric* symmetry operations. The concept of a symmetry needs to be enlarged to include non-geometric operations.

Colour symmetries

In general, an operation is described as a symmetry of a system if no change can be detected in the system after the operation has been performed. The rotations which carry a polyhedron onto itself are symmetries under this definition since they carry the polyhedron to a position indistinguishable from its original position. They are geometric symmetries of a geometric object.

Examples of non-geometric symmetries appear in physics. When measuring certain quantities in physical systems, the choice of the zero-level on the gauge or scale is not important. The origin can be placed anywhere and the scale can be translated. A physical system which is invariant under a translation of a measuring scale is said to have a *gauge* symmetry. An example is the scale of voltages. The choice of where to place zero volts on the scale does not matter since only differences in voltage levels can be measured. The equations of electrical systems deal with voltage differences and they have a gauge symmetry. This kind of non-geometric symmetry has far-reaching consequences: in the case of voltage, it leads to the law of conservation of electric charge.

In order to improve our mathematical description of the symmetrical nature of a coloured object or pattern, we need a better understanding of the way in which the eye–brain system interprets coloured images. This will lead to another kind of non-geometric symmetry called a colour symmetry.

When faced with a coloured scene, one of the things the mind does is to study each colour separately against the background formed by the others. This gives an impression of the distribution of each colour within the scene. Attention can be switched from colour to colour quite rapidly. Painters make use of this response to link areas of a picture together: areas of the same colour become associated in the mind. In view of this, it seems reasonable to assume that when the brain is processing the image formed by looking at a coloured polyhedron, the following properties of the colouring play a role in determining the brain's assessment of the degree of structure in the pattern formed by the colours:

(*i*) the structure of the monochromatic patterns formed by each colour taken individually,

(*ii*) the variety of these structures and their interrelationships,

(*iii*) the positions of the colours in relation to one another.

To some extent, the first of these properties is described mathematically by the colour-preserving symmetries. A colour-preserving symmetry carries each face to one of like colour so it is also a symmetry of each of the monochromatic patterns. However, the set of colour-preserving symmetries may not be the whole set of symmetries which preserve a pattern of a particular colour. For example, if a cube is properly coloured with three colours so that opposite faces have the same colour then the group of colour-preserving symmetries is D_2 comprising the 2-fold rotations about each of the three axes joining opposite face centres. The symmetries which preserve the faces of a given colour, red (say), also include 4-fold rotations so the red-preserving group is D_4.

The second property concerns the variety and interrelationships of these single-colour structures. In the coloured models which we intuitively regard as highly ordered or symmetrical (such as the compounds), all of the monochromatic structures are the same. After the eye has picked out one component (being all of one colour) it then moves on to another colour and identifies another copy of the same thing. Whilst scanning the patterns formed by each colour individually, the structure abstracted from the pattern does not change. This may be why the brain interprets these colourings as having a high degree of symmetry. If this property can be described mathematically, it may capture the intuitive idea of what we regard as a symmetrical colouring.

The fact that there is something here which is the same in different situations suggests that a symmetry could be used to describe it. (Recall that, when used in a technical sense, a symmetry is an operation which cannot be detected or which leaves something unchanged.) When observing a coloured compound, the structure of the pattern formed by any one colour does not depend on the colour you choose to look at. Even though your attention shifts from one colour to another, from the point of view of their structure, the patterns are indistinguishable since all the components are the same.

A symmetry operation can be defined which models this consistency of structure to some degree. It is analogous to the way that the structure in the individual monochromatic patterns is partially modelled by the colour-preserving symmetries of a polyhedron. Like the colour-preserving symmetries, it is related to the geometric symmetries of the (uncoloured) polyhedron, and a restriction is placed on the way in which it can affect the colouring: the colours can be reallocated but the overall pattern must not change. For example, a rotation of a polyhedron may result in every red face being carried onto a green face, every green face to a blue face, and every blue face to a red face. This rearrangement of the colour scheme is called a *permutation* of the colours. This particular permutation can be written

$$(\text{red} \to \text{green}, \text{green} \to \text{blue}, \text{blue} \to \text{red}).$$

The rotation is said to have *induced* the permutation of the colours. Clearly, for this to happen, the structure of the monochromatic patterns formed by the red, green, and blue faces individually must be the same since the rotation carries one pattern onto another. Such an operation is called a colour symmetry.

Definition.   A *colour symmetry* of a coloured polyhedron is a geometric symmetry of the (uncoloured) polyhedron which induces a permutation of the colours.

The set of colour symmetries of a polyhedron forms a group which is called the colour symmetry group or simply the *colour group*. The colour-preserving symmetries carry faces to ones of like colour so they induce the identity permutation of the colours. Therefore, they can also be interpreted as colour symmetries and a copy of the colour-preserving group is contained in the colour group.

An example which illustrates the differences between the three kinds of symmetry is provided by the following colouring of a regular octahedron. Regard the octahedron as a triangular-based antiprism and colour the base and the top red, the faces adjacent to the top white, and those adjacent to the base black. This results in a proper 3-colouring of the octahedron. The rotational symmetry group of the underlying polyhedron is the octahedral group consisting of 24 rotational symmetries. The colour-preserving symmetries are the rotations about the axis joining the centres of the red faces, so the colour-preserving group is C_3. The colour symmetries include the three colour-preserving symmetries and also three others: these are the 2-fold rotations about axes which join the midpoints of opposite edges and which are parallel to the base. They all induce the same permutation of the colours:

$$(\text{red} \to \text{red}, \text{black} \to \text{white}, \text{white} \to \text{black}).$$

Therefore, the colour group contains six colour symmetries and is D_3. (There is a subtle change of interpretation hidden in the last sentence. In Chapter 8, we defined D_3 to be a group of rotational symmetries. In the current context, the elements of the group are realised as colour symmetries, not geometric ones.)

The third property which may influence the brain's interpretation of an arrangement of colours is the way in which the monochromatic patterns are physically related. This has more to do with the local properties of the colouring than with the global symmetry of the pattern. The jump between cases (*ii*) and (*iii*) is larger than that between the first two cases, both perceptually and mathematically. This is reflected by the fact that mathematicians have only recently begun to study the problem of when two colourings have the same 'pattern type'. The following example illustrates that the above groups of symmetries are inadequate to give a complete account of a colouring. Figure 10.31 shows two identical pyramids with irregular hexagons for bases viewed from above. Although they are coloured differently, their colour groups are the same, as are their colour-preserving groups and symmetry groups. And in both cases, a clockwise rotation of 120° induces the same permutation:

$$(\text{black} \rightarrow \text{white}, \text{white} \rightarrow \text{grey}, \text{grey} \rightarrow \text{black}).$$

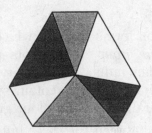 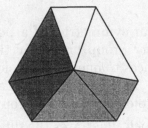

Figure 10.31.

Perfect colourings

A colouring of a polyhedron is called a *perfect* colouring if every (rotational) symmetry of the underlying polyhedron is also a colour symmetry. The colouring of the octahedron in the previous example is not perfect since its colour group (D_3) and symmetry group (O) are different. A tetrahedron which is properly coloured (in four colours) is perfectly coloured even though its colour-preserving group is trivial. The coloured compounds are also examples of perfectly coloured polyhedra. This confirms that colour symmetries capture something of our intuitive idea of a systematic pattern.

The coloured polyhedra described in Examples 1–5 above provide further examples of perfectly coloured polyhedra. Earlier, these examples were used to illustrate the kinds of colour-preserving group that can occur and their relationships to the rotational symmetry groups. As all the colourings are perfect, these examples also show the variety of ways that a group of colour-preserving symmetries can sit inside a colour group. Besides the cases where the colour-preserving group is either trivial or the whole of the colour group, the following relationships appeared in these examples:

$$T < O, \quad D_2 < T, \quad D_2 < O, \quad C_n < C_{np}.$$

The exercise on colouring a dipyramid gave examples of two more relationships:

$$C_n < D_{np}, \quad D_n < D_{np}.$$

Examining a large number of coloured polyhedra leads to the observation that these seem to be the only relationships which occur. The colour-preserving group always sits inside the colour group in one of these ways.

One thing to notice about this list is the absence of the icosahedral group. This seems strange since, for all the other types of groups (the octahedral, tetrahedral, dihedral and some cyclic groups), examples of coloured polyhedra can be found where the colour-preserving group is non-trivial and yet neither is it the complete set of colour symmetries. But for the icosahedral group, no such intermediate groups occur. For any coloured polyhedron with icosahedral colour group, the group of colour-preserving symmetries is either trivial (the compound of five cubes) or else it is the whole colour group (the polyhedron in Example 2).

This phenomenon is a consequence of the internal structure of the icosahedral group. On one hand, as the icosahedral group is large (having 60 elements), one might expect that it has the potential to contain a lot of subgroups that can arise as groups of colour-preserving symmetries. On the other hand, there is a tendency for some kinds of systems to become more complex as they become larger. When there is sufficient space within a system, its elements can become tangled and intertwined to produce a complicated structure. The icosahedral group fits into the latter category. When the members of the icosahedral group are interpreted as colour symmetries, its internal structure is such that none of its proper subgroups can occur as groups of colour-preserving symmetries. This property of the subgroups of the icosahedral group can be identified abstractly and can be demonstrated to be an algebraic property of the group structure itself.

(†) The reader who is familiar with elementary group theory will have realised that the structure being referred to is a lack of normal (or invariant) subgroups. In fact, the above list of colour-preserving groups is based on an enumeration of the normal subgroups of the rotation groups. In the examples, all of the colour-preserving groups are normal subgroups of the symmetry group of the underlying polyhedron. This happened because all of the colourings are perfect and because of the following fact:

Theorem. A colour-preserving group is always a normal subgroup of the colour group. To see this let G be the colour group and let S_n be the group of permutations of n objects. There is a natural homomorphism from G to S_n. The kernel of this map, the group of symmetries which map onto the identity permutation, is the colour-preserving group. A fundamental theorem in group theory states that the kernel of a homomorphism is a normal subgroup. It is also well-known that the only normal subgroups of the icosahedral group are itself and the trivial group.

The solution of fifth degree equations

The complexity of the structure of the icosahedral group has deep implications in another area of mathematics, namely the theory of equations. This branch of mathematics is concerned with finding solutions of polynomials.

Methods for solving polynomial equations date back to antiquity. The Greeks were able to solve quadratic equations using geometrical constructions. Before them, the Babylonians (and possibly the Egyptians too) could solve problems requiring the solution of a quadratic equation, although they did not have the notation to express their method. An algebraic formulation did not appear until after the first century AD. The Chinese also studied the theory of equations and by the thirteenth century their techniques for finding solutions were way ahead of those in the West.

We can find the two solutions of the quadratic equation $ax^2 + bx + c = 0$ by using the formula

$$x = \frac{-b \pm \sqrt{b^2 - 4ac}}{2a}.$$

A similar method for solving cubic equations was unknown in Europe until the sixteenth century. At the end of his *Summa de Arithmetica* (1494) Luca Pacioli remarked that solving the equations $x^3 + cx = d$ and $x^3 + d = cx$ with the current state of knowledge was as impossible as squaring the circle. Yet within fifty years, the cubic could be solved. In 1545, the physician Girolamo Cardano published his *Ars Magna* in which he gave a complete discussion of the solution of cubic equations. The methods are ascribed to Niccolò Fontana (also known by his nickname 'Tartaglia', the stammerer) who had discovered them at least ten years earlier. He had refused to reveal the details and was persuaded to divulge them to Cardano only after having sworn him to secrecy. The *Ars Magna* also contains a method of reducing equations of degree four to cubics which is due to Ludovico Ferrari.

The methods resulted in formulae similar to that above for the quadratic. The solutions are expressed in terms of the coefficients of the equations combined using only the four basic operations of arithmetic (addition, subtraction, multiplication and division) and the extraction of roots. All polynomials of degree less than five could now be solved by formulae of this kind, and the search started for a formula which would solve quintic (degree five) equations.

Despite great effort no progress was made, and by the nineteenth century people were beginning to speculate that a formula might not exist. Joseph-Louis

Lagrange (1736–1813) observed that the method for solving a cubic equation is to reduce the problem to that of solving a quadratic, and that solving a quartic proceeds using essentially the same idea to reduce the problem to a cubic. However, applying the trick to quintic equations did not reduce the scale of the problem, but produced a polynomial of degree six.

The first proof that a formula for finding the solutions of a general equation of degree five does not exist was given in 1799 by the Italian physician Paolo Ruffini (1765–1822). However, some people remained unconvinced by Ruffini's arguments, especially those professional mathematicians who had devoted a lot of time to the problem. The discussion was ended in 1824 when the Norwegian Niels Henrik Abel (1802–1829) gave a conclusive proof.

After it was known that a general formula for solving all quintic equations was not possible, attention shifted to the problem of finding necessary and sufficient conditions for determining whether or not a formula for the solutions of a particular equation existed. Underlying the arguments used by Ruffini and Abel is a relationship between the radicals (n^{th} roots) and certain subgroups of permutation groups. At that time, the group concept had not been isolated and the ideas involved were not sufficiently well understood for a general theory that applied to arbitrary equations to be worked out. A major step in this direction was taken by a young Frenchman named Evariste Galois (1811–1832). He also realised the importance of 'normal subgroups' and this enabled him to answer completely the problem of which equations can be solved by a formula.

Galois' idea was to associate a group with the equation which described the way that its solutions are related to each other. The properties of this group of 'algebraic symmetries' reflected the properties of the solutions. In particular, if a formula for finding the solutions of the equation exists then the group can be decomposed into smaller groups in a special way. The structures of the Galois groups associated with the equations of degree less than five can be decomposed in the required way, and knowledge of these decompositions allows the formulae to be reconstructed. However, the group associated with most equations of degree five contains the icosahedral group and the structure of this group is too complicated to be split up in the required way. So there is no formula to solve equations of degree five.

In the sixteenth century François Viète had shown that cubic equations can be reduced to a form where only one of the coefficients is unknown. This enabled him to express its solutions in terms of trigonometric functions. In 1786 E. S. Bring showed how to reduce quintic equations to a similar form, but his work went unnoticed for fifty years. The reduction of a general quintic equation to one having a single parameter meant that its solutions could be found by applying a technique analogous to Viète's solution of cubics. This was achieved in 1858 by Charles Hermite (better known for proving the transcendence of e). Following a hint of Galois, he showed that the solutions of a quintic could be found using elliptic functions in place of trigonometric ones.

These many strands—the symmetries of polyhedra, group theory, the solution of equations, and elliptic functions—were all woven together by Felix Klein in his *Lectures on the Icosahedron and the Solution of Equations of the Fifth Degree*. This illustrates the remarkable unity of mathematics. On some occasions similar ideas will occur to different groups of people working in different disciplines. For a while they develop in parallel, then unexpected connections are found; eventually the unifying theme is identified and abstracted. The group concept, for instance, arose in algebraic equations and geometric symmetries. At other times, many pieces of the patchwork of mathematics are brought together and produce a coherent and beautifully interwoven whole. In the prologue to his famous lecture of 23 problems David Hilbert commented on this unexpected unity and used Klein's work as an illustration:

> *It often happens* ··· *that the same problem finds application in the most unlike branches of mathematical knowledge.* ···
> *How convincingly has F. Klein, in his work on the icosahedron, pictured the significance which attaches to the problem of the regular polyhedra in elementary geometry, in group theory, in the theory of equations and in that of linear differential equations.*[c]

Like the components of a compound, the branches of mathematics interlock in intriguing ways.

Appendix I

Because symmetry groups have been studied in so many areas of mathematics, and also in crystallography, several notations have been developed to denote them. Each notation has its own rules and internal consistency which comes from the kind of analysis used to study the symmetry groups. Some are compared in the table below.

C_1		$[1]^+$	1	11	C_1
C_s		$[1]$	m	$*11$	C_2C_1
C_i		$[2^+, 2^+]$	$\tilde{2}$	$1\times$	$\overline{C}_1$
C_n	$n \geqslant 2$	$[n]^+$	n	nn	C_n
C_{nv}	$n \geqslant 2$	$[n]$	$n \cdot m$	$*nn$	D_nC_n
C_{nh}	$n \geqslant 2$	$[2, n^+]$	$n : m$	$n*$	$\begin{cases} C_{2n}C_n & \text{if } n \text{ odd} \\ \overline{C}_n & \text{if } n \text{ even} \end{cases}$
S_{2n}	$n \geqslant 2$	$[2^+, 2n^+]$	$\widetilde{2n}$	$n\times$	$\begin{cases} \overline{C}_n & \text{if } n \text{ odd} \\ C_{2n}C_n & \text{if } n \text{ even} \end{cases}$
D_n	$n \geqslant 2$	$[2, n]^+$	$n : 2$	$22n$	D_n
D_{nv}	$n \geqslant 2$	$[2^+, 2n]$	$\widetilde{2n} \cdot m$	$2*n$	$\begin{cases} \overline{D}_n & \text{if } n \text{ odd} \\ D_{2n}D_n & \text{if } n \text{ even} \end{cases}$
D_{nh}	$n \geqslant 2$	$[2, n]$	$m \cdot n : m$	$*22n$	$\begin{cases} D_{2n}D_n & \text{if } n \text{ odd} \\ \overline{D}_n & \text{if } n \text{ even} \end{cases}$
T		$[3, 3]^+$	$3/2$	332	T
T_d		$[3, 3]$	$3/\tilde{4}$	$*332$	OT
T_h		$[3^+, 4]$	$\tilde{6}/2$	$3*2$	$\overline{T}$
O		$[3, 4]^+$	$3/4$	432	O
O_h		$[3, 4]$	$\tilde{6}/4$	$*432$	$\overline{O}$
I		$[3, 5]^+$	$3/5$	532	I
I_h		$[3, 5]$	$3/\widetilde{10}$	$*532$	$\overline{I}$

In the first column is the notation used in this book (described in Chapter 8). It is essentially that used by Arthur Schoenflies in his *Kristallsysteme und Krystallstructur* and described by Harold Hilton in his *Mathematical Crystallography and the Theory of Groups of Movements*. Its labels are derived from the kinds of symmetry elements (axes and mirror-planes) which are present.

The second column lists the notation used by H. S. M. Coxeter and W. O. J. Moser in their book *Generators and Relations for Discrete Groups*. Their labels indicate how the groups can be generated as subgroups of reflection groups. A plus sign (+) indicates that only the rotational symmetries contributed by certain generators are to be included.

The labels in the third column are described by A. V. Shubnikov and V. A. Koptsik in *Symmetry in Science and Art*. As with Schoenflies' notation, the symbols indicate some of the symmetry elements which are present. The letter m denotes the presence of a mirror-plane, a single dot ($\cdot$) means that it contains a rotation axis, and two dots (:) means that it is normal to a rotation axis. The tilde symbol ($\sim$) over a number indicates an axis of rotation-reflection.

The notation in the fourth column was derived by John Conway and William Thurston from the topological features of quotient orbifolds. Numbers following an asterisk ($*$) refer to the angles at the corners of the orbifold's boundary, other numbers indicate the kinds of cone-points, and a cross ($\times$) indicates a cross-cap. For a more complete explanation see Conway's account: 'The Orbifold Notation for Surface Groups' published in the *Proceedings of the 1990 Durham Conference on Groups, Combinatorics and Geometry*, Cambridge Univ. Press 1992.

The final column lists the notation used by L. Fejes Toth in his book *Regular Figures*. Rather than concentrate on reflections as the previous notations do, the structure underlying this notation comes from rotations and inversion. This causes a discrepancy in the cases of some cyclic and dihedral groups as to when two groups belong to the same kind of system. This shows up in the table as odd and even cases of n being listed separately.

Appendix II

Chapters 9 and 10 made use of a group theoretic result known as the Orbit–Stabiliser theorem. It was quoted in the form:

$$\begin{pmatrix} \text{the number of} \\ \text{equivalent objects} \end{pmatrix} \times \begin{pmatrix} \text{the number of symmetries} \\ \text{of each object} \end{pmatrix}$$

$$= \begin{pmatrix} \text{the number of symmetries} \\ \text{of the polyhedron} \end{pmatrix}$$

In Chapter 9 the objects were coloured polyhedra and we were counting equivalent colourings; in Chapter 10 the objects were faces, edges or vertices, and the right hand term (the number of symmetries of each object) was called the stabiliser. It will be no surprise to learn that the other term (number of equivalent objects) is called an orbit. The purpose of this appendix is to give a proof of this theorem.

Suppose that we have a polyhedron and we wish to study some component or property of it which we will denote by A. Thus A might be a particular face or a colouring. Let G be the symmetry group of the polyhedron.

The *stabiliser* of A, written $\mathrm{stab}_G(A)$, is the set of all symmetries of the polyhedron which carry A to itself:

$$\mathrm{stab}_G(A) \;=\; \{\, g \text{ in } G \text{ such that } g(A) = A \,\}.$$

In the earlier examples, the stabiliser was the set of symmetries which fixed a face, or the (colour-preserving) symmetries of a coloured polyhedron. The stabiliser is a subgroup of G. For convenience we shall write H for $\mathrm{stab}_G(A)$ and suppose that it contains m symmetries:

$$H \;=\; \{\, h_1, h_2, \cdots, h_m \,\}.$$

One of these must be the identity and we can choose $h_1 = 1$.

The *orbit* of A, written $\mathrm{orbit}_G(A)$, is the set of all the places A is carried to by the symmetries of the polyhedron:

$$\mathrm{orbit}_G(A) \;=\; \{\, g(A) \text{ for all } g \text{ in } G \,\}.$$

In the colouring example, the orbit consists of the different orientations of a coloured polyhedron. In the transitivity examples, all the faces of a face-transitive polyhedron form a single orbit. Suppose that there are n different things in the orbit of A:

$$\text{orbit}_G(A) \ = \ \{\, A_1, A_2, \cdots, A_n \,\}.$$

One of these must be A itself so we can choose $A_1 = A$.

Choose n symmetries $g_1, g_2, \cdots, g_n$ in G such that $g_i(A) = A_i$. There may be several symmetries to choose from in each case. For example, any element of $\text{stab}_G(A)$ can be chosen as g_1.

Writing $|X|$ to denote the number of members of the set X, we can state the Orbit–Stabiliser theorem as follows.

Theorem.

$$|G| \ = \ |\,\text{orbit}_G(A)\,| \ \times \ |\,\text{stab}_G(A)\,|.$$

PROOF: To prove the theorem we split up the symmetries of the polyhedron into several sets. We can form a new set (denoted gH) of elements of G as follows:

$$gH \ = \ \{\, gh_1, gh_2, \cdots, gh_m \,\}.$$

All the products $g{\cdot}h_i$ are distinct because the h_i are distinct.

Consider the following n such sets:

$$
\begin{aligned}
g_1 H \ &= \ \{\, g_1 h_1, g_1 h_2, \cdots, g_1 h_m \,\} \\
g_2 H \ &= \ \{\, g_2 h_1, g_2 h_2, \cdots, g_2 h_m \,\} \\
&\ \vdots \qquad\qquad \vdots \\
g_n H \ &= \ \{\, g_n h_1, g_n h_2, \cdots, g_n h_m \,\}
\end{aligned}
$$

These n sets are called *cosets* of H. The first one is H itself since g_1 is in H. If we can show that all these sets are disjoint and that every element of G appears in some coset then we will have proved $|G| = n \times m$ which is what we need.

STEP 1. All elements of G can be written in the form $g_i h_j$.

Given any symmetry g in G we consider $g(A)$ in $\text{orbit}_G(A)$. Now $g(A)$ must equal A_i for some i. Hence

$$
\begin{aligned}
g(A) \ &= \ g_i(A) && \text{for some } i \\
\Rightarrow \qquad g_i^{-1} g(A) \ &= \ A \\
\Rightarrow \qquad g_i^{-1} g \ &= \ h && \text{for some } h \text{ in } H \\
\Rightarrow \qquad g \ &= \ g_i h.
\end{aligned}
$$

We have written g in the required form.

STEP 2. The sets $g_i H$ are disjoint.

Suppose some element is in two different sets so that $g_i x = g_j y$, where x and y are some elements of H. Then $g_j^{-1} g_i = y x^{-1}$. Now $y x^{-1}$ is a product of elements of H and is therefore also in H. So $g_j^{-1} g_i = h$ for some h in H and thus $g_i = g_j h$.

We now consider the action of these elements on A.

$$
\begin{aligned}
g_i(A) &= g_j h(A) \\
\Rightarrow \qquad g_i(A) &= g_j(A) \qquad \text{because } h \text{ is in stab}_G(A) \\
\Rightarrow \qquad A_i &= A_j
\end{aligned}
$$

But this is a contradiction since the g_i's were chosen so that this could not happen. ∎

Sources of Quotations

Introduction

a) J. D. Barrow and F. J. Tipler, *The Anthropic Cosmological Principle*, Oxford Univ. Press 1986, quoted on p79

Chapter 1

a) E. Maor, *To Infinity and Beyond: a Cultural History of the Infinite*, Birkhauser 1986, quoted on p179 and p226

b) R. J. Gillings, *Mathematics in the Time of the Pharaohs*, M. I. T. Press 1972, p185

c) *ibid.*, p188

d) *ibid.*, p234

e) B. L. van der Waerden, *Geometry and Algebra in Ancient Civilizations*, Springer-Verlag (1983) p*xi*

f) K. von Fritz, 'The Discovery of Incommensurability by Hippasus of Metapontum', *Annals of Math.* **46** (1945) p256

g) J. Needham, *Science and Civilisation in China*, Cambridge Univ. Press 1959, volume 3, p92

h) Li Yan and Du Shiran, *Chinese Mathematics: a Concise History*, translated by J. N. Crossley and A. W.-C. Lun, Clarendon Press, Oxford 1987, p21

i) E. Maor, *op. cit.*, quoted on p224

j) E. J. Dijksterhuis, *Archimedes*, translated by C. Dikshoorn, Princeton Univ. Press 1987, p314

k) T. L. Heath, *The Thirteen Books of Euclid's Elements*, Cambridge Univ. Press 1908, volume 3, p368

l) D. B. Wagner, 'An Early Chinese Derivation of the Volume of a Pyramid: Liu Hui, 3rd Century AD', *Historia Math.* **6** (1979) pp178–179

m) *ibid.*, p181

n) Li Yan and Du Shiran, *op. cit.*, p69

o) Lao Tzu, *Tao Te Ching*, translated by Gia-fu Feng, Wildwood House Ltd, London 1972, chapter 14 *p*) T. L. Heath, *op. cit.*, p365

q) D. Hilbert, 'Mathematical Problems'. English translation by M. W. Newson reprinted in *Mathematical Developments Arising from Hilbert Problems*, edited by F. E. Browder, Proc. of Symposia in Pure Math. 23, American Math. Soc., Providence Rhode Island 1976, pp10–11

Chapter 2

a) W. C. Waterhouse, 'The Discovery of the Regular Solids', *Archive for History of Exact Sciences* **9** (1972) p214

b) Plato, *Timaeus*, §54. Taken from H. D. P. Lee, *Timeaus and Critias*, Penguin Books 1977, pp75–76

c) *ibid.*, §55. pp76–78

d) Plutarch, *Platonicae Quaestiones*, question 5 part 1. Taken from *Plutarch's Moralia*, volume 13 part 1, translated by H. Cherniss, Loeb Classical Library, Harvard Univ. Press 1976, p53

e) J. V. Field, 'Kepler's Star Polyhedra', *Vistas in Astronomy* **23** (1979) p123

f) *ibid.*, p124

g) Plato, *Republic*, §510. Quoted in M. Kline, *Mathematical Thought from Ancient to Modern Times*, Oxford Univ. Press 1972, p44

h) B. L. van der Waerden, *Science Awakening*, translated by Arnold Dresden, P. Noordhoff Ltd, Groningen, Holland 1954, volume 1, p173

i) W. C. Waterhouse, *op. cit.*, p216

j) G. J. Allman, *Greek Geometry from Thales to Euclid*, Dublin 1889, p211

k) H. S. M. Coxeter, 'Regular Skew Polyhedra in Three and Four Dimensions, and Their Topological Analogues', *Proc. London Math. Soc.* (series 2) **43** (1937) pp33–34

l) Pappus, *Mathematical Collection*. Taken from *Selections Illustrating the History of Greek Mathematics*, translated by I. Thomas, Loeb Classical Library, Harvard Univ. Press 1939, volume 2, p195

m) J. V. Field, *op. cit.*, p139

Chapter 3

a) T. L. Heath, *The Thirteen Books of Euclid's Elements*, Cambridge Univ. Press 1908, volume 3, p512

b) S. J. Edgerton, *The Renaissance Discovery of Linear Perspective*, Basic Books Inc., New York 1975, pp79–80

c) G. Vasari, *The Lives of the Painters, Sculptors and Architects*, translated by A. B. Hinds, Everyman's Library, Dent, London 1963, volume 1, p233

d) *ibid.*, p335

Chapter 4

a) H. S. M. Coxeter, 'Kepler and Mathematics', *Vistas in Astronomy* **18** (1975) p661

b) O. Gingerich, 'Kepler'. Article in the *Dictionary of Scientific Biography*, volume 7, p292

c) D. Pedoe, *Geometry and the Liberal Arts*, Penguin Books 1976, p267

d) J. V. Field, 'Kepler's Star Polyhedra', *Vistas in Astronomy* **23** (1979) p114

e) *ibid.*, p115

f) C. Hardie, *The Six-cornered Snowflake*, Oxford Univ. Press 1966, p11

g) J. V. Field, *op. cit.*, p114

h) *ibid.*, p134

i) *ibid.*, p115

j) *ibid.*, p136

k) *ibid.*, p135

l) *ibid.*, p133

m) *ibid.*, p133

Chapter 5

a) P. J. Federico, *Descartes on Polyhedra: a Study of the 'De Solidorum Elementis'*, Springer-Verlag 1982, quoted on p71

b) *ibid.*, quoted on p63

c) *ibid.*, pp43–44

d) N. L. Biggs, E. K. Lloyd and R. J. Wilson, *Graph Theory 1736–1936*, Clarendon Press, Oxford 1976, p76

e) *ibid.*, p77

f) P. J. Federico, *op. cit.*, p66

g) H. Lebesgue, 'Remarques sur les Deux Premières Démonstrations du Théorème d'Euler, Relatif aux Polyèdres', *Bull. Soc. Math. France* **52** (1924) p316

h) A. L. Cauchy, 'Recherches sur les Polyèdres (first memoire)', *J. École Polytechnique* **9** (1813) p77

i) C. L. Dodgson, 'Through the Looking Glass', p136; in *The Complete Illustrated Works of Lewis Carroll*, edited by Edward Guiliano, 1982, pp81–176

j) L. Poinsot, 'Note sur la Théorie des Polyèdres', *Comptes Rendus des Séances de l'Académie des Sciences* **50** (1860) p70

k) P. J. Federico, *op. cit.*, p66

l) R. Hoppe, 'Ergänzung des Euler'schen Lehrsatze von Polyedern', *Archiv der Mathematik und Physik* **63** (1879) p102. Quoted in I. Lakatos, *Proofs and Refutations: the Logic of Mathematical Discovery*, Cambridge Univ. Press 1976, p78

m) P. J. Federico, *op. cit.*, pp54, 57

Chapter 6

a) H. S. M. Coxeter, *Introduction to Geometry*, Wiley and Sons 1961, p5

b) Proclus, *Commentary on the First Book of Euclid's Elements*, translated by G. L. Morrow. Princeton Univ. Press 1970, pp187–88

c) T. L. Heath, *The Thirteen Books of Euclid's Elements*, Cambridge Univ. Press 1908, volume 3, p226

d) H. Freudenthal, 'Cauchy'. Article in the *Dictionary of Scientific Biography*, volume 3, p143

e) B. Belhoste, *Augustin-Louis Cauchy: a Biography*, (translated by F. Ragland), Springer-Verlag 1991, p29

f) H. Gluck, 'Almost All Simply-connected Closed Surfaces are Rigid', *Lecture Notes in Math.* **438**, 'Geometric Topology', Springer-Verlag (1975), quoted on p225

g) *ibid.*, p225

Chapter 7

a) L. Poinsot, 'Mémoire sur les Polygones et Polyèdres', *J. École Polytechnique* **10** (1810) pp34–35

b) *ibid.*, p42

c) B. Grünbaum, 'Polyhedra with Hollow Faces', *Proc. NATO-ASI Conference on Polytopes: Abstract, Convex and Computational* (Toronto 1993), edited by T. Bisztriczky, P. McMullen, R. Schneider and A. Ivic'Weiss, Kluwer Academic Publ. Dortrecht 1994, pp43–70

Chapter 8

a) A. Badoureau, 'Mémoire sur les Figures Isosceles', *J. École Polytechnique* **49** (1881) p51

b) F. E. Browder and S. MacLane, 'The Relevance of Mathematics', p339; in *Mathematics Today: Twelve Informal Essays*, edited by L. A. Steen, Springer-Verlag 1978, pp323–350

c) H. Hilton, *Mathematical Crystallography and the Theory of Groups of Movements*, Clarendon Press, Oxford 1903, p259

Chapter 9

a) G. H. Hardy, *A Mathematician's Apology*, §10. Canto edition, Cambridge Univ. Press 1992, pp84–85

b) A. B. Kempe, 'How to Colour a Map with Four Colours', *Nature* **21** (26th February, 1880) p399

Chapter 10

a) Aristotle, *Metaphysics* 10f.1045a

b) I. Stewart, *New Scientist* (4th November, 1989) p42

c) D. Hilbert, 'Mathematical Problems'. English translation by M. W. Newson reprinted in *Mathematical Developments Arising from Hilbert Problems*, edited by F. E. Browder, Proc. of Symposia in Pure Math. 23, American Math. Soc., Providence Rhode Island 1976, pp2–3

Bibliography

The main part of the bibliography is arranged according to chapter. However, the books in the first section are general references, mainly on the history of mathematics.

W. W. R. Ball,
  1922 *A Short Account of the History of Mathematics*, MacMillan, London
M. Brückner,
  1900 *Vielecke und Vielflache, Theorie und Geschichte*, Teubner, Leipzig
H. S. M. Coxeter,
  1969 *Introduction to Geometry* (second edition), Wiley, New York
  1973 *Regular Polytopes* (third edition), Dover, New York
  1974 *Regular Complex Polytopes*, Cambridge Univ. Press
H. T. Croft, K. J. Falconer and R. K. Guy,
  1991 *Unsolved Problems in Geometry*, Springer-Verlag
H. M. Cundy and A. P. Rollett,
  1961 *Mathematical Models* (second edition), Oxford Univ. Press
H. Eves,
  1983 *An Introduction to the History of Mathematics* (fifth edition), Saunders
    College Publishing, Philadelphia
J. Fauvel and J. Gray (editors),
  1987 *The History of Mathematics: a Reader*, MacMillan
L. Fejes Tóth,
  1964 *Regular Figures*, International Series of Monographs in Pure and Applied
    Math. 48, Pergamon, Oxford
C. C. Gillispie (editor),
  1974 *Dictionary of Scientific Biography*, Charles Scribner's Sons, New York
B. Grünbaum,
  1967 *Convex Polytopes*, Wiley
M. Kline,
  1972 *Mathematical Thought from Ancient to Modern Times*, Oxford Univ. Press
  1982 *Mathematics: the Loss of Certainty*, Oxford Univ. Press

J. Malkevitch,

  1988 'Milestones in the History of Polyhedra', in M. Senechal and G. Fleck [1988] pp80–92

D. E. Rowie and J. McCleary (editors),

  1988 *The History of Modern Mathematics* (2 volumes), Academic Press

M. Senechal and G. Fleck (editors),

  1988 *Shaping Space—a Polyhedral Approach*, Birkhauser, Basel

G. C. Shephard,

  1968 'Twenty Problems on Convex Polyhedra', *Math. Gazette* **52** (1968) pp136–147, 359–367

D. E. Smith,

  1923 *History of Mathematics volume 1: General Survey of the History of Elementary Mathematics*, Athenaeum Press, New York

E. Steinitz,

  1916 'Polyeder und Raumenteilungen', *Encyklopaedie der Mathematischen Wissenschaften* **3** (1922), 'Geometrie', part 3AB12 pp1–139

J. Stillwell,

  1989 *Mathematics and its History*, Springer-Verlag

C. Wiener,

  1864 *Über Vielecke und Vielflache*, Leipzig

L. Young,

  1981 *Mathematicians and Their Times*, Mathematics Studies 48, North-Holland

Introduction

B. Artmann,

  1993 'Roman Dodecahedra', *Math. Intelligencer* **15** no. 2 (1993) pp52–53

  1993 Response to I. Hargittai [1993], *Math. Intelligencer* **15** no. 4 (1993) p3

E. Haeckel,

  1887 *Report on the Radiolaria*, Report on the Scientific Results of the Voyage of H. M. S. Challenger During Years 1873–1876. Zoology (Edinburgh) 18

I. Hargittai,

  1993 'Imperial Cuboctahedron', *Math. Intelligencer* **15** no. 1 (1993) pp58–59

P. Hoffman,

  1988 *Archimedes' Revenge: the Joys and Perils of Mathematics*, Penguin

K. Miyazaki,

  1993 'The Cuboctahedron in the Past of Japan', *Math. Intelligencer* **15** no. 3 (1993) pp54–55

E. L. Muetterties and W. H. Knoth,

  1968 *Polyhedral Boranes*, Marcel Dekker Inc. New York

L. S. Seiden,

   1989 *Buckminster Fuller's Universe: an Appreciation*, Plenum Press

M. J. Wenninger,

   1971 *Polyhedron Models*, Cambridge Univ. Press

Chapter 1

R. C. Archibald,

   1930 'Mathematics Before the Greeks', *Science* **71** (1930) pp109–121, 342; *Science*
      72 (1930) p36

F. Bagemihl,

   1948 'On Indecomposable Polyhedra', *American Math. Monthly* **55** (1948)
      pp411–413

V. G. Boltianskii,

   1978 *Hilbert's Third Problem*, (translated from the Russian by R. A. Silverman),
      Scripta Technica Inc.

W. Bolyai,

   1832 *Tentamen. Juventutem Studiosam in Elementa Matheseos Purae, Elementaris*
      ac Sublimioris, Methodo Intuitiva Evidentiaque Huic Propria, Introducendi,
      Marosvasarhely

R. Bricard,

   1896 'Sur une Question de Géométrie Relative aux Polyèdres', *Nouvelles Annales de*
      Math. **15** (1896) pp331–334

P. Cartier,

   1985 'Décomposition des Polyèdres: le Point sur le Troisième Problème de Hilbert',
      Séminaire Bourbaki 1984/85 Exposés 633–650, *Astérisque* **133–134** (1986)
      pp261–288

J. L. Cathelineau,

   1992 'Quelques Aspects du Troisième Problème de Hilbert', *Gazette des*
      Mathematiciens **52** (1992) pp45–71

A. B. Chace, L. Bull, H. P. Manning and R. C. Archibald (editors),

   1927 *The Rhind Mathematical Papyrus* (2 volumes), Oberlin, Ohio 8 (1927–29)

M. Dehn,

   1900 'Über raumgleiche Polyeder', *Nachrichten von der Konigl. Gesellschaft der*
      Wissenschaften zu Gottingen Mathematisch-Physikalischen Klasse, pp345–354

E. J. Dijksterhuis,

   1987 *Archimedes*, (translated from the Dutch by C. Dikshoorn), Princeton Univ.
      Press

I. E. S. Edwards,

   1988 *The Pyramids of Egypt* (revised edition), Penguin Books

K. von Fritz,

1945 'The Discovery of Incommensurability by Hippasus of Metapontum', *Annals of Math.* **46** (1945) pp242–264

R. J. Gillings,

1972 *Mathematics in the Time of the Pharaohs*, M. I. T. Press

M. Goldberg,

1958 'Tetrahedra Equivalent to Cubes by Dissection', *Elemente der Mathematik* **13** (1958) pp107–109

1969 'Two More Tetrahedra Equivalent to Cubes by Dissection', *Elemente der Mathematik* **24** (1969) pp130–132; Correction: *ibid.* **25** (1970) p48

1974 'New Rectifiable Tetrahedra', *Elemente der Mathematik* **29** (1974) pp85–89

D. Hilbert,

1902 'Mathematical Problems—a Lecture Delivered Before the International Congress of Mathematicians in Paris in 1900'. English translation by M. W. Newson, *Bull. American Math. Soc.* **8** (1902) pp437–479

M. J. M. Hill,

1896 'Determination of the Volumes of Certain Species of Tetrahedra Without Employing the Method of Limits', *Proc. London Math. Soc.* **27** (1896) pp39–53

W. H. Jackson,

1912 'Wallace's Theorem Concerning Plane Polygons of the Same Area', *American J. Math.* **34** (1912) pp383–390

V. F. Kagan,

1903 'Über die Transformation der Polyeder', *Math. Ann.* **57** (1903) pp421–424

W. R. Knorr,

1975 *The Evolution of the Euclidean Elements: a Study of the Theory of Incommensurable Magnitudes and its Significance for Early Greek Geometry*, D. Reidel Publ. Co., Dordrecht

H. Lebesgue,

1945 'Sur l'Équivalence des Polyèdres, en Particulier des Polyèdres Réguliers, et sur la Dissection des Polyèdres Réguliers en Polyèdres Réguliers', *Annales de la Soc. Polonaise de Math.* **17** (1938) pp193-226; **18** pp1–3

Li Yan and Du Shiran,

1987 *Chinese Mathematics: a Concise History*, (translated from the Chinese by J. N. Crossley and A. W.-C Lun), Clarendon Press, Oxford

H. Lindgren,

1964 *Geometric Dissections*, D. van Nostrand

J. Needham,

1959 *Science and Civilisation in China volume 3: Mathematics and the Sciences of the Heavens and the Earth*, Cambridge Univ. Press

O. Neugebauer,

1951 *The Exact Sciences in Antiquity*, Oxford Univ. Press

O. Neugebauer and A. Sachs,

  1945 *Mathematical Cuneiform Texts*, American Oriental Series 29

T. E. Peet (editor),

  1923 *The Rhind Mathematical Papyrus*, London

Plutarch,

  1976 *De Communibus Notitiis Adversus Stoicos*. English translation by H. Cherniss
     in 'Plutarch's Moralia' volume 13 part 2, Loeb Classical Library, Harvard
     Univ. Press. pp622–873

C. H. Sah,

  1979 *Hilbert's Third Problem: Scissors Congruence*, Pitman

A. Seidenberg,

  1978 'The Origin of Mathematics', *Archive for History of Exact Sciences* **18** (1978)
     pp301–342

W. W. Struve,

  1930 'Mathematischer Papyrus des Staatlichen Museums der Schönen Künste in
     Moskau', *Quellen und Studien zur Geschichte der Mathematik*, part A, volume
     1, Berlin, 1930

J. P. Sydler,

  1943 'Sur la Décomposition des Polyèdres', *Commentarii Math. Helvetici* **16**
     (1943–44) pp266–273

  1965 'Conditions Nécessaires et Suffisantes pour l'Équivalence des Polyèdres de
     l'Espace Euclidien à Trois Dimensions', *Commentarii Math. Helvetici* **40**
     (1965) pp43–80

A. Szabó,

  1978 *The Beginnings of Greek Mathematics*, (translated from the German by A. M.
     Ungar), D. Reidel Publ. Co.

B. L. van der Waerden,

  1954 *Science Awakening*, (translated from the Dutch by Arnold Dresden),
     P. Noordhoff Ltd, Groningen, Holland

  1980 'On Pre-Babylonian Mathematics I and II', *Archive for History of Exact
     Sciences* **23** (1980) pp1–25, 27–46

  1983 *Geometry and Algebra in Ancient Civilisations*, Springer-Verlag

D. B. Wagner,

  1979 'An Early Chinese Derivation of the Volume of a Pyramid: Liu Hui, 3rd
     Century AD', *Historia Mathematica* **6** (1979) pp164–188

Chapter 2

G. J. Allman,

  1889 *Greek Geometry from Thales to Euclid*, Dublin

M. Berman,

  1979 'Regular-faced Convex Polyhedra', *J. Franklin Institute* **291** (1979) pp321–352 plus 7 pages of photographs

H. S. M. Coxeter,

  1937 'Regular Skew Polyhedra in Three and Four dimensions, and Their Topological Analogues', *Proc. London Math. Soc.* (series 2) **43** (1937) pp33–62

H. M. Cundy,

  1955 'Deltahedra', *Math. Gazette* **39** (1955) pp263–266

H. Freudenthal and B. L. van der Waerden,

  1947 'Over een Bewering van Euclides', *Simon Stevin* **25** (1947) pp115–128

J. Gow,

  1884 *History of Greek Mathematics*, Cambridge Univ. Press

B. Grünbaum and N. W. Johnson,

  1965 'The Faces of a Regular-faced Polyhedron', *J. London Math. Soc.* **40** (1965) pp577–586

T. L. Heath,

  1921 *A History of Greek Mathematics*, Clarendon Press, Oxford

  1926 *The Thirteen Books of Euclid's Elements* (3 volumes), Cambridge Univ. Press

N. W. Johnson,

  1966 'Convex Polyhedra with Regular Faces', *Canadian J. Math.* **18** (1966) pp169–200

K. Lamotke,

  1986 *Regular Solids and Isolated Singularities*, Advanced Lectures in Mathematics, Friedr. Vieweg & Sohn Braunschweig/Wiesbaden

C. Lanczos,

  1970 *Space Through the Ages: the Evolution of Geometrical Ideas from Pythagoras to Hilbert and Einstein*, Academic Press, London

F. Lasserre,

  1964 *The Birth of Mathematics in the Age of Plato*, Hutchinson, London

H. D. P. Lee,

  1977 *Timaeus and Critias*, Penguin Books

G. L. Morrow,

  1970 *Proclus: a Commentary on the First Book of Euclid's Elements*, Princeton Univ. Press

Plutarch,

  1976 *Platonicae Quaestiones*. English translation by H. Cherniss in 'Plutarch's Moralia' volume 13 part 1, Loeb Classical Library, Harvard Univ. Press. pp18–129

E. Sachs,

  1917 *Die Fünf Platonischen Körper*, Weidmann, Berlin. Reprinted by Arno Press
        (1976) New York

T. Smith,

  1902 *Euclid—His Life and System*, Clark, Edinburgh

I. B. Thomas,

  1939 *Selections Illustrating the History of Greek Mathematics: Translations of
        Greek Sources* (2 volumes), Loeb Classical Library, Harvard Univ. Press

W. C. Waterhouse,

  1972 'The Discovery of the Regular Solids', *Archive for History of Exact Sciences* **9**
        (1972) pp212–221

V. A. Zalgaller,

  1966 *Convex Polyhedra with Regular Faces* (in Russian). Seminars in
        Mathematics 2, Steklov Institute, Leningrad, Nauka. English translation
        published by Consultants Bureau, New York 1969

Chapter 3

L. B. Alberti,

  1436 *Della Pittura*. English translation in J. R. Spencer [1956]

F. Anzelewsky,

  1980 *Dürer: His Art and Life*, Chartwell Books

M. Baxandall,

  1985 *Painting and Experience in Fifteenth Century Italy*, Oxford Univ. Press

O. Benesch,

  1965 *The Art of the Renaissance in Northern Europe: its Relation to the
        Contemporary Spiritual and Intellectual Movements*, Phaidon, London

J. L. Berggren,

  1986 *Episodes in the Mathematics of Medieval Islam*, Springer-Verlag

A. Chastel,

  1963 *The Age of Humanism—Europe 1480–1530*, Thames and Hudson

L. Cheles,

  1981 *The Inlaid Decorations of Federigo da Montefeltro's Urbino 'Studiolo': An
        Iconographical Study*, Mitteilungen des Kunsthistorischen Instituts in
        Florenz 25

A. Cobban (editor),

  1969 *The Eighteenth Century: Europe in the Age of the Enlightenment*, Thames
        and Hudson, London

W. M. Conway,

  1958 *The Writings of Albrecht Dürer*, Philosophical Library, New York

J. L. Coolidge,

  1990 *The Mathematics of Great Amateurs* (second edition), Clarendon Press, Oxford

J. Cousin,

  1560 *Livre de Perspective*, Paris

M. Daly Davis,

  1977 *Piero della Francesca's Mathematical Treatises*, Longo Editore, Ravenna

  1980 'Carpaccio and the Perspective of Regular Bodies', in M. D. Emiliani [1980] pp183–200

A. Dürer,

  1525 *Underweysung der Messung*,

S. Y. Edgerton,

  1975 *The Renaissance Discovery of Linear Perspective*, Basic Books Inc., New York

M. D. Emiliani (editor),

  1980 *La Prospettiva Rinascimentale: Codificazioni e Trasgressioni*, Florence

J. V. Field,

  1985 'Giovanni Battista Benedetti on the Mathematics of Linear Perspective', *J. Warburg and Courtauld Institutes* **48** pp71–99

  1988 'Perspective and the Mathematicians: Alberti to Desargues', in C. Hay [1988] pp236–263

E. Grant and J. E. Murdoch (editors),

  1987 *Mathematics and its Applications to Science and Natural Philosophy in the Middle Ages*, Cambridge Univ. Press

N. L. W. A. Gravelaar,

  1902 'Stevin's Problemata Geometrica', *Nieuw Archief voor Wiskunde* (2) **5** (1902)

F. Hartt,

  1970 *A History of Italian Renaissance Art; Painting, Sculpture, Architecture*, Thames and Hudson

C. Hay (editor),

  1988 *Mathematics from Manuscript to Print: 1300–1600*, Clarendon Press, Oxford

J. P. Hogendijk,

  1984 'Greek and Arabic Constructions of the Regular Heptagon', *Archive for History of Exact Sciences* **30** (1984) pp197–330

W. Jamnitzer,

  1568 *Perspectiva Corporum Regularium*, Nürnberg

M. Kemp,

  1981 *Leonardo da Vinci: The Marvellous Works of Nature and Man*, Dent

  1990 *The Science of Art: Optical Themes in Western Art from Brunelleschi to Seurat*, Yale Univ. Press

J. A. Levenson (editor),

    1991 *Circa 1492—Art in the Age of Exploration*, National Gallery of Art,
        Washington DC, Yale Univ. Press

N. MacKinnon,

    1993 'The Portrait of Fra Luca Pacioli', *Math. Gazette* **77** (1993) pp130–219

N. H. Nasr,

    1976 *Islamic Science: an Illustrated Study*, World of Islam Festival Publ. Co.

L. Pacioli,

    1494 *Summa de Arithmetica, Geometria, Proportioni et Proportionalità*, Venice

    1509 *Divina Proportione*, Venice

E. Panofsky,

    1955 *The Life and Art of Albrecht Dürer*, Princeton Univ. Press

D. Pedoe,

    1976 *Geometry and the Liberal Arts*, Penguin Books

P. Rotondi,

    1960 *Il Palazzo Ducale di Urbino* (2 volumes), Urbino

J. C. Smith,

    1983 *Nuremberg: a Renaissance City, 1500–1618*, Univ. of Texas Press

J. R. Spencer,

    1956 *Leon Battista Alberti: On Painting*, Yale Univ. Press

S. Stevin,

    1583 *Problematum Geometricorum Libri V*, Antwerp

R. E. Taylor,

    1942 *No Royal Road: Luca Pacioli and his Times*, Chapel Hill N. C.

G. Vasari,

    1550 *The Lives of the Painters, Sculptors and Architects*. English translation by
        A. B. Hinds, Everyman's Library, Dent, London 1927

K. H. Veltman,

    1980 'Ptolemy and the Origins of Linear Perspective', in M. D. Emiliani [1980]
        pp403–407

J. White,

    1967 *The Birth and Rebirth of Pictorial Space* (second edition), Faber and Faber,
        London

Chapter 4

A. Badoureau,

    1881 'Mémoire sur les Figures Isosceles', *J. École Polytechnique* **49** (1881) pp47–172

A. Beer and P. Beer (editors),

    1975 'Kepler—400 years, Proceedings of Conferences Held in Honour of Johannes
        Kepler', *Vistas in Astronomy* **18** (1975)

S. Bilinski,

   1960 'Über die Rhombenisoeder', *Glasnek Mat. Fiz. Astronom. Društvo Mat. Fiz. Hrvatske* (series 2) **15** (1960) pp251–262

M. Caspar (editor),

   1938 *Johannes Kepler Gesammelte Werke*, Beck, Munich

E. C. Catalan,

   1865 'Mémoire sur la Théorie des Polyèdres', *J. École Polytechnique* **24** (1865) pp1–71 .

H. S. M. Coxeter,

   1975 'Kepler and Mathematics', *Vistas in Astronomy* **18** (1975) pp661–670

H. S. M. Coxeter, M. S. Longuet-Higgins and J. C. P Miller,

   1953 'Uniform Polyhedra', *Philosophical Trans. Royal Soc. London* (series A) **246** (1953) pp401–450

P. R. Cromwell,

   1995 'Kepler and Polyhedra', *Math. Intelligencer* **17** no. 3 (1995) pp23–33

J. V. Field,

   1979 'Kepler's Star Polyhedra', *Vistas in Astronomy* **23** (1979) pp109–141

   1988 *Kepler's Geometrical Cosmology*, Univ. Chicago Press

C. Hardie,

   1966 *The Six-cornered Snowflake*, Oxford Univ. Press

I. Hargittai,

   1996 'Sacred Star Dodecahedron', *Math. Intelligencer* **18** no. 3 (1996) pp52–54

J. Kepler,

   1595 *Mysterium Cosmographicum*, Tubingen. Also in M. Caspar [1938] volume 1

   1611 *De Nive Sexangula*, Prague. English translation in C. Hardie [1966]

   1619 *Harmonices Mundi Libri V*, Linz. Also in M. Caspar [1938] volume 6. English translation of book 2 in J. V. Field [1979]

A. Koestler,

   1986 *The Sleepwalkers: a History of Man's Changing Vision of the Universe*, Penguin Books

J. Lesavre and R. Mercier,

   1947 'Dix Nouveaux Polyèdres Semi-régulièrs sans Plan de Symétrie', *Comptes Rendus des Séances de l'Académie des Sciences* **224** (1947) pp785–786

J. P. Phillips,

   1965 'Kepler's Echinus', *Isis* **56** (1965) pp196–200

J. Pitsch,

   1881 'Uber Halbregulare Sternpolyeder', *Zeitschrift für das Real Schulwesen von Kolbe, Wien* **6** (1881) pp9–24, 72–89, 216

J. Skilling,

    1975 'The Complete Set of Uniform Polyhedra', *Philosophical Trans. Royal Soc. London* (series A) **278** (1975) pp111–135

S. P. Sopov,

    1970 'Proof of the Completeness of the Enumeration of Uniform Polyhedra', *Ukrain. Geom. Sbornik* **8** (1970) pp139–156 (1975) pp111–135

B. Stephenson,

    1987 *Kepler's Physical Astronomy*, Springer-Verlag

M. J. Wenninger,

    1971 *Polyhedron Models*, Cambridge Univ. Press

Chapter 5

J. Bertrand,

    1860 'Note on Preceding Memoire (E. Prouhet [1860])', *Comptes Rendus des Séances de l'Académie des Sciences* **50** (1860) pp781–782

N. L. Biggs, E. K. Lloyd and R. J. Wilson,

    1976 *Graph Theory 1736–1936*, Clarendon Press, Oxford

O. Bonnet,

    1848 'Mémoire sur la Théorie Générale des Surfaces', *J. École Polytechnique* **19** (1848) pp1–146

F. Cajori,

    1928 *A History of Mathematical Notations*, Open Court Publ. Co.

A. L. Cauchy,

    1813 'Recherches sur les Polyèdres (first memoire, part 2)', *J. École Polytechnique* **9** (1813) pp68–86

S. S. Chern,

    1979 'From Triangles to Manifolds', *American Math. Monthly* **86** (1979) pp339–349

A. Crum Brown,

    1864 'On the Theory of Isomeric Compounds', *Trans. Royal Soc. Edinburgh* **23** (1864) pp707–719

R. Descartes,

    1630 *De Solidorum Elementis*, (unpublished). English translation in P. J. Federico [1982]

P. Dombrowski,

    1979 'Differential Geometry—150 Years after Carl Friedrich Gauss' "Disquisitiones Generales circa Superficies Curvas"', *Astérisque* **62** (1979) pp97–153

L. Euler,

    1758 'Elementa Doctrinae Solidorum', *Novi Commentarii Academiae Scientiarum Petropolitanae* **4** (1752/53) pp109–140

1758 'Demonstratio Nonnullarum Insignium Proprietatum Quibus Solida Hedris Planis Inclusa Sunt Praedita', *Novi Commentarii Academiae Scientiarum Petropolitanae* **4** (1752/53) pp140–160

P. J. Federico,

1982 *Descartes on Polyhedra: a Study of the 'De Solidorum Elementis'*, Springer-Verlag

C. F. Gauss,

1828 'Disquisitiones Generales circa Superficies Curvas', *Commentationes Societatis Regiae Scientiarum Gottingensis Recentiones* **6** (1828). Reprinted with parallel English translation in *Astérisque* **62** (1979) pp1–81

A. Girard,

1629 *Invention Nouvelle en l'Algebre*, Amsterdam

B. Grünbaum and G. C. Shephard,

1994 'A New Look at Euler's Theorem for Polyhedra', *American Math. Monthly* **101** (1994) pp109–128

J. C. F. Hessel,

1832 'Nachtrag zu dem Euler'schen Lehrsatze von Polyedern', *J. für die Reine und Angewandte Mathematik* **8** (1832) pp13–20

P. Hilton and J. Pederson,

1981 'Descartes, Euler, Poincare, Pòlya—and Polyhedra', *L'Enseignement Math.* (2) **27** (1981) pp327–343

R. Hoppe,

1879 'Ergänzung des Eulerschen Satzes von den Polyedern', *Archiv der Mathematik und Physik* **63** (1879) pp100–103

E. de Jonquières,

1890 'Note sur un Point Fondamental de la Théorie des Polyèdres', *Comptes Rendus des Séances de l'Académie des Sciences* **110** (1890) pp110–115

1890 'Note sur le Théorème d'Euler dans la Théorie des Polyèdres', *Comptes Rendus des Séances de l'Académie des Sciences* **110** (1890) pp169–173

1890 'Note sur un Mémoire de Descartes Longtemps Inédit et sur les Titres de son Auteur a la Priorité d'une Decouverte dans la Théorie des Polyèdres', *Comptes Rendus des Séances de l'Académie des Sciences* **110** (1890) pp261–266

1890 'Écrit Posthume de Descartes sur les Polyèdres', *Comptes Rendus des Séances de l'Académie des Sciences* **110** (1890) pp315–317

S. A. J. L'Huilier,

1811 'Démonstration Immédiate d'un Théorème Fondamental d'Euler sur les Polyhèdres, et Exceptions dont ce Théorème est Susceptible', *Mémoires de l'Académie Imperiale de Saint Petersbourg* **4** (1811) pp271–301

1812 'Mémoire sur la Polyèdrométrie', *Annales de Math., Pures et Appliquées* **3** (1812/13) pp168–191

S. A. J. L'Huilier (*continued*),

1812 'Mémoire sur les Solides Réguliers', *Annales de Math., Pures et Appliquées* **3** (1812/13) pp233–237

I. Lakatos,

1976 *Proofs and Refutations: the Logic of Mathematical Discovery*, Cambridge Univ. Press

H. Lebesgue,

1924 'Remarques sur les Deux Premières Démonstrations du Théorème d'Euler, Rélatif aux Polyèdres', *Bull. Soc. Math. France* **52** (1924) pp315–336

A. M. Legendre,

1794 *Éléments de Géométrie*, Paris

J. B. Listing,

1848 *Vorstudien zur Topologie*

1862 'Der Census Räumlicher Complexe oder Verallgemeinerung des Euler'schen Satzes von den Polyedren', *Abhandlungen der Königlichen Gesellschaft der Wissenschaften zu Gottingen* **10** (1862) pp97–180

A. F. Möbius,

1865 'Uber die Bestimmung des Inhaltes eines Polyeders', *Königlich-Sächsischen Gesellschaft der Wissenschaften, Mathematisch-Physikalische Klasse* **17** pp31–68

B. O'Neill,

1966 *Elementary Differential Geometry*, Academic Press

L. Poinsot,

1858 'Note sur la Theorie des Polyèdres', *Comptes Rendus des Séances de l'Académie des Sciences* **46** (1858) pp65–79

E. Prouhet,

1860 'Remarques sur un Passage des Ouevres Inedites de Descartes', *Comptes Rendus des Seances de l'Academie des Sciences* **50** (1860) pp779-781

G. Pólya,

1954 *Mathematics and Plausible Reasoning, volume 1: Induction and Analogy in Mathematics*, Oxford Univ. Press

K. G. C. von Staudt,

1847 *Geometrie der Lage*, Nürnberg

Chapter 6

A. D. Alexandroff,

1950 *Convex Polyhedra* (in Russian). German translation: *Konvexe Polyeder*, Berlin (1958)

B. Belhoste,

1991 *Augustin-Louis Cauchy: a Biography*, (translated from the French by F. Ragland), Springer-Verlag

G. T. Bennett,

1912 'Deformable Octahedra', *Proc. London Math. Soc.* (series 2) **10** (1912)
pp309–343

R. Bricard,

1895 'Reponse a Question 376 (C. Stephanos [1894])', *L'Intermédiaire des
Mathématiciens* **2** (1895) p243

1897 'Mémoire sur la Théorie de l'Octaèdre Articulé', *J. de Math., Pures et
Appliquées* (series 5) **3** (1897) pp113–148

A. L. Cauchy,

1813 'Sur les Polygones et Polyèdres (second memoire)', *J. École Polytechnique* **9**
(1813) pp87–98

R. Connelly,

1978 'A Counter-example to the Rigidity Conjecture for Polyhedra', *Publications
Math. de l'Institute des Hautes Études Scientifiques* **47** pp333–338

1978 'A Flexible Sphere', *Math. Intelligencer* **1** (1978) pp130–131

1978 'Conjectures and Open Questions in Rigidity', *Proc. International Congress
Math.*, (Helsinki) volume 1 pp407–414

1979 'The Rigidity of Polyhedral Surfaces', *Math. Magazine* **52** (1979) pp275–283

1981 'Flexing Surfaces', *The Mathematical Gardner*, edited by D. A. Klarner,
Wadsworth International pp79–89

H. Gluck,

1975 'Almost All Simply Connected Closed Surfaces are Rigid', *Lecture Notes in
Math.* **438**, 'Geometric Topology', Springer-Verlag pp225–239

M. Goldberg,

1978 'Unstable Polyhedral Structures', *Math. Magazine* **51** (1978) pp165–170

B. Grünbaum and G. C. Shephard,

1975 *Lectures on Lost Mathematics*, unpublished notes (1975). Revised version
(1978)

J. Hadamard,

1907 'Erreurs de Mathematicians', *L'Intermédiaire des Mathématiciens* **14** (1907)
p31

N. H. Kuiper,

1978 'Spheres Polyhedriques Flexibles dans E³, d'apres Robert Connelly', *Lecture
Notes in Math.* **710** 'Séminaire Bourbaki Exposés 507–524 (1977/78)'
Springer-Verlag (1979) pp514.01–514.22

H. Lebesgue,

1909 'Démonstration Complète du Théorème de Cauchy sur l'Égalité des Polyèdres
Convexes', *L'Intermédiaire des Mathématiciens* **16** (1909) pp113–120

1967 'Octaèdres Articulé du Bricard', *L'Enseignement Math.* (2) **13** (1967)
pp175–185

L. A. Lyusternik,

  1956 *Convex Figures and Polyhedra* (in Russian). English translations by T.
     Jefferson Smith, Dover 1963; and D. L. Barnette, Heath, Boston 1966

I. G. Maksimov,

  1995 *Bendable Polyhedra and Riemannian Surfaces*, Uspekhi Matematicheskikh
     Nauk 50 (1995)

B. Roth,

  1981 'Rigid and Flexible Frameworks', *American Math. Monthly* **88** (1981) pp6–20

I. J. Schoenberg and S. K. Zaremba,

  1967 'On Cauchy's Lemma Concerning Convex Polygons', *Canadian J. Math.* **19**
     (1967) pp1062–1071

E. Steinitz and H. Rademacher,

  1934 *Vorlesungen über die Theorie der Polyedern*, Springer, Berlin

C. Stephanos,

  1894 'Question 376', *L'Intermédiaire des Mathématiciens* **1** (1894) p228

J. J. Stoker,

  1968 'Geometrical Problems Concerning Polyhedra in the Large', *Communications
     on Pure and Applied Math.* **21** (1968) pp119–168

W. Whiteley,

  1984 'Infinitesimally Rigid Polyhedra I: Statics of Frameworks', *Trans. American
     Math. Soc.* **285** (1984) pp431–465

  1988 'Infinitesimally Rigid Polyhedra II: Modified Spherical Frameworks', *Trans.
     American Math. Soc.* **306** (1988) pp115–139

W. Wunderlich,

  1965 'Starre, Kippende, Wackelige und Bewegliche Achtflache', *Elemente der
     Mathematik* **20** (1965) pp25–32

W. Wunderlich and C. Schwabe,

  1986 'Eine Familie von Geschlossen Gleichflachigen Polyedern, die Fast Beweglich
     Sind', *Elemente der Mathematik* **41** (1986) pp88–98

Chapter 7

J. Bertrand,

  1858 'Note sur la Théorie des Polyèdres Réguliers', *Comptes Rendus des Séances de
     l'Académie des Sciences* **46** (1858) pp79–82

A. L. Cauchy,

  1813 'Recherches sur les Polyédres (first memoire, part 1)', *J. École Polytechnique* **9**
     (1813) pp68–86

A. Cayley,

  1859 'On Poinsot's Four New Regular Solids', *Philosophical Mag.* **17** (1859)
     pp123–128

1859 'Second Note on Poinsot's Four New Polyhedra', *Philosophical Mag.* **17** (1859) pp209–210

H. S. M. Coxeter, P. Du Val, H. T. Flather and J. F. Petrie,

1951 *The Fifty-nine Icosahedra*, Univ. of Toronto

A. W. M. Dress,

1981 'A Combinatorial Theory of Grünbaum's New Regular Polyhedra I: Grünbaum's New Regular Polyhedra and their Automorphism Group', *Aequationes Math.* **23** (1981) pp252–265

1985 'A Combinatorial Theory of Grünbaum's New Regular Polyhedra II: Complete Enumeration', *Aequationes Math.* **29** (1985) pp222–243

B. Grünbaum,

1977 'Regular Polyhedra—Old and New', *Aequationes Math.* **16** (1977) pp1–20

1993 'Regular Polyhedra', *Companion Encyclopaedia of the History and Philosophy of the Mathematical Sciences volume 2*, edited by I. Grattan-Guiness, Routledge, London, pp866–876

1994 'Polyhedra with Hollow Faces', *Proc. NATO-ASI Conference on Polytopes: Abstract, Convex and Computational*, (Toronto 1993), edited by T. Bisztriczky, P. McMullen, R. Schneider and A. Ivic'Weiss, Kluwer Academic Publ. Dortrecht pp43–70

1997 'Isogonal Prismatoids', *Discrete and Computational Geometry* **18** (1997) pp13–52

R. Haussner,

1906 *Abhandlung über die Regelmässigen Sternkörper von L. Poinsot (1809), A. L. Cauchy (1811), J. Bertrand (1858), A. Cayley (1859)*, Leipzig

J. L. Hudson and J. G. Kingston,

1988 'Stellating Polyhedra', *Math. Intelligencer* **10** no 3. (1988) pp50–61

T. Hugel,

1876 *Die Regulären und Halbregulären Polyeder*, Neustadt a. d. Halle

D. Luke,

1957 'Stellations of the Rhombic Dodecahedron', *Math. Gazette* **41** (1957) pp189–194

L. Poinsot,

1810 'Mémoire sur les Polygones et les Polyèdres', *J. École Polytechnique* **10** (1810) pp16–48

O. Terquem,

1849 'Sur les Polygones et les Polyèdres Étoilés, Polygones Funiculaires; d'apres Poinsot', *Nouvelles Annales de Math.* **8** (1849) pp68–74

1849 'Polyèdres Régulier Ordinaires et Polyèdres Régulier Étoilé; d'apres M. Poinsot', *Nouvelles Annales de Math.* **8** (1849) pp132–139

A. H. Wheeler,

 1924 'Certain Forms of the Icosahedron and a Method for Deriving and Designating
      Higher Polyhedra', *Proc. International Math. Congress*, (Toronto) pp701–708

Chapter 8

A. Bravais,

 1849 'Mémoire sur les Polyhèdres de Forme Symetrique', *Extrait J. Math., Pures et
      Appliquées* **14** (1849) pp141–180

 1850 'Mémoire sur les Systèmes Formés par des Points Distribués Regulièrement sur
      un Plan ou dans l'Espace', *J. École Polytechnique* **19** (1850) pp1–128

 1851 'Etudes Cristallographiques', *J. École Polytechnique* **20** (1851) pp101–278

M. J. Buerger,

 1971 *Introduction to Crystal Geometry*, McGraw-Hill, New York

P. Curie,

 1885 'Sur les Répétitions et la Symétrie', *Compte Rendus des Séances de
      l'Académie des Sciences* **100** (1885) pp1393–1396

W. Dyck,

 1882 'Gruppentheoretischen Studien', *Math. Ann.* **20** (1882) pp1–44

 1883 'Gruppentheoretischen Studien II', *Math. Ann.* **22** (1883) pp70–108

E. S. Fedorov,

 1885 *Elements of the Theory of Figures* (in Russian). English translation by
      D. Harker and K. Harker, American Crystallographic Association, monograph
      7 (1971)

W. R. Hamilton,

 1856 'Memorandum Respecting a New System of Roots of Unity', *Philosophical
      Mag.* **12** (1856) p446

I. Hargittai (editor),

 1986 *Symmetry: Unifying Human Understanding*, Pergamon, New York

I. Hargittai and E. Y. Rodin (editors),

 1989 'Symmetry II: Unifying Human Understanding (part 2)', *Intern. J. Computers
      and Math. with Applications* **17**

R. J. Haüy,

 1801 *Traité de Minéralogie*, Delance, Paris

 1822 *Traité de Cristallographie*, Bachelier et Huzard, Paris

J. F. C. Hessel,

 1830 *Krystallometrie oder Krystallonomie und Krystallographie*, Leipzig (1830).
      Reprinted in Ostwald's 'Klassiker der Exakten Wissenschaften', Engelmann,
      Leipzig (1897)

H. Hilton,

 1903 *Mathematical Crystallography and the Theory of Groups of Movements*,
      Clarendon Press, Oxford

R. Hooke,

1665 *Micrographia, or some Physiological Descriptions of Minute Bodies*, London

C. Huygens,

1690 *Traité de la Lumière*, Leyden

C. Jordan,

1866 'Recherches sur les Polyèdres', *Comptes Rendus des Séances de l'Académie des Sciences* **62** (1866) pp1339–1341

1867 'Sur les Groupes de Mouvements', *Comptes Rendus des Séances de l'Académie des Sciences* **65** (1867) pp229–232

1869 'Mémoire sur les Groupes de Mouvements', *Annali di Matematica* **2** (1869) pp167–215, 322–345

J. Lima-de-Faria (editor),

1990 *Historical Atlas of Crystallography*, Kluwer Academic

A. F. Möbius,

1849 'Ueber das Gesetz der Symmetrie der Krystalle und die Anwendung dieses Gesetzes auf die Eintheilung der Krystalle in Systeme', *Königlich-Sächsischen Gesellschaft der Wissenschaften, Mathematisch-Physikalische Klasse* **1** (1849) pp65–75

1851 'Ueber Symmetrische Figuren', *Königlich-Sächsischen Gesellschaft der Wissenschaften, Mathematisch-Physikalische Klasse* **3** (1851) pp19–28

F. C. Phillips,

1956 *An Introduction to Crystallography*, Longman

L. Poinsot,

1851 'Théorie Nouvelle de la Rotation des Corps', *J. de Math., Pures et Appliquées* **16** (1851) pp9–129, 289–336

J. B. L. Romé de l'Isle,

1772 *Essai de Crystallographie, ou Description des Figures Géométriques, Propres à Differens Corpes du Règne Minéral, Connus Vulgairement sous le nom de Cristaux*, Paris

1783 *Crystallographie, ou Description des Formes Propres à Tous les Corps du Regne Minéral*, Paris

A. M. Schoenflies,

1891 *Krystallsysteme und Krystallstruktur*, Teubner, Leipzig

E. Scholz,

1989 'Crystallographic Symmetry Concepts and Group Theory (1850–1880)', *The History of Modern Mathematics volume 2*, edited by D. E. Rowie and J. McCleary, Academic Press pp3–28

M. Senechal,

1990 'Brief History of Geometrical Crystallography', in J. Lima-de-Faria [1990] pp43–59

A. V. Shubnikov and V. A. Koptsik,

   1972 *Symmetry in Science and Art* (in Russian). English translation by G. D.
        Archard (1974) Plenum, New York

L. Sohncke,

   1874 'Die Regelmässigen Ebenen Punktsysteme von Unbegrenzter Ausdehnung',
        J. für die Reine und Angewandte Mathematik **77** (1874) pp47–101

   1879 *Entwickelung einer Theorie der Krystallstruktur*, Teubner, Leipzig

H. Weyl,

   1952 *Symmetry*, Princeton Univ. Press

F. Klein,

   1956 *Lectures on the Icosahedron and the Solution of Equations of the Fifth Degree*
        (second revised edition), Dover

Chapter 9

V. A. Aksionov and L. S. Mel'nikov,

   1980 'Some Counter-examples Associated with the Three-color Problem',
        J. Combinatorial Theory (series B) **28** (1980) pp1–9

K. Appel and W. Haken,

   1977 'Every Planar Map is Four Colorable I: Discharging', *Illinois J. Math.* **21**
        (1977) pp429–490

   1978 'The Four-color Problem', *Mathematics Today: Twelve Informal Essays*, edited
        by L. A. Steen, Springer-Verlag pp153–180

   1986 'The Four Color Proof Suffices', *Math. Intelligencer* **8** no. 1 (1986) pp10–20

   1989 *Every Planar Map is Four Colorable*, American Math. Soc. Contemporary
        Math. Series, 98

K. Appel, W. Haken and J. Koch,

   1977 'Every Planar Map is Four Colorable II: Reducibility', *Illinois J. Math.* **21**
        (1977) pp491–567

W. W. R. Ball and H. S. M. Coxeter,

   1987 *Mathematical Recreations and Essays* (thirteenth edition), Dover, New York

D. W. Barnette,

   1983 *Map Coloring, Polyhedra, and the Four-color Problem*, Math. Association of
        America

K. A. Berman,

   1981 'Three-colouring of Planar 4-valent Maps', *J. Combinatorial Theory* (series B)
        30 (1981) pp82–88

N. L. Biggs,

   1983 'De Morgan on Map Colouring and the Separation Axiom', *Archive for
        History of Exact Sciences* **28** (1983) pp165–170

G. D. Birkhoff,

   1913 'The Reducibility of Maps', *American J. Math.* **35** (1913) pp114–128

R. L. Brooks,

1941 'On Colouring the Nodes of a Network', *Proc. Cambridge Philosophical Soc.* **37** (1941) pp194–197

W. Burnside,

1911 *Theory of Groups of Finite Order*, Cambridge Univ. Press (1911). Second edition, Dover (1955)

A. Cayley,

1866 'Notes on Polyhedra', *Quarterly J. of Pure and Applied Math.* **7** (1866) pp304–316

1879 'On the Colouring of Maps', *Proc. Royal Geog. Soc.* (new series) **1** (1879) pp259–261

A. Ehrenfeucht,

1964 *The Cube Made Interesting*, (translated from the Polish by W. Zawadowski), Pergamon Press

J. C. Fournier,

1977 'Le Théorème du Coloriage des Cartes', *Lecture Notes in Math.* **710**, 'Séminaire Bourbaki Exposés 507–524 (1977/78)', Springer-Verlag (1979) pp509.01–509.24

P. Franklin,

1922 'The Four Colour Problem', *American J. Math.* **44** (1922) pp225–236

1938 'Note on the Four Colour Problem', *J. Math. and Phys.* **16** (1938) p172

G. Frobenius,

1887 'Ueber die Congruenz Nach einem Aus Zwei Endlichen Gruppen Gebildeten Doppelmodul', *J. für die Reine und Angewandte Mathematik* **101** (1887) pp273–299

H. Grötzsch,

1958 'Zur Theorie der Diskreten Gebilde VII: ein Dreifarbensatz für Dreikreisfrei Netze auf der Kugel', *Wiss. Z. Martin Luther Univ. Halle-Wittenberg Math.-Naturw. Reihe* **8** (1958/59) pp109–120

B. Grünbaum,

1963 'Grötzsch's Theorem on 3-colorings', *Michigan Math. J.* **10** (1963) pp303–310

F. Guthrie,

1880 'Note on the Colouring of Maps', *Proc. Royal Soc. Edinburgh* **10** (1880) pp727–728

P. J. Heawood,

1890 'Map Colour Theorem', *Quarterly J. Math.* **24** (1890) pp332–338

H. Heesch,

1969 *Untersuchungen zum Vierfarbenproblem*, B-I-Hochschulskripten 810/810a/810b, Bibliographisches Institute, Mannheim/Vienna/Zurich

A. B. Kempe,

1878 (untitled abstract), *Proc. London Math. Soc.* **10** (1878) pp229–231

A. B. Kempe (*continued*),

 1879 'On the Geographical Problem of the Four Colours', *American J. Math.* **2** (1879) pp193–200

 1880 'How to Colour a Map with Four Colours', *Nature* **21** (26th February, 1880) pp399–400

A. Kotzig,

 1965 'Coloring of Trivalent Polyhedra', *Canadian J. Math.* **17** (1965) pp659–664

K. O. May,

 1965 'The Origin of the Four-colour Conjecture', *Isis* **56** (1965) pp346–348

O. Ore,

 1967 *The Four Color Problem*, Academic Press, New York

O. Ore and J. Stemple,

 1968 'On the Four Colour Problem', *Notices American Math. Soc.* **15** (1968) p196

C. N. Reynolds,

 1927 'On the Problem of Colouring Maps in Four Colours', *Annals of Math.* (2) **28** (1927) pp1–15

G. Ringel,

 1974 *Map Color Theorem*, Springer-Verlag

T. L. Saaty and P. C. Kainen,

 1977 *The Four Colour Problem: Assaults and Conquest*, McGraw-Hill

W. Stromquist,

 1975 'The Four-color Theorem for Small Maps', *J. Combinatorial Theory* (series B) **19** (1975) pp256–268

P. G. Tait,

 1880 'Remarks on the Colouring of Maps', *Proc. Royal Soc. Edinburgh* **10** (1880) p729

H. Whitney,

 1932 'Congruent Graphs and the Connectivity of Graphs', *American J. Math.* **54** (1932) pp150–168

C. E. Winn,

 1940 'On the Minimum Number of Polygons in an Irreducible Map', *American J. Math.* **62** (1940) pp406–416

E. M. Wright,

 1981 'Burnside's Lemma: a Historical Note', *J. Combinatorial Theory* (series B) **30** (1981) pp89–90

Chapter 10

N. H. Abel,

 1826 'Beweis der Unmöglichkeit Algebraische Gleichungen von Höheren Graden als dem Vierten Allgemein Aufzulösen', *J. für die Reine und Angewandte Mathematik* **1** (1826) pp65–84

T. Bakos,

1959 'Octahedra Inscribed Inside a Cube', *Math. Gazette* **43** (1959) pp17–20

M. Brückner,

1905 'Uber die Diskontinuierlichen und Nicht-konvexen Gleicheckig-gleichflächigen Polyeder', *Proc. Third International Congress Math.*, (Heidelberg), Teubner, Leipzig pp707–713

1906 'Uber die Gleicheckig-gleichflächigen, Diskontinuirlichen und Nichkonvexen Polyeder', *Nova Acta, Abhandlungen der Kaiserlichen Leopoldinisch-Carolinischen Deutschen Akademie der Naturforscher* **86** (1906) pp1–348

1907 'Zur Geschichte der Theorie der Gleichechik-gleichflächigen Polyeder', *Unterrichtblätter für Mathematik und Naturwissenschaften* **13** (1907) pp104–110, 121–127

G. Cardano,

1545 *Ars Magna*. English translation in T. R. Witner [1968]

M. Dedò,

1994 'Topologia delle Forme di Poliedri', *L'Insegnamento della Matematica e delle Scienze Integrate* **17B** no. 2 (1994) pp149–192

B. Grünbaum and G. C. Shephard,

1984 'Polyhedra with Transitivity Properties', *Comptes Rendus Math. Reports of the Academy of Science, Canada* **6** (1984) pp61–66

1998 'Isohedra with Non-convex Faces', *J. Geometry* **63** (1998) pp76–96

C. Hermite,

1858 'Sur la Résolution de l'Équation du Cinquième Degré', *Comptes Rendus des Séances de l'Académie des Sciences* **46** (1858) pp508–515

E. Hess,

1875 'Uber Zwei Erweiterungen des Begriffs der Regelmässigen Körper', *Sitzungsberichte der Gesellschaft zur Beförderung der gesammten Naturwissenschaften zu Marburg*, pp1–20

1876 'Zugleich Gleicheckigen und Gleichflächigen Polyeder', *Schriften der Gesellschaft zur Beförderung der Gesammten Naturwissenschaften zu Marburg* **11** (1876) pp5–97

E. Hess and M. Brückner,

1910 'Über die Gleicheckigen und Gleichflächigen, Diskontinuierlichen und Nichtkonvexen Polyeder', *Nova Acta, Abhandlungen der Kaiserlichen Leopoldinisch-Carolinischen Deutschen Akademie der Naturforscher* **93** (1910)

J. L. Lagrange,

1771 *Refléxions sur la Résolution Algébrique des Équations*, Nouv. Mém. Acad. Berlin (1770/71)

A. Rosenthal,
  1910 'Untersuchungen über Gleichflächige Polyeder', *Nova Acta, Abhandlungen der Kaiserlichen Leopoldinisch-Carolinischen Deutschen Akademie der Naturforscher* **93** (1910) pp45–192

P. Ruffini,
  1799 *Teoria Generale delle Equazioni in cui si Dimostra Impossibile la Soluzione Algebrica delle Equazioni Generali di Grado Superiore al Quarto*, Bologna

S. A. Robertson and S. Carter,
  1970 'On the Platonic and Archimedean Solids', *J. London Math. Soc.* (series 3) **2** (1970) pp125–130

S. A. Robertson, S. Carter and H. R. Morton,
  1970 'Finite Orthogonal Symmetry', *Topology* **9** (1970) pp79–95

J. Skilling,
  1976 'Uniform Compounds of Uniform Polyhedra', *Math. Proc. Cambridge Philosophical Soc.* **79** (1976) pp447–457

I. Stewart,
  1973 *Galois Theory*, Chapman and Hall

M. J. Wenninger,
  1968 'Some Interesting Octahedral Compounds', *Math. Gazette* **52** (1968) pp16–23

T. R. Witner,
  1968 *The Great Art (or the Rules of Algebra)*, M. I. T. Press

Name Index

Read the symbol ▶ as 'see'.

Subject Index

Read the symbols ▶ and ▷ as 'see' and 'see also' respectively.